The

VICTORIAN MAPS OF DEVON

PRINTED MAPS 1838-1901

The VICTORIAN MAPS OF DEVON

PRINTED MAPS 1838-1901

KIT BATTEN

FRANCIS BENNETT

DEVON BOOKS

First Published in Great Britain in 2000 by Devon Books

British Library Cataloguing in Publication Data

Data for this publication is available from the British Library

ISBN 1 85522 756 8

DEVON BOOKS
Official Publisher to Devon County Council

Halsgrove House
Lower Moor Way
Tiverton
Devon EX16 6SS
Tel: 01884 243242
Fax 01884 243325
www.halsgrove.com

Printed and bound in Great Britain by
Bookcraft Ltd, Midsomer Norton

Contents

FOREWORD

In one of her full-length portraits Elizabeth 1 is shown standing on the map of England produced by Christopher Saxton. She was monarch of all that he had surveyed and the royal support of the production of the first maps of Devon and other counties in the 1570s was a significant political statement. Saxton's mapping formed the basis of most county maps for more than a century and only gradually did other surveys, notably those of Donn in 1765 and the Ordnance Survey in 1809, provide a more accurate image of the county as maps became ever more widely used in an increasing range of activities. The plates resulting from those early surveys and their numerous derivatives were meticulously chronicled by the present authors up to 1837 in their companion volume.

Now they continue the story into the relatively uncharted waters of the Victorian era, a period when new methods of printing add more complexity to the identification of different plates and their variants. By providing an illustration of at least one state of each plate the authors make the volume not only more attractive but infinitely more useful to the librarian, collector or researcher who wishes to identify individual items. In the nineteenth century printing plates could be modified more easily and so kept up to date, sometimes too much so, as witness the depiction of railway lines projected but never constructed. This volume guides the unwary past these and other pitfalls in a period when maps proliferated not only as separate publications but also in guidebooks, directories, atlases and official reports.

The county of Devon has made a major contribution to the study of cartography. The work of scholars such as JB Harley on the Ordnance Survey, W Ravenhill on early surveyors and Roger Kain on the tithe surveys is being continued by Landmark's Exeter based activities digitising early Ordnance Survey mapping of the whole country and the work of Mary Ravenhill and Margery Rowe's listing of the more than 1,000 manuscript surveys of the county prior to the tithe maps of *c.*1840. The present authors are also working on the early printed town and district maps of Devon, so it can be seen that this volume takes an important place in a much larger and more comprehensive endeavour. As a librarian who has custody of one of the largest public collections of Devon maps, I welcome this valuable information resource which will make my work very much easier.

Ian Maxted
County Local Studies Librarian

PREFACE

The first printed map of Devonshire appeared in 1575 and, in this age of multi-media and the supposedly paperless office, it will be interesting to see when the last printed map will be published. It will very probably not be in our lifetimes. In the meantime we can begin to catalogue those maps that are available to us. The first part of this work has already been done. The *Printed Maps of Devonshire 1575 - 1837* appeared in 1996 and now with this volume we have brought the cartobibliography of the county up to the beginning of the twentieth century - just in time for the twenty-first.

We decided to complete volume one with the occasion of Queen Victoria's accession to the throne in 1837: a date which comes after the Reform Bill but before the advent of the railways and one that (nearly) avoids the multi-reproductions possible through lithography. In this present volume the progress from engraving to lithography is completed and the course of the railways in Devon almost completed

In the 64 years covered in this present volume the authors have identified some 65 separate plates, listed nearly 260 states and referenced nearly 600 sources of atlases and maps. The first volume covered a period almost four times longer. Between 1575 and 1837 some 117 plates were identified, (and one more has since been found, see Appendix I); between 1838 and 1901 although only 65 new plates have been discovered the number of states is comparatively higher than in the previous years. With the advent of photolithography it became much easier to amend plates without spoiling the originally prepared material, hence the lifespans of many maps were extended.

We would once again like to take this opportunity to thank all those without whose help this cartobibliography could not have been completed: firstly those individuals who have so generously given us access to their collections or who have, so often, resolved difficult problems. Thanks above all to Ian Maxted of the Exeter Westcountry Studies Library; to Francis Herbert of the Royal Geographic Society; to Eugene Burden for access to his collection and many letters; to David Smith who made available much unpublished material; the librarians and curators who have helped us to examine their collections or answered our queries (see p. xxxiii); and finally to Devon Books and especially to Karen Binaccioni for her patience and understanding even with our last minute changes.

Thanks must also go to Joan for her support and editing and Karin, Katrina and Kim for their encouragement.

In the course of our researches we may have omitted something, overlooked a source or falsely interpreted our notes. We apologise for any errors and welcome correspondence from fellow enthusiasts and collectors.

October 2000

Francis Bennett
Menryn
Newton Ferrers
Devon PL8 1BW

Kit Batten
Auerhahnweg 7
70499 Stuttgart
Germany

ACKNOWLEDGEMENTS

We would like to thank the following for their permission to use illustrations making it possible for this to be a fully-illustrated county bibliography (numbers refer to maps, not page numbers):

133, 139, 148, 152, 159, 160, 177 are reproduced by permission of the Exeter Westcountry Studies Library
145, 158 are reproduced by permission of the Bodleian Library, Oxford
173, 180 are reproduced by permission of the Trustees of the National Library of Scotland
125 is reproduced by permission of the Devon & Exeter Insitute, Exeter
162 is reproduced by permission of Glasgow University Library
163, 167, 174, 179 are reproduced by permission of Eugene Burden

All other illustrations are taken from maps in possession of the authors.

The major portions of the collections of Kit Batten and Francis Bennett (pre-1837) were acquired by the Westcountry Studies Library in Exeter in spring 1997. The Westcountry Studies Library now possesses one of the country's best collections of a county's maps. A provisional list of the Exeter collection is available from the Library.

Frotispiece vignette from Black's Guide to Devon *(1.42).*

INTRODUCTION

The history of county map production before 1837 can be seen as a series of phases. From 1575, with the production of the first printed map of Devon by Christopher Saxton, until 1645 and the publication of Joan Blaeu's county map, one has what might be termed the *Romantic Age* of map production. This period is marked by the production of four of the most attractive and ornamental county maps. The work of Christopher Saxton was not surpassed by any later surveyor until Benjamin Donn in 1765, John Cary at the end of the eighteenth century and the Ordnance Survey at the beginning of the nineteenth. Saxton's surveys were not bettered but his work was copied, and in some artistic respects improved, by John Speed, Jan Jansson and Joan Blaeu[1]. The quality of these maps, their artistic flourishes and popular appeal made them the forerunners for over a hundred years. During the next seventy-five years only twelve printed maps were produced and all were based on Saxton's survey.

The *Age of Reason* lasted from 1650 until 1780 with the rise of a Middle Class. News sheets, part works and magazine publications brought new ideas and information to a rising class of affluent well-educated people eager to find out about the world at large. Over this period of a hundred and thirty years more than forty new maps of the county were printed. In 1765 the famous large-scale map of Benjamin Donn appeared. Produced on twelve sheets it was the first surveyed and detailed map of the county to appear since Saxton and would remain the best depiction until the Ordnance Survey in 1809.

The next major period covers the years from 1787 to 1837 and the advent of Victoria. This *Age of Variety* was the busiest period of county mapping. In fifty years a total of seventy-one different printed maps of Devon were published. The variety is awesome. We see simple maps such as that produced by William Cobbett - a map of Devon that looks more like a representation of a kidney - to the highly detailed and superbly crafted engravings of John Cary. It was Cary's efforts at depicting a true representation of the county, in the same manner as Benjamin Donn before him, that led to the clear and correct mapping of the county. During this period we also see some of the most decorative of county maps. Walker, Dix & Darton, Langley & Belch, Schmollinger and Moule introduced colourful scenes to complement the map detail. But of paramount importance was the advent of the Ordnance Survey. As Saxton had been the benchmark for the preceding two hundred and forty years so the Ordnance Survey would become the benchmark for the future and would often be copied, though seldom if ever improved.

The Victorian era could be called the *Age of Precision.* It marked the end of romantic mapping and the adoption of the precision of John Cary and the accuracy of the Ordnance Survey. Between 1837, when Victoria ascended to the throne and 1901 when she died, a total of sixty-five different printed maps of the county were produced. In comparison with the previous years it might seem that fewer maps were printed. In reality the new techniques of lithography and chromolithography made it much easier to change a plate to meet new demands and to allow the production of many more maps per print run, as well as many more print runs per plate. It was also the era of the railway and the advance of the lines led to more and more revisions to the plates.

Of the many printing and publishing companies that had produced maps before the accession of Victoria, only two would be of importance in the following seventy years. These were the companies established by John Cary and the two Walker brothers, John and Charles. John Cary died in 1835 leaving the company to two relatives, George and John, who continued to publish maps for a number of years. However, they do not appear to have produced any new work and seem to have discontinued and sold their business *c.*1844. But the map plates they still held at the end were important throughout the century and into the twentieth century. The plates passed to G F Cruchley who exploited them as much as possible. He printed and published a vast amount of atlases and folding maps from plates that were already over half a century old. The plates were then bought by Gall & Inglis who continued to print from them until the end of the nineteenth century. The Walker brothers printed and published their *Royal Atlas* in 1836 when Victoria was still a princess (the atlas was dedicated to *their Royal Highnesses the Duchess of Kent and the Princess Victoria*). The maps were printed almost to the end of the century. They were included in guide books and directories[2] and became the famous fox hunting maps of Hobson and later of the Walkers themselves. But it was a handful of new firms that would dominate the market for maps and atlases in the nineteenth century. The Victorian age was strongly influenced by the efforts of the family firms of A & C Black, the Philips and the Bartholomews, sometimes independently and sometimes cooperating. Together with the companies of G W Bacon, W & A K Johnston, Cassells and Collins they printed and published by far the greatest share of the growing atlas market in Great Britain.

1. K Batten: *Saxton's Survey of Devon;* unpublished manuscript - a copy is available at the Exeter Westcountry Studies Library.
2. Walker's maps appeared in editions of Murray's Devon and Cornwall (both combined and individual county editions) and in *White's Directory* of 1878/9.

This was an age of education and new technology and these companies used both to produce vast numbers of maps, wall-maps, folding maps, guide book maps as well as the thousands of assorted atlases covering every part of the world and every feature in the new science of thematic mapping. The history of many of these companies during this time is one of ever-increasing production and most had to move premises several times. The Blacks took over a two-acre site for printing and production in the 1870s to meet the growing educational market's demands. The Philips established their company in Paradise Street, Liverpool in 1834 but before long had added a factory to manufacture stationery and carry out letter-press printing concentrating on educational works. By 1859 they had opened their Caxton Works; 'a purpose-built, multi-storied, spacious, integrated, technologically-advanced printing works. The latest state-of-the-art equipment.' At one time they employed over eighty colourists.[1]

W & A K Johnston moved to new premises in 1855 and began to exploit lithography and chromo-lithography. The Bartholomews, who had begun as engravers, were not slow to follow suit and in 1859 they embarked on a programme of expansion, establishing their own printing works. They moved to new premises and trebled the number of steam presses at their disposal. They were able to compete in every stage of map production from first draft to final printing.[2] By the outbreak of the First World War Bartholomews was one of the leading map companies even challenging the Ordnance Survey's position. Such was their output that an edition of their successful *Citizen's Atlas* could sell one hundred and forty one thousand copies.

When William Collins, son of the Scottish firm's founder, visited London he took time to visit the publishing offices of Clowes on 13th July 1837 and reported 'This forenoon we visited Messrs. Clowes' printing office, which is the largest in London - indeed I should suppose the largest in the world. We were first conducted through the machine-rooms, in which were about 20 machines, throwing off at the rate of 750 sheets, printed on both sides, per hour. These machines were driven by a steam engine'. By the middle 'fifties the Collins firm was operating ten single-cylinder steam printing machines, five older style printing presses and several litho presses in their works. However, expansion came so quickly that in 1869 the book-printing plant consisted of sixteen single and double cylinder printing machines, seven litho presses and a number of smaller presses. In the same year the records show an output of over a million printed and bound works. By 1895 one thousand nine hundred workers printed and bound two and a half million works using a printing plant of forty cylinder machines, twenty five commercial jobbing machines, and twenty cylinder litho presses.[3]

This mammoth rise in production, spurred on by increasing demand and sated by cheaper means of production, meant that Gall and Inglis, the heirs to the Cruchley and Cary plates, could be regarded as one of the most important suppliers of the mass map market towards the end of the century and were mentioned in committees investigating sales of OS maps.[4] And this even though they were churning out reprints of maps one hundred years old!

One of the factors in the growth of all of these companies was that during the eighteen-seventies, the Education Acts of 1870 (England) and 1872 (Scotland) were passed and publishers were not slow to see the potential. Demand was so large that already in 1875 there were nine hundred and twenty schoolbooks in the Collins' catalogue, and the number of employees had risen to more than twelve hundred.[3]

Another factor that undoubtedly led to increased demand was the steady decrease in the number of hours being worked. The Bank Holidays Act was first passed in 1871. Before then, although some seasonal workers might migrate looking for work, the working classes rarely travelled much further than to the nearest market town. Employers were not known for being generous in providing leisure for their employees: Sunday was still very much a day of rest, and Christmas and Good Friday were both legal holidays. However, Boxing Day, Easter Monday and Whit Monday were only holidays according to the whim of the employer. Sir John Lubbock promoted the Bank Holidays Act of 1871, hence they were sometimes known as St Lubbock's Days, but the immediate impact was slight. The Act was extended by the Holidays Extension Act of 1875. Lubbock's activities led to the Shops Hours Regulation Act in 1886, which restricted the working hours to seventy four a week for shop assistants under eighteen.[5] All these measures led to increased leisure hours for all, and not just for the working classes.

Here again, William Collins seems to have been a leader in the field: having already reduced the weekly working hours of his printers from sixty-six to sixty hours a week – an example quickly followed by the other master printers of Glasgow. Already by the end of 1870 he had reduced them again to fifty-seven – an announcement which his workers greeted, said a contemporary report, with 'immense applause'. By 1895 his workers were working fifty two and a half hours a week.[3]

With reduced working hours came more leisure time. The end of the nineteenth century saw a tremendous rise in the popularity of cycling and with it an enormous output of mapping material associated with the new sport. Many maps that had originally been produced for guide books or for information purposes were adapted to show information vital to the cyclist such as repair stations and accommodation or the gradients of hills, eg Bartholomew's map for Pattisons (**175**), or Gall & Inglis' Contour Road Maps (**182**). So popular was the sport that printing runs of 60,000 cycling maps were not uncommon.[6]

1. D Smith; Map Publishers of Victorian Britain - The Philip Family Firm; *The Map Collector* 38; 1987.
2. D Smith; The Business of the Bartholomew Family Firm; *IMCoS Journal* 75; 1998.
3. David Keir; 1952; pp. 121, 167, 176, 178, 204.
4. D Smith; Gall and Inglis; *IMCoS Journal* 73; 1998.
5. John Vaughan; 1974; p. 23.
6. D Smith; The Cartography of the Bartholomew Family Firm; *IMCoS Journal* 76; 1998.

Printing

Although the first printed maps had used the woodcut, a relief process where the back is cut away leaving the design raised, this technique was quickly superseded by copper plate engraving. The only county maps where wood was used were for Joshua Archer's plates in Pinnock's *Guide to Knowledge* series and Seeley & Jackson's and J P's children's puzzles. The former maps were engravings and the result was white on black (**108**, 1833). The latter were small county maps embedded in a children's rebus (**147** and **158**). Wood would remain an important medium for illustrations and early guides sometimes included attractive woodcuts.

Copper plate engraving dominated the production of maps for some three hundred years until the middle of the 19th century. The design was cut into the metal in reverse, the plate was then inked and wiped clean leaving the ink within the incised lines. Passing the plate through rollers under pressure 'lifted' the lines onto the paper, so that they stood proud of the surface. This technique had many advantages over the woodcut; the engraver could work much faster and could use many other techniques, dots, pecked lines, stipples and lettering of greater fluency. Changes were also fairly easy to carry out; lines could be burred or hammered out and re-engraved. Copper engraving had the major disadvantage of eventually wearing down so that strengthening of incised lines was needed or even a new plate, as can be seen with some of John Cary's maps. Cary's *New and Correct English Atlas,* 1787 (**51**) was re-engraved in 1809 (**73**), and his *Traveller's Companion* of 1789 (**55**) was re-engraved twice in 1806 and 1822 (**69** and **92**).

It was not until the 1800s that it was possible to engrave on steel. The use of steel allowed a longer print run, hence maps engraved on steel are more common. James Pigot's series for his *Directories* from the mid-1820s were the first to employ this technique in 1829 (**98**) and Henry Fisher's maps were *Engraved on Steel by the Omnigraph, F P Becker* in the early 1840s (**120**).

Acid etching was also introduced in the 1800s. In this technique the design is cut through a wax coating applied to the plate. The plate is then immersed in acid which eats into the exposed copper, creating the etched image. A noticeable difference in the technique is that etched lines tend to end square or blunt, whereas engraved lines taper to a point. Although giving the artist much greater flexibility, the etching technique was not used for any county map series.

However, the biggest breakthrough was the invention of lithography. This was invented in 1798 and patented a year later. Alois Senefelder, a German playwright (1771-1834) who published his own plays, found that by drawing with special greasy ink or crayon on a flat limestone slab the grease was absorbed and the image would then accept printer's ink which was repelled by the rest of the stone, provided the surface was kept moistened. Senefelder went on to experiment with colour and constructed various printing machines. The technique was already being used by the 1820s but was not adopted by the map trade until much later, although some maps were produced at this time. It is surprising how open Alois Senefelder and his brother, Karl Friedrich Matthias Senefelder, were about publicising the new development and encouraging students to come to Munich to learn about the system. However, their reasons were not completely altruistic: Carlos Gimbernat, Deputy Director of the Real Gabinete de Historia Natural, was the first Spaniard to practise the process of lithography. He made contact with Karl Senefelder and for the princely sum of a thousand florins signed a contract on 24 March 1806, whereby the German undertook to teach him everything about lithography but he was not to communicate its secrets to anyone for three months afterwards. As early as 1807 Gimbernat drew on stone the illustrations for *Manual del Soldado español en Alemania* (Munich, 1807), which was printed by F Hubschman. This book included a map of the North Sea coast drawn on stone by Alois Senefelder which, according to Gimbernat, was the first example of lithography for geographical work.[1]

Other early examples of maps are those of Capri and the Pianta del Real Orto Botanicodi Napoli, both from 1818,[2] and possibly the first American lithographed map was published in October 1821.[3] The development of lithography revolutionised the map world enabling quick and accurate reproduction, changes and ultimately cheap and numerous copies. The Germans were quicker to spot the potential of lithography in connection with maps: Senefelder was put in charge of the cadastral surveys undertaken by the Steuer Kataster Commission in Munich between 1809 and 1827. In his *Complete Course of Lithography* (London, 1819) Senefelder wrote about the benefits of lithographic engraving for the production of maps, 'This Manner [drawing on stone] is one of the most useful in Lithography, and is nearly equal to the best copper-plate engraving'.[4] Manuals of the time

1. Jesusa Vega; Lithography and Spain: the difficult beginnings of a new art; *Journal of the Printing Historical Society No.27*; 1998; pp. 33-34.
2. Vladimiro Valerio; Patrelli, Müller and the Officio Topografico: the beginnings of lithography in Naples; in the *J.P.H.S. No.27*; 1998. The maps are illustrated on pages 11 and 17.
3. Philip J Weimerskirch; The Beginnings of Lithography in America; *J.P.H.S. No.27*; 1998. The map is illustrated on page 58. It appeared as a folding plate in *The American Journal of Science* (1822).
4. Vladimiro Valerio; Patrelli, Müller and the Officio Topografico: the beginnings of lithography in Naples; *J.P.H.S. No.27*; 1998, p. 11.

describing the techniques sometimes added maps as examples. Other German agencies were quick to adopt Senefelder's methods.[1]

Charles Joseph Hullmandel (1789-1850) was London's first successful lithographer and he too had learnt the technique from Alois Senefelder. Hullmandel made his own improvements and 'did more than any man to foster lithography in England and particularly the topographical lithograph'.[2] Examples of his work can be found in some early guide books such as W T Moncrieff's *The Visitor's New Guide to the Spa of Leamington Priors* (3rd ed, 1824) or Mary Southall's *A Description of Malvern*, 2nd ed (1825). The first real competition to Hullmandel came from the firm of Day & Haghe, successively lithographers to William IV and to Queen Victoria.[3]

One of the major problems facing the lithographer at the time was the quality of stone available. At first only certain stones could be used but gradually the heavy limestone blocks that were originally used gave way to thin metal plates. These were usually made of zinc (first used in 1824) or aluminium specially prepared with a finely grained porous surface which enabled them to function like the porous limestone. They were lighter and easier to use as well as being cheaper and easier to produce. When a drawing had been completed it was covered with a solution of water and gum arabic. When the gum was washed away sufficient remained in the non-printing area to maintain a clean water-attractive ink-repellant surface. Also the grease penetrated the 'stone' so well that even when the design was apparently washed off sufficient remained to attract a new application of ink. A succession of prints could then be made by alternately dampening the 'stone' and inking the design and rolling the applied paper.

It was further discovered that drawings could be made on paper having a special coating and sufficiently smooth for pens to be used. The drawing could now be transferred to the 'stone'. Fresh impressions from lithographs or from engravings printed in greasy ink onto litho-transfer paper could be transferred. This technique enabled many of the earlier copper engravings to be transferred and subsequently modified. The whole cartographic technique was revolutionised enabling large numbers of maps and variations to be produced. This transfer system was the beginning of off-set printing as we know it today. As mentioned, a number of maps originally engraved were later reproduced as lithographs. Some of Cary's maps first appearing in 1809 and 1822 (**73** and **92**) were later republished by G F Cruchley. Later still G W Bacon and Gall & Inglis both obtained copper plates from auctions of Cary's property and proceeded to produce lithographic copies (see **Appendix II**)

Chromolithography followed, enabling maps to be reproduced in colour. Each colour had a separate plate transferred from the same base plate, with only the colour area being treated. The map would be overprinted taking care for alignment and using the black plate last.

In 1859 a successful method was evolved to enable the reproduction of line drawings by photolithography. Paper coated in gelatine was sensitized by being immersed in a bichromate solution and then dried. The paper was then exposed under a photographic negative until an image appeared. The surface was covered with a film of greasy ink and the paper soaked. Unexposed gelatine areas absorbed the water and were wiped off. The retained inked design was then transferred in the usual way enabling great accuracy. This technique combined with zincography was popular with the Ordnance Survey, and many later copies of the Mudge survey of 1809 (**74**), as well as new maps prepared 1885-1892 (**162, 164** and **169**), were noted as *Zincographed* or *Photozincographed at the Ordnance Survey Office.*

At first printing was done by hand. The paper was laid on to the prepared and inked surface and pressed by a boxwood 'scraper'. In 1860 self-acting presses were introduced. The 'stone' moved to and fro beneath a cylinder and two sets of rollers, one for dampening and the other for inking. The paper rotated with the cylinder pressing it into contact with the plate. This technique, used until *c.*1895, was gradually replaced by the 'direct' and then the 'offset' rotary presses. Here the lithograph is the right way round on the first roller, reversed on the second and printed from the third or 'impression' roller.

Although at first sight similar, the lithograph can be distinguished from an engraving by the absence of raised ink lines and the plate block depression in the paper, as less force was needed to transfer the image. The Reynolds series (**123**) was one of the last to be engraved first and later lithographed. Larger paper copies of the pre-1860 maps sometimes have traces of the impression in the paper near the edges, whereas all later copies are smooth and flat over the whole surface.

1. Ian Mumford; Lithography for Maps; in the *Journal of the Printing Historical Society No.27;* 1998 writes that authors of technical manuals on lithography often used a small map alongside other examples of reproductive capability. Senefelder's manual in both the 1818 German and 1819 (quarto) French edition included an example of a map. The map in the French edition is a negligible example of lithographic engraving of a sparse fictitious map. The German supplement has a more splendid topographic map of the environs of Munich engraved on stone by L. Zertahelly. Quoted from page 70 and see pp71 and 74.
2. M Twyman; *Lithography 1800-1850: the techniques of drawing on stone in England and France and their application in works of topography;* 1970; p. 70.
3. John Vaughan; 1974; p. 111.

Fig. 1: Ordnance Survey Box-set Label

Topography

A map is a topographical drawing and, while being drawn to an accurate horizontal scale, should still show those features which cannot be seen from the vertical viewpoint. From the first the map-maker introduced or copied conventional signs, pictures and symbols and by the nineteenth century these had become firmly established.

While not all maps had the detailed geographical or geological information of Henry De la Beche's work (**118**),[1] later to be included in Ordnance Survey maps, most had the sort of reference material included by Joshua Archer in his series of maps for James Dugdale's *Curiosities of Great Britain.* This included different styles of lettering for various sizes of localities, roads, railways and canals as well as including Polling Places and Parliamentary representation.

Hill shading was, however, still a problem. Saxton's 1575 (**1**) sugar-loaves, shaded on the east side, gave way to alpine mountains for Jansson in 1644 (**11**), clefts for Blome in 1673 (**14**) and joined-in lines for Thomas Kitchin in 1750 (**34**). Benjamin Donn who probably - and correctly - did not like the idea of the multi-molehills omitted them entirely in 1765 (**44, 45**). In 1801 Charles Smith's *New English Atlas* (**61**) introduced hachuring which, when used for larger-scale maps, would become fully conventionalised and improved by the Ordnance Survey. Even so, De la Beche, who worked for the OS before becoming its Director of the Geological Survey, ignored hills while Archer's hills were carefully hachured. Unfortunately, hachuring meant that other information at a hill point would not show up. It was the advent of colour printing that would ultimately lead to the inclusion of hills on most, but certainly not all, maps used for travelling.

Colouring was originally only undertaken at the purchaser's request. For shields standard conventions were used, either letters or hatching or both. For example 0 stood for Gold or Yellow and the engraver would ensure the appropriate area was dotted and the colourist would proceed to colour them correctly. For the map itself there were also standards: boundary lines were shown in different colours on each side, sometimes with areas colour-washed; woods and seats were green; rivers and the sea, or just the sea-coast blue; towns were red. All else was left to the colourist's imagination. These conventions were still being employed until the middle of the nineteenth century. The Archer maps from early editions of *Curiosities* show outline colouring on the engraved issues.

One of the earliest pioneers of lithography also experimented with colouring. Lt. Franz von Hauslab, an instructor at the Engineers Academy in Vienna (from 1819), taught the use of heighted contours as the basis for plan drawing, especially of fortifications drawn at large scales. By 1825 he had developed more elaborate ideas for presenting terrain details at the smaller scales used for topographic maps. His ideas were implemented in a collection of plates reproduced lithographically, some of which are in colour, with the title *Versuche über die Anwendung der Lithographie für die Situations-Zeichnung.* It is not known whether this printed work was intended for publication or whether it was simply a class demonstration of his methods.[2]

1. The Royal Agricultural Society of England included a series of maps at infrequent intervals in its Journal between 1843 and 1870 published by Murray. There might have been the intention to publish a map of each county but Devon and 11 other counties were not included; a map of Exeter was, however, included. The maps are geological in nature showing various rocks and soil strata.
2. This paragraph is adapted from Ian Mumford; Lithography for Maps; J.P.H.S. *No.27*; 1998; p. 69.

It was a very successful early example of the innovative use of colour in lithographic printing, but it went unnoticed and others strove to find a way of printing in colour and of representing heights effectively. One early system for colour printing was patented by George Baxter (1804-1867). He patented his method in 1835, using oil colours and several blocks, and licensed others to use the technique. Although W Dickes, illustrator and publisher who prepared some county maps, used this technique for landscapes, it was never used for mapping. Charles Knight, an influential printer and publisher, attempted to produce the first series of printed books with coloured county maps but only succeeded in producing Berks, Derby and Hants in 1840. The fourth in the series (Kent) was coloured by hand in the time-honoured tradition and no more were produced.

It was the efforts of the family firms such as A & C Black, Philips, W & A K Johnston and the Bartholomews, who dominated the Victorian map market, that led to colour printing in Britain. The Johnston brothers were keen innovators and early exploiters of lithography and chromolithography. Great supporters of thematic maps, they exploited the new technology to produce various maps from one plate. Their *National Atlas of Historical, Commercial and Political Geography* published in 1843 included maps by Heinrich Berghaus (1797-1884) and the work on isotherms of Alexander von Humboldt (1769-1859). It used tonal progression and shading to show the geographical distribution of food plants.[1] The 1849 edition of the *National Atlas* was the first general atlas to be colour printed by lithography and in 1852 Blackwoods published *Johnston's Atlas of General and Descriptive Geography*. This was notable for the use of chromolithography and the twenty two maps were printed from five plates. The firm is also credited with the first London map with contours (1851).[2]

One of the first county atlases to be produced with machine applied colour was George Philips' *County Atlas* which first appeared in 1865, exploiting maps engraved up to three years earlier. This firm led the field in pioneering the mechanical colouring of maps and its huge staff was replaced by the latest Senefelder machines.[3]

John Bartholomew senior (1805-1861) was sceptical of the technique of lithography and it was John Bartholomew II (1831-1893) who introduced it to the firm, which in the 1860s advertised itself as the Edinburgh Engraving & Lithographic Establishment.[4] When the *Encyclopaedia Britannica* county maps first started to appear in 1877 these were produced lithographically with colour printing, first from Bartholomew's presses, later from Johnston's. John junior not only overcame family resistance to introduce lithography he also invented layer colouring, first exhibited in his maps at the Paris Exhibition in 1878. He had not been satisfied with the system of hachuring then being used as it either obscured, or left less room for, other details. Consequently he attempted to improve height representation, first through experimenting with hill shading in brown rather than black and later by a system of layer colouring. The 1 inch Ordnance Survey had introduced contours in 1846[5] but it was John George Bartholomew who was responsible for the refinement of layer colouring and contours as used for the first time in the *Thorough Guides* published by Dulau and Co. and largely written by Baddeley and Ward[6] (**157**): Baddeley's guide to the *English Lake District* (1880) being the first in the series.[7] Consequently, it was not long before Bartholomew was claiming to use 'revolutionary production methods'. Layer colouring[8] was first used for county maps *c.*1895 as seen in Bartholomew's large sheet Devon maps (**174**) and the sectional maps for the Murray guides of Devon (**150.14**).

Colour printing could always be added to maps from plates produced much earlier. Hence, G W Bacon took over the Edward Weller *Dispatch Atlas* plates in 1869 and although these had been used fairly frequently since 1858 it was Bacon who added overprinted colour and successfully exploited the plates until well into the twentieth century. Gall and Inglis carried out similar updating to their stock of Cruchley maps and plates, substituting printed colour for the hand-colouring originally employed and introducing coloured road classification to denote quality of road for cycling or motoring.

Other machines for automation were also invented during this period. F P Becker's famous *Omnigraph* (a versatile ruling machine) was designed to apply commonly occuring symbols onto maps automatically saving the engraver time and effort. 'The ultimate application of this development was undoubtedly Francis Paul's Omnigraph, invented in the early 1840s as a means of speeding-up engraving. The copper plate was passed through a machine comprising a series of punches for each character - which could be impressed into the plate surface using an automatic hammer. Becker sold his 'Omnigraph' to the Ordnance Survey, claiming that it would save £5000 a year on the staff of forty five engravers. It was used until 1875 when the survey decided that

1. Royal Scottish Geographical Society; *The Early Maps of Scotland*; RSGS; Edinburgh; 1973; p.132.
2. D Smith; The Cartography of W & A K Johnston; IMCoS Journal 82; 2000.
3. D Smith; Map Publishers of Victorian Britain - The Philip Family Firm; *The Map Collector* 38; 1987.
4. L Gardiner; 1976.
5. Lancashire was the first county 'contoured'
6. L Gardiner; 1976; p.32.
7. Some of the series was still being offered for sale by Ward & Lock in the 1950s!
8. This was first used in Britain in 1845 by Thomas Larcon, but had been used earlier in Germany (see above).

it was actually more costly than hand production because of the great amount of manual retouching required by 'Omnigraph'-produced signs'.[1]

Cassell, Petter and Galpin, better known for their publishing were also innovators. They not only printed but also manufactured their own paper and produced printing machinery. One of their Belle Sauvage staff, Samuel Bremner, designed a horizontal, single-cylinder gripper, one-and-two-colour, two-feeder printing machine. This won honourable mention at the 1862 exhibition and was popular for some time after. It claimed to print eight hundred sheets an hour by hand or two thousand by steam if it could be fed fast enough.[2]

Railways

The earliest railway in Devon was probably John Smeaton's track in Mill Bay, Plymouth (1756-9). Too short to appear on any map, it was fully described in Smeaton's book of the Eddystone.[3] He devised his 'rail road' to facilitate moving the one-ton blocks from mason to mason, then to the test bed and on to the boats. It was a simple narrow gauge track based on the colliery designs of the time and presumably used man or horse power for propulsion.

John Rennie, the well-known civil engineer, used a 3ft 6in railway when he was constructing the Plymouth Breakwater (1812-1844). The track was used to bring stone from the quarry at Pomphlett down to the quay at Oreston, all three and a half million tons of it.[4]

The first tramway was on the 'inclined plane' from the Tavistock canal to Morwellham opened in 1817. It was worked by a water wheel supplied from the canal itself. Morwellham Quay largely served the Devon Consols mine. The area both east and west of the Tamar at this point was to become the most important mining area in the United Kingdom and for some minerals the world's leading source. By 1900 some seven hundred and fifty thousand tons of copper alone had been exported through Morwellham Quay from the Devon Consols Mine.[5]

The first real 'line' was the Heytor Granite Tramway which opened in 1820 and connected the quarries at Heytor to the Stover canal at Ventiford. The track was constructed with two granite 'rails' which lay inside the wagon's wheels preventing the horses from deviating left or right. Trains were made up and were pulled by as many as twelve horses. The tramway was shown on many county maps, for example Greenwood in 1827 and Archer/Dugdale in 1842. The line closed in 1858 and although then omitted from most maps the hard granite tracks can still be seen today.[6]

In 1820 William Stuart, then the Admiralty engineer with Rennie at Plymouth, was appointed as the part time engineer for the *Plymouth & Dartmoor Railway*.[7] This second tramway ran from Princetown to Plymouth and was incorporated in an act of 1819 for the line as far as Crabtree, and again in 1820 extending the line to Sutton Pool. It was opened in September 1823 and was the first line to use the 4ft 6in., or Dartmoor, gauge to be followed by the Lee Moor line in 1854. The line was the brainchild of Sir Thomas Tyrwhitt, an eccentric yet successful politician, who lived at Tor Royal, near Princetown. He published a pamphlet in 1818 in an attempt to persuade the Plymouth *Chamber of Commerce* to join him and construct the new line. He believed passionately that his railroad would 'gratify the lover of his country; reward the capitalist; promote agricultural, mechanic and commercial arts'. Although he spent large sums to improve the moor the only real use made of the line was to export granite from the quarries, near Princetown, down to the Cattewater. A portion of the line was on a gentle downhill slope where the full waggons ran under their own weight, otherwise the locomotion was provided by horses. The line was first shown on the Greenwoods' map of 1827 (**96**) along with Heytor and was usually shown on the later county maps. At first it had nothing to do with the prison which had been closed in 1816 at the end of the war, and was not re-opened until 1850. The route above Yelverton was bought by the GWR and redeveloped as a separate line in 1883. It was on this line that in 1891, the year of the great blizzard, a train was stranded for two long nights. The lower length was used occasionally until 1900. A branch line was added in 1833 to run to the blue slate quarry in Cann Wood owned by the Earl of Morley, in return for his earlier agreement for the Plymouth & Dartmoor line. To facilitate the development of the china clay works at Lee Moor a further branch line was proposed and this was built as far as Plympton in 1834. However, local opposition stopped further progress and it was not until 1854 that the Lee Moor railway was built from the clay

1. Quoted by David Smith; 1985; p. 17.
2. S Nowell-Smith; 1958; p. 73 and the machine is illustrated on page 74.
3. J Smeaton; *A Narrative of the building of the Eddystone Lighthouse;* London; 1791.
4. Martin Smith; 1993; p. 105.
5. F Booker; 1967 (1974).
6. Helen Harris; *The Haytor Tramway & Stover Canal;* Peninsula Press; 1994.
7. David St John Thomas; *Regional History of the Railway -The West Country;* David & Charles; 1960.

works to the branch line at Cann Wood. The line again used horses but included two incline systems. In 1899 steam locomotives replaced the horses at the Moor end.[1]

In 1831 the merchants of Bideford and Okehampton raised funds for a report on the feasibility of a railway between the two towns. In October Roger Hopkins completed a survey and produced plans, books of reference, costs and a full report for lines from Okehampton to Torrington and Bideford. His scheme proposed a narrow gauge track and the use of two steam engines. This would have been the third line in the country, following the diminutive Stockton & Darlington and the Manchester-Liverpool lines of George Stephenson in 1830. In 1831 a House of Commons Committee reported in favour of the use of steam engines on roads and completely ignored the railway. The Bideford to Okehampton line was not to be and by 1836 some of its promoters were diverted to 'Stephenson's line', the L&SWR line from a point between Basingstoke and Winchester to Exeter and thence to Plymouth and Falmouth. Part of the Hopkins' scheme was reintroduced in the proposals of 1845 when his sons proposed a line from Tavistock (the South Devon Railway from Plymouth) to Okehampton and then with a branch to both Bideford and Crediton (the Exeter & Crediton railway). But this too failed when the South Devon Railway decided not to extend their line to Tavistock.[2] Although these schemes always included full drawings the lines apparently did not appear on any county map, probably because no act was proposed in parliament.

The development of steam railways in the West Country almost exactly coincides with the reign of Queen Victoria. The line to Bristol was operational in 1841 and by 1901 only four small branch lines were still to be opened in Devon. In just two generations the life of the county was altered and the effect of the railway on travel to and within the West Country was dramatic.

In 1780 it took nearly two days to travel from Exeter to London. By 1826, with the advent of the post roads, the stage coach *Telegraph* took seventeen hours; and the *Quicksilver* sixteen and a half hours on its way to Plymouth, which was reached some four and three quarter hours later.[3] There were only two fast coaches a day and each could carry only ten passengers, four of whom were outside!

But with the coming of the railway travel changed for ever. On 1 May 1844 the first through train reached Exeter. 'We had a special train with a large party from London to go down to the opening. A great dinner was given in the Goods shed at Exeter Station. I worked the train with *Actaeon* engine, one of our 7-ft class, with six carriages. We left Exeter at 5.20 pm and stopped at Paddington platform at 10. Sir Thomas Acland, who was with us, went at once to House of Commons, and by 10.30 got up and told the House he had been in Exeter at 5.20.'[4]

The most dramatic account of the change comes from the notes in Cecil Torr's book of Wreyland; 'On 19 March 1841 my father started from Piccadilly in the *Defiance* coach at half past four, stopped at Andover for supper and at Ilminster for breakfast, and reached Exeter at half past ten' (eighteen hours later). 'On 10 October 1842 he started from Paddington by the mail train at 8.55 pm reached Taunton at 2.55 am and came on by the mail coach stopping at Exeter at 6.15' (nine hours and forty minutes later). '20 March 1845 coming on the same train he reached Exeter at 4.25 (seven and a half hours later). On 8 August 1846 he came by the express train, 9.45 am to 2.15pm' (four and a half hours later).[5]

However, not everyone was so enthusiastic about the advance of the railways. William Collins, the publisher, is known to have chaired a demonstration in Glasgow in November, 1840[6] protesting against Sunday trains. Many landowners were opponents of railway development. Lord Fortescue objected to a branch from Umberleigh Bridge to meet the Taunton-Barnstaple line at South Molton and the line was never built. Sir William Williams objected that the Ilfracombe line, passing between his house and the river, would damage his property and it was not until 1870 that he withdrew his objections and a year later the line was built.[7] Hutchings, writing about Barnstaple, complained that: 'Just below the hideous bridge which carries the South Western line across the Taw is the Quay ... and on the right bank ... is the North Walk, now unhappily cut up for the purposes of the new railway from Lynton'.[8] But perhaps the worst comment was written by J Lloyd Page: 'Within the last four or five years the repose of this valley has been broken by the locomotive ... He is noisy; he is obtrusive. Where he gets his iron foot romance dies. And from Dulverton to Exeter he has spoilt the Exe Valley'.[9]

1. Bryan Gibson; Plymouth Railway Circle; 1997
2. Sir R. Lethbride; *The Bideford & Okehampton Railway;* Devon Association transactions XXXIV-1902.
3. C G Harper; *The Exeter Road*; Chapman & Hall; 1899.
4. Daniel Gooch, later knighted, a locomotive engineer and designer who joined Brunel in 1837. He also wrote in his diary that his back was so sore from working on the footplate that he could hardly walk the next day.
5. Cecil Torr; *Small Talk at Wreyland 1st series;* Cambridge; 1918.
6. David Keir; 1952; p. 137.
7. Victor Thompson; 1983; pp. 18 and 30. The South Molton line was shown on Archer/Dugdale (**119.5**).
8. W W Hutchings; in *The Rivers of Great Britain*. See entry **176**.
9. J Lloyd Page; in *The Rivers Of Devon*. See entry **170**.

But the railway was not to be stopped. The first proposals for a line from London to Bristol had been made in 1824 yet little was done until March 1833 when Isambard Kingdom Brunel was appointed. He not only surveyed the route, he designed the line. Some idea of the undertaking can be appreciated from his diary and correspondence. His own duty of superintendence severely taxed his great powers of work. He spent several weeks travelling from place to place by night and riding about the country by day, directing his assistants and endeavouring, very frequently without success, to conciliate the landowners in whose properties he proposed to trespass. His diary shows that when he halted at an inn for the night little time was spent in rest; he often sat up writing letters and reports until it was almost time for his horse to come round to take him for the day's work. 'Between ourselves', he wrote to Hammond his assistant, 'it is harder work than I like. I'm rarely much under 20 hours a day at it'.[1] A great example of his surveying ability must be the Box Tunnel. This is straight for 3193 yards at a constant fall of 1:100 to the west and when clear of smoke one can see through the entire length. It is said that on or about 9th April the sun is visible from the west end before it rises over Box Hill.

The costs were large, not only the compensations for land and the actual construction of line and rolling stock but also for the bills to pass through parliament. The first Bill for the *Great Western Railway* approved by Parliament cost the company £87,197 or about £775 per mile of line constructed.[2]

Initially there was also the problem of time. The GWR timetable of 30 July 1841 contained the disconcerting statement that 'London time is about four minutes earlier than Reading time, seven and a half minutes earlier than Cirencester and fourteen minutes before Bridgewater'. Although by 1848 most lines had adopted Greenwich time,[3] this was very much a local problem and was not finally resolved until 1852 when Greenwich time was adopted in the West of England. The Mayor of Exeter issued a notice on the 28th October 'That upon and after Tuesday the Second Day of November next the Cathedral Clock, and other Public clocks in the city, will be set to and indicate Greenwich Time'.

The GWR scheme was completed, approved by parliament and received Royal Assent in 1835. The actual construction was much quicker and by then Brunel was also working for the *Bristol & Exeter Railway.* In 1841 the line was open as far as Bridgewater.[4] By the 1st of May 1843 it had reached Beambridge, just west of Wellington, then through the Whiteball Tunnel into Devon and, exactly a year later, Exeter: 'The opening day was kept as a general holiday, all business was suspended, a splendid *dejeuner a la fourchette* took place at the Railway Station, and vast numbers from all parts of the County came to witness the arrival of the first day's train'.[5] A local newspaper journalist was even more enthusiastic: 'intercourse with the more populous districts of England cannot but prove highly advantageous to the fair and lovely spinsters of Devon'.[6]

Two routes were suggested for the line to Plymouth; the coastal line through the South Hams; the inland route through Crediton and Tavistock. In 1840 Plymouth merchants proposed a third line, a high-moor railway through Dunsford, Chagford and Yelverton.[7] The route was surveyed by Nathaniel Beardmore, an apprentice of J M Rendel, and a final route plan was prepared.[8] The estimated cost was £770,780 including the branch to Tavistock, a saving, so it was claimed, of £1,000,000 over the coastal route. But it was too late. Business interests in both cities were concerned over the *waste land* and that mines and quarries would not provide sufficient income.[9]

As early as 1836 Brunel had surveyed the route to Plymouth and the *South Devon Railway* bill received Royal Assent in July 1844. In 1846 trains ran to Teignmouth and the sea. So great was the excitement on that Whit weekend when Exeter went to the coast, that 'upwards of 1,500 persons went down and at Dawlish and Teignmouth bands of music were stationed and flags flew from every tower'.[10]

The route to Plymouth was designed to use the newly patented atmospheric power.[11] Stationary engines pumped the air out of an iron pipe laid between the rails, creating a vacuum ahead of the carriage's piston which was propelled by the air behind it. It was claimed to be cheaper and quieter to operate with heavier trains. The first atmospheric train to Teignmouth ran in September 1847 and to Newton Abbot four months later. 'Many prefer the noiseless track to the long drawn sighs of Puffing Billy'.[12] After the first teething troubles were overcome the system ran well but sadly the valves proved difficult to maintain. The scheme was abandoned in September 1848 and the route changed back to using ordinary steam locomotives.

1. Isabel Brunel; *Life of I K Brunel*; Longmans Green; 1870.
2. J S Jeans; *Railway Problems* .
3. P S Bagwell; *The Transport Revolution from 1770;* 1974.
4. The journey took 4 hours to Bristol and 1 1/2 more to Bridgewater.
5. Alfred Wheaton; *The Handbook of Exeter;* (1845).
6. Thomas Latimer writing in *The Western Times.*
7. R S Gregory; *South Devon Railway.*
8. Deposited Plan 148 - Devon Records Office.
9. J Dilley; *The Devon Historian;* Oct. 1988.
10. Reported in the *Exeter Flying Post.*
11. In 1838 by Messrs Clegg, Samuda & Samuda.
12. Newspaper article of the time.

The line had reached Totnes by February 1848 and to Laira, on the outskirts of Plymouth, by the 5th of May. On that Friday 'By common consent all business in Plymouth was suspended, banks, shops, and all other places of trade, with but very few exceptions, being closed ... and tens of thousands of persons assembled to greet the arrival of the opening train, gaily decorated with flags, (which had) completed the distance from Totnes in 42 minutes.'[1]

The line from Newton Abbot to Torre followed in December of the same year but work on the branch line was slow. It took eleven years to reach Torquay and Paignton, where a party was held with a pudding weighing over a ton, and it took a further five years to reach Kingswear, opposite Dartmouth. Although South Devon was open it was not until May 1859, just four months before his death at the age of 53, that Brunel completed the Royal Albert Bridge, and the railway line entered Cornwall. Nevertheless, the south coast was quick to exploit the fact that two main lines came into the county. Exmouth trains ran from 1861 and the neighbouring towns of Seaton and Sidmouth saw their first trains in 1868 and 1874 respectively.

At the same time branch lines continued to be built and to be absorbed. Most of these lines were built by local private subscription and were run as separate companies with the South Devon Railway. There was great hope that the lines would open up the county for both passengers and trade. The *Torbay and Brixham branch* was opened in 1868 to the benefit of the fishing fleet. The south of Dartmoor was opened up to much acclaim. High days and holidays greeted the first trains with celebrations and even tea for the one thousand five hundred residents of Ashburton. The twelve mile line to Moreton Hampstead was opened in July 1866, but not without troubles. The engines on this branch were quite unequal to their work, and there were then no effective brakes. Coming down the incline the trains often passed the station and passengers had to walk from where the train stopped. The last branch of the *South Devon Railway* was the Totnes-Ashburton line which opened in 1872. This nine and a half mile stretch followed the river (whereas an earlier suggestion - directly from Newton - would have involved some very steep gradients). Although the once flourishing wool industry was in decline the area was responsible for half of Devon's serge trade and Buckfastleigh station was often busier than Newton Abbot.[2] Plans were made for a line from Totnes to Exeter via Ashton and the first stretch was completed in 1882, providing access to the Teign House Races at Christow. It was not until 1903 that the line finally reached Exeter.

North Devon was not so fortunate. Though the *Taw Vale Railway & Dock Company* was incorporated in 1838, to run a line from Fremington to Barnstaple, nothing happened and in 1845 the extension to Crediton was abandoned as a result of the controversy over the gauge.

This problem affected all the early lines in Devon. Brunel proposed the famous broad-gauge of 7' 0¼" for the *Great Western Railway* and this was adopted by all the early west country companies including the line from Exeter to Crediton.[3] Most other railways in Britain, including the *London & South Western Railway* were using the narrow gauge of 4' 8½" and this was favoured by the *Taw Vale.*

The two track systems were very different in their construction. The narrow gauge had 'conventional' cross sleepers while the broad gauge used longitudinal timbers. Brunel had designed for speed and for comfort. By 1845 the London train to Exeter (194 miles) took only four and a half hours and the luxurious, ten feet by seven feet, first class compartments were designed for only eight passengers. But it was not until the L&SWR pushed westward that the problems arose, and nowhere was the conflict more real and more damaging than on the line from Exeter to Barnstaple.

In 1845 the Exeter and Crediton line was incorporated and by 1847 their broad gauge line was almost ready. But also in 1845 the Taw Vale line from Fremington to Barnstaple, enacted in 1838 proposed, with *London & South Western Railway* interests, an extension to Crediton and in August 1847 the line obtained approval but only with broad gauge support. However, by then the L&SWR had bought considerable stock in both lines and their nominees effectively stopped the Crediton line from opening; moreover they proceeded to narrow the track leaving the broad lines to rust.

In February 1848 the Railway Commissioners ruled that the line from Barnstaple to Crediton, the *Taw Vale Extension,* should be broad. The L&SWR cancelled the opening and halted the construction of the new line. In 1851 the *North Devon Railway* took over both the Crediton and the Taw Vale lines and leased them to the *Bristol and Exeter.* The Crediton line was re-converted back to broad gauge and was opened in May. Work northward commenced in February 1852. Finally, regular service to Barnstaple commenced in August 1854, but the extension to Bideford beyond Fremington was only opened to passengers in November 1855. The time, the cost, and the frustration were enormous. The north of the county had remained virtually isolated while to the south trains had reached Plymouth six years earlier.

1. Newspaper article of the time.
2. Martin Smith; 1993; p. 31.
3. The Exeter and Crediton Act of 1845.

The gauge dispute continued until the end of the century,[1] aggravated in Devon with the arrival of the London & South Western Railway, whose narrow-gauge lines from Salisbury to Exeter, either by a central or coastal route, received assent in July 1848. After many years of confusion and debate the Salisbury-Yeovil line was agreed and the line to Exeter was opened in July 1860. The Exeter/Crediton line, the Bideford Extension and the North Devon leases were all taken over in 1862 and amalgamated with the L&SWR in 1865. The gauge of this line was mixed and broad-gauge goods trains still visited Bideford as late as 1872.

The county was in a sense split. The North East and South West were developed by the *Great Western Railway* (God's Wonderful Railway) and the North West and South East by the L&SWR, crossing at Exeter. There were two anomalies. Both companies were to construct lines to Launceston, the GWR from Plymouth via Tavistock and Lydford and the L&SWR from Halwill Junction, and both had lines to Plymouth when the L&SWR extended its line from Okehampton through Tavistock and Bere Alston. The Plymouth line introduced a change-over in the town with a stretch of mixed gauge track and a new station at North Road. Shortly after this the GWR took over the SDR and dissolved it altogether in 1878. There were also two other variations. Brunel's philosophy for the GWR was for directness. If a town was not in line, so be it. Tiverton was an early example with the town served by the Tiverton Road Station five miles away. On the other hand, the L&SWR were content to have branch lines totally isolated from the main system. A good example was the Bodmin & Wadebridge line in Cornwall which was taken over in 1846, fifty years before it was connected.

By 1865 Devon and the West Country were fully linked, if confusedly, with the rest of England. Tourist tickets from London to the north coast resorts were issued in two parts: passengers changed to horse-drawn coaches at Taunton for Williton, Lynmouth, Lynton and Ilfracombe. The tickets included the coach fares and all fees to Coachmen and Guards. First class tickets entitled the holders to either inside or outside places, but second class were for outside only. Lynton coaches left Williton station (just south of Watchet) daily, except on Sundays, at 2.50 pm after the arrival of the 9.15 am Express from Paddington. Passengers for Ilfracombe broke their journey at Lynton to continue the next day at 7.00 am.

Trains ran regularly from Paddington to Exeter (four hours and forty five minutes); Torquay (six hours and thirty eight minutes); Kingswear (seven hours and twelve minutes); Plymouth (seven and a half hours); Barnstaple (eight hours and thirteen minutes). Journey times up or down were much the same. Fares were cheap compared to the £3 single to Exeter on the *Defiance* coach: Paddington to Lynton cost 65s & 46s (1st & 2nd return); Ilfracombe 79s & 56s; Torquay 50s & 37s; Plymouth 58s & 42s.[2]

Times improved and in 1887 the *Jubilee Express* took only six hours and fifty minutes to reach Plymouth. Service facilities took far longer to improve. Until 1882 there were no toilet facilities on any trains in the country, nor dining arragements until 1893, and it was not until 1887 that a West Country train carried third class passengers.

For a long time the north west of Devon remained untouched by the railway although the Bideford to Torrington stretch had been completed by 18th July 1872. Before the Exeter-Barnstaple line was started the *Taw Vale Railway Company* had built a stretch for horse wagons from Fremington to Barnstaple to bring cargoes to the railway station (1848),[3] but few people. In 1880 the *North Devon Clay Company* laid a three foot gauge from Marland to Torrington for the goods and workers of the mineral companies at Marland and Meeth, but it would be the twentieth century before a passenger line would be completed to Halwill. In 1879 a line was built from Meldon, near Okehampton, through Halwill to reach Holsworthy. Here for years it rested, while the company decided where to go next. Then instead of turning north to Torrington the line headed west to the sea and Bude, being opened in August 1898. In the meantime the L&SWR completed their connection from Launceston to Halwill in 1886.

The *Somerset and Devon Railway* (known as the Slow and Dirty) proposed a shorter link from Taunton to Barnstaple which the L&SWR instantly opposed. However, the first turf was lifted at South Molton in 1864 and by 1866 the section to Barnstaple was making good progress. The line opened in sections (Norton to Wiveliscombe in June 1871) but it was not until August 1873 that directors were able to travel on the first train from Bristol to Barnstaple via the new line.[4] Planners obviously saw the chance to provide Tiverton with a through railway at last, and during the years 1884-85 a line connecting Exeter with Tiverton and then Morebath (Bampton) was opened in two stages. Also in the east the branch line to Hemyock was built by the *Culm Valley Light Railway* to serve Coldharbour Halt, Uffculme, Culmstock Halt, Whitehall Halt and Hemyock, and opened 29th May 1876.

One of the last lines to be sponsored by the two railway giants was the second Tavistock to Plymouth line. The

1. The GWR abandoned the broad-gauge for all main lines on 20th May 1892.
2. *Time Tables Great Western and other railways in connection Bristol And Exeter Railway To Plymouth; South Devon Railway To Plymouth; Cornwall Railway To Falmouth; West Cornwall Railway To Penzance*. June 1865. Price One Penny. London: Henry Tuck. f/s 1971.
3. Victor Thompson; 1983; p. 36.
4. Victor Thompson; 1983; pp. 30-33.

L&SWR had been paying high fees to use the GWR line linking the two lines. They were reluctant to build the new line themselves but they provided the funds to the *Plymouth, Devon and South Western Junction Railway* to build and operate it. Opened in June 1890, it connected Tavistock with Devonport via Bere Alston and, at last, the L&SWR had their own route linking London to Plymouth. As a consequence it was in a good position to steal the lucrative fruit and flower trade of the Tamar valley from the GWR. The line improved official L&SWR times from London to Plymouth by thirty to forty minutes for the 229 mile line: but as the trains were frequently late, this was mainly academic.[1]

One of the later small companies was the *Lynton & Barnstaple.* It used the narrowest gauge in the county for passenger trains, a mere one foot and eleven and a half inches, and was sponsored almost entirely by local people. Sir George Newnes, head of a famous publishing house (he published Bartholomew's *Royal Atlas c.*1900) was one of the most supportive advocates of the new line. It was his wife, Lady Newnes, who cut the first turf at the projected site of Lynton station in September 1895.[2] The line was not a success, partly due to the terminus at Lynton being several hundred feet above the village, and it was sold at a loss to the *Southern Railway* in 1921.

By the beginning of the twentieth century the network was almost complete. In the south the long-awaited route to Kingsbridge was opened in 1894 - it had been anticipated from at least 1872 - and in 1898 trains ran from Plymouth to Yealmpton. On the North coast the Westward Ho! line was opened in 1901 although it had been planned as early as 1860 and in the centre Bude was finally linked to Holsworthy.

In 1903 lines connecting Exmouth to Budleigh Salterton, Axminster to Lyme Regis and the Ashton to Exeter route were completed. In 1908 the Westward Ho! line was extended to Appledore, only to be closed in 1917. This line was not connected to the main Bideford line - one walked over the bridge. The only time that trains crossed the river was in 1917 after the closure, when rails were laid down to transport the rolling stock to its new home. This railway had a remarkable time-table stating that 'the published time tables ... are only intended to fix the time before which the train will not start, and the Company do not undertake that the trains shall start or arrive at the times specified'.[3] When in 1915 the army in France began to construct narrow gauge supply lines the tracks of the Appledore railway were taken up and shipped out of Portishead to be sunk by a U-boat off Padstow.[4]

The year 1908 also saw the line from Bere Alston to Callington officially opened on 2nd March.[5] The Callington-Calstock Railway was founded in 1869, became the *East Cornwall Mineral Railway* two years later, and began goods traffic on 7th May 1872. This connected Kelly Bray (just north of Kit Hill and Callington) and Gunnislake with Calstock on the east bank of the Tamar and was vital to the mineral trade in the area. From *c.*1844 to 1870 both banks of the river experienced a major boom in copper production especially through Morwellham Quay. The value of ore shipped out, both copper and the seventy two thousand tons of arsenic, which took over in importance from 1860 to 1890, is estimated to have been three and a half million pounds. The lodes extended east and west of the Tamar and suggestions were constantly put forward that the ECMR be linked with the *Plymouth, Devonport and South Western Junction Railway* (the L&SWR subsidiary responsible for the Plymouth-Tavistock link) but by the time the lines were connected via an impressive viaduct the mineral trade had disappeared. Although copper and arsenic remained the chief exports of the area a number of other minerals [6] were mined and quarried including small amounts of gold.[7]

In 1925 the last line was built, Torrington to Halwill Junction, though projected routes for this line had appeared on maps at least 60 years before a rail was laid! Bacon's *County Atlas* of 1869 seems to have been the first to indicate a line south of Torrington. This was soon followed by representations of lines on Heydon's large map of 1872 and on Philips *County Atlas* map of 1874. The Heydon map, produced by Bartholomew, was perhaps the most influential, being used by W H Smith, Black's guides and Murray's handbooks. It showed the completed line to Bideford and a proposed line from Bideford to Torrington. From there the line was drawn continuing to Little Torrington, skirting Hatherleigh to the southwest before travelling eastwards to meet the (projected) L&SWR line Exeter to Tavistock at Greenslade, just south of Sampford Courtenay. A branch line runs from Hele Bridge (just north of Hatherleigh) due west to Holsworthy and thence to Bude. James Jervis produced a map of the proposed railway *c.*1897 which varies from Heydon's in that the line continued south after Hatherleigh, linking with the L&SWR line at Okehampton (and ignoring the branch to Holsworthy).

Colonel H Stephens planned the *Devon and Cornwall Junction Light Railway* to bridge the Torrington gap and

1. F Booker; 1967 (1974); pp. 204-209.
2. Victor Thompson; 1983; p. 42. Sir George Newnes also supported and financed the Lynton Cliff Railway.
3. Martin Smith; 1993. p. 64.
4. Victor Thompson; 1983; p. 55.
5. The following account is taken from F Booker; 1974; especially pp. 22-23.
6. Besides copper and arsenic, tin, silver, lead, wolfram, and pyrites were extracted. The line was also used for the transport of bricks, tiles and granite.
7. In 1999 exploration took place at a recently discovered site in Devon to ascertain whether gold could be mined in viable quantities. See Andrew Mosley; The Quest for Gold; in *Devon Today*; Devonshire Press; Torquay; January 1999.

partly follow the track of the old Marland light railway; indeed the prospect of carrying clay, cattle and Welsh coal as well as tourists was a reason for its construction.[1] As built the line varied slightly as it meandered from Petrockstow to Meeth (Halt) before arriving in Hatherleigh. From there it progressed in a south-westerly direction to Hole before joining the Okehampton-Bude line at Halwill Junction. It finally opened on 27th July 1925.

The railway network in Devon was complete to operate as such for only thirty eight years when the Beeching plan closed many of the branch lines. During the 1960s alone Devon lost some forty stations from its total of nearly one hundred and twenty; a loss of one-third!

In addition to the established lines other railroads and tramways, operated by independent companies, appear on some maps: the Zeal Tor Tramway to Shipley Bridge (1854-1890), and the Lee Moor Tramway to Plymouth (1854-1960), both built for the extraction of minerals from Dartmoor (shown for example on the Bartholomew/Black maps *c.*1862). There are also others which did not appear such as the Great Consols line west of Tavistock, which ended in the inclined plane down to the quays at Morwellham on the Tamar. This first operated in 1856 and was fully completed in 1859. The line included a loop to Wheal Emma (named in honour of the widow of William Morris, who had been one of the main shareholders).[2] Another line which did not appear on county maps was the Exeter Railway, though incorporated in 1883, it was only completed from Teign House to Exeter in 1903.

The progress of the railways is reflected in the county maps but to use them as a guide to the dates of various states requires care. Some publishers noted the applications to Parliament for new lines and showed them falsely in anticipation. Examples of these are the line from Newton Abbot directly to Ashburton, and not via Totnes as it was later built (Archer/Dugdale 1858, Wood 1855); the incorrect eastern route from Barnstaple to Ilfracombe (Besley 1866, Murray 1865, Bartholomew 1878); a line from the south west to South Molton (Archer/Dugdale); and the line right round the town of Tiverton (Stanford 1878).

As a good example of the confusion one need look no further than the Archer/Dugdale series (**119**). The Bristol & Exeter Railroad to Exeter, appears on maps in editions usually dated *c.*1842, although it was not to be opened until 1844. In amended maps, published *c.*1850, the South Devon Railway is shown to Devonport, the Taw Vale Railway to Barnstaple, the Exmouth Railway to Exmouth and the Dartmouth line to Kingswear, to be opened respectively in 1848, 1854, 1861 and 1864. Another, later, 1850 state showed lines to South Molton from Umberleigh Bridge, and to Ashburton from Newton Bushell, neither of which was built and both were to be erased on later editions. The surveyor or engraver was probably aware of the various Acts of Parliament or delays or cancellation due to lack of funding and adjusted the maps accordingly. Another example was the line to Kingswear which was taken to Torre in 1848, then suffered hold-ups during the building of the Greenway Tunnel only reaching Paignton in 1859, Churston in 1861, Kingswear in 1864 and with a spur to Brixham built in 1868. To some extent the states reflected the position. For example, in 1858 the Dartmouth line was shortened to Torquay, with roads and estuary redrawn and the Credition-Bideford line was erased only to be reinstated to Barnstaple two years later. The final state of the map, a later 1860 state, showed the line to Tavistock and the line into Cornwall, both opened in 1859, and the removal of uncompleted lines. Joseph Archer had state and date at last in concert. Look also at Black's 1855 map which incorrectly shows the line from Exeter to Exmouth not to be completed until 1861 and also shows two lines from Exeter to Crediton, the projected line never to be built (**130**)

The Becker/Kelly series (**132**) is a good example of erroneous railway information. The early maps showed a line to Ashburton via Newton Abbot, but later states corrected this to the correct line via Totnes. Two Somerset lines, never completed, were shown from Bridgewater to Stolford in 1860 and Watchet to Dulverton in 1873.

But perhaps the most interesting example of confusion was the line from Barnstaple to Ilfracombe. The Besley map of North Devon was first issued in 1857 (**134**) and showed with a pecked line the mapmaker's expected route. This showed the line running north through Pilton and Bittadon to just south of Berry Narbor before approaching Ilfracombe from the east. A later state shows the line completed but still wrong. It is now shown running west along the estuary to Heanton Punchardon before turning north to Wrafton and Braunton and passing two miles east of Mortehoe to approach Ilfracombe from the west! Traces of the earlier proposed line can still be seen on this state. Finally we have the actual line through Wrafton which stopped at Braunton and with a noticeable S-bend ran into Mortehoe before reaching Ilfracombe.

Later editions of Philip's *Handy Atlas* have an interesting alteration to the Kingsbridge line. Although not completed until 1894 the line was shown in 1873 as closely following the river and it was not until 1895 that the line was redrawn correctly, some distance east of the river, and to include Gara Bridge and Doddiswell stations.

As an aid to dating the various states the map and following table show the lines, the companies and their opening dates.

1. Victor Thompson; 1983; p. 37.
2. D B Barton; *The Mines of East Cornwall and West Devon*; Truro;1964.

No.	Year	Line
1	1843	Taunton/Beambridge
	1844	Beambridge/ Exeter
	1846	Exeter/Newton Abbot
	1847	Newton Abbot/Totnes
	1848	Totnes/Plymouth
	1859	Plymouth/Saltash
2	1848	Tiverton Jn./Tiverton
2a	1876	Tiverton Jn./Hemyock
3	1848	Newton Abbot/Torre
	1859	Torre/Paignton
	1861	Paignton/Churston
	1864	Churston/Kingswear
3a	1868	Churston/Brixham
4	1851	Exeter/Crediton
	1854	Crediton/Barnstaple
	1855	Barnstaple/Bideford
	1872	Bideford/Torrington
4a	1925	Torrington/Halwill
5	1859	Plymouth/Tavistock
	1865	Lydford/Launceston
6	1860	Axminster/Exeter (L&SWR)
7	1861	Exeter/Exmouth
8	1865	Crediton/North Tawton
	1867	N.Tawton/Okehampton
	1871	Okehampton
	1874	Okehampton/Lydford
	1876	Lydford/Tavistock
	1890	Tavistock/Plymouth (L&SWR)
9	1866	Newton Abbot/Moreton Hampstead
10	1868	Seaton Jn./Seaton
11	1872	Totnes/Ashburton
12	1873	Barnstaple/Dulverton/Taunton
13	1874	Barnstaple/Ilfracombe
14	1874	Sidmouth Jn /Sidmouth
14a	1897	Sidmouth /Budleigh Salterton
15	1879	Halwill Jn./Holsworthy
	1886	Launceston/Halwill Jn.
	1898	Holsworthy/Bude
16	1882	Heathfield/Ashton
16a	1903	Ashton/Exeter
17	1883	Yelverton/Princetown
18	1885	Exeter/Tiverton
	1886	Tiverton/Bampton
19	1890	Bere Alston/Callington
20	1891	Plymouth/Millbay
22	1892	Plymouth/Plymstock
	1898	Plymstock/Yealmpton
23	1893	Brent/Kingsbridge
24	1897	Plymouth/Turnchapel
25	1898	Barnstaple/Lynton
26	1901	Bideford/Westward Ho!
26a	1908	Northam/Appledore
27	1903	Axminster/Lyme Regis

Table 1. *Showing development of the railway lines*

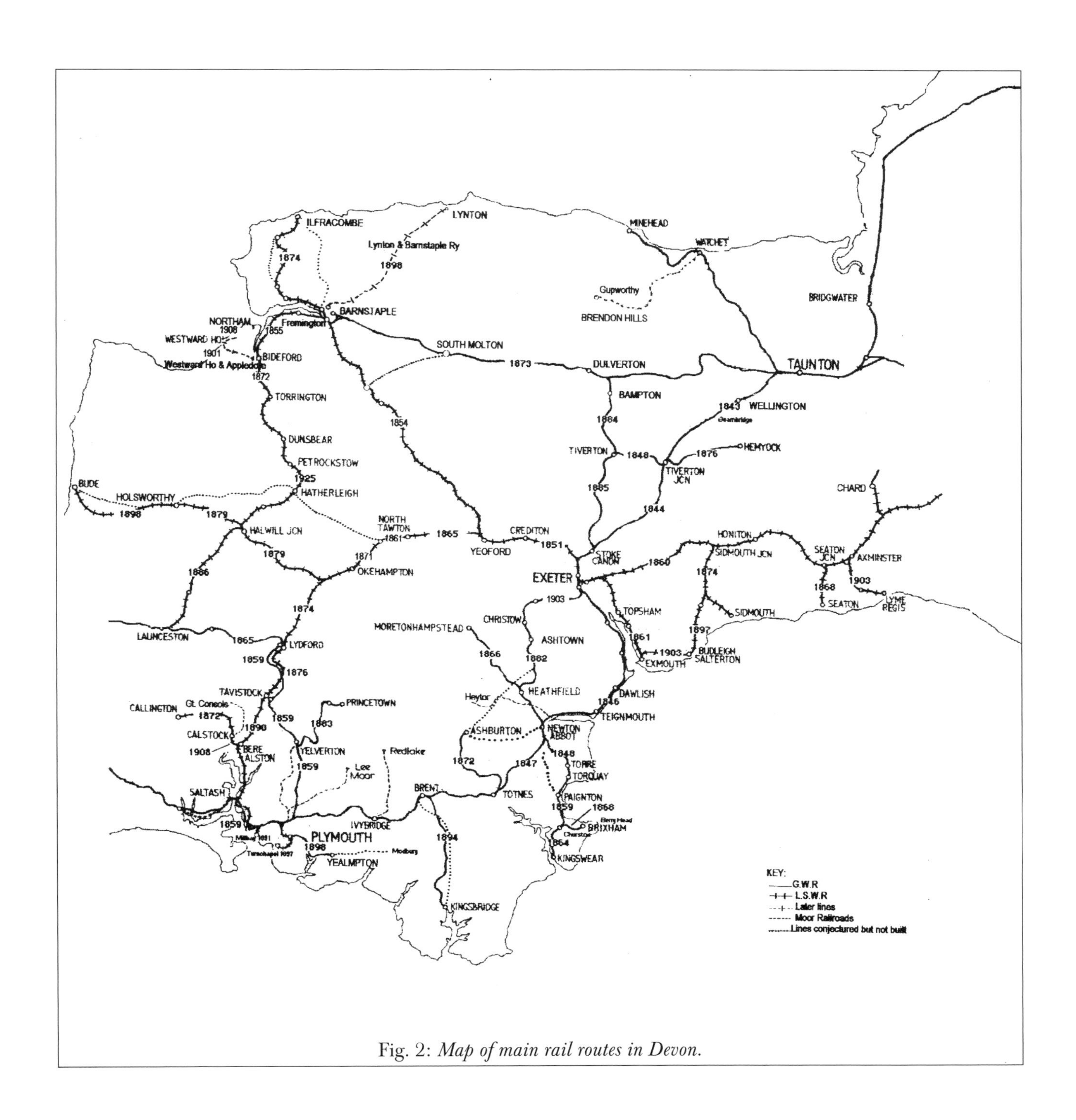

Fig. 2: *Map of main rail routes in Devon.*

The Victorian Guide Book

One of the interesting developments of the Victorian age was that of the guide book. Facets of today's guide books had appeared in England much earlier, indeed John Norden's triangular table of distances for each county – a must for many a modern guide book – first appeared in 1625, to be reprinted complete with county maps by Matthew Simmons in 1636 (**9**). Many of the county books produced before 1770 were topographical with accounts of the geography and history of the area and often with their text taken from Camden's *Britannia,* published in 1586 with but one map[1]. However none were arranged in the manner we would expect to find in a typical guide book; an up-to-date map of the area and of towns along a recommended route with information on things to see and places worth visiting.

The earlier road books of the seventeenth and eighteenth century copied John Ogilby (1675) showing only the single road in strip form with its junctions, bridges and hills. Owen and Bowen in 1720 took the strip form a step further when they added county maps (**24**) and descriptions of towns and places. One or two other writers began to introduce places to visit; for example, Francis Grose in his 1787 *Antiquities* gave a list of those 'in this county worth notice' but the maps, reprints of John Seller's 1694 originals (**20**), were not exactly up-to-date and had no roads.

It was not until 1771 with the publication of Daniel Paterson's *New and Accurate Description of the Direct and Principal Cross Roads in England and Wales* that a book described the actual route. He followed this in 1785 with his *British Itinerary* which included a full set of road strip maps (to be revived by Gall & Inglis in the late nineteenth century). In 1789 Cary's *Traveller's Companion* (**55**) was published and this included for the first time county maps which showed all the main coach and post roads. John Cary followed his *Companion* in 1798 with his *New Itinerary*, ideally to be kept, if not bound, together.

These two Itineraries were sophisticated road books. Although very dry for today's readers, each combined volume included a map of Great Britain, an index of all main routes from London with branch roads and cross roads, lists of coach routes, inns and staging points on the way, giving the traveller a complete picture of the route, local estates and houses, with descriptions of the principal towns and local history. Bulky works by anyone's standards, Paterson's and Cary's works were colossal achievements compared to their forerunners. Even so they remained very much a route book, one where the route to be taken is the focal point, rather than the sightseeing along the way.

Not until 1805, with G A Cooke's *Topographical Survey* (**68**), was it possible to purchase handy, pocket-sized guides incorporating a road map of the county. His County series was available as individual volumes or in groups of adjoining counties. The *Topographical and Statistical Description of the County of Devon* included an itinerary of the roads, a description of the rivers, agriculture etc, accounts of the main towns and a small pocket-sized map of the county. This was the first guide book with the emphasis on scenery rather than the road. It was followed by a number of dictionaries or gazetteers which, in the manner of an encyclopedia, listed all towns and villages in alphabetical order, often including maps of each county; Fullarton's *A New and Comprehensive Gazetteer* of 1833 with Scott's map of Devon (**107**) or Lewis' *A Topographical Dictionary of England* of 1835 with maps by Creighton (**113**) are good examples.

As John Vaughan writes[2]: the period of the formation of the English guide book, roughly 1780 to 1870, coincided with the revitalising of cartography in England. The Ordnance Survey was established in 1791 and in 1801 produced the first of its one inch county maps, a series completed by 1867. Some guide book proprietors saw the advantage of the work of the Ordnance Survey and sought to use Survey material in their own publications.[3]

This is perhaps too general. The earlier Cooke's maps were specially drawn and engraved and Scott and Creighton produced county maps for their gazetteer and dictionary respectively. In the tradition of cartography many of these maps were copies of earlier work and not of new material. But both Paterson and Cary surveyed the roads and Cary's maps and itinerary were commissioned by the Post Master General and were original maps. We must remember, too, that Cary's *Traveller's Companion* was in its second form in 1806 (**69**) three years before the second ordnance survey map, Devon, was issued. Yet, true to form, when and if maps contained new and up-to-date information it was highly likely that they had used the Ordnance Survey as source material.

Among the earliest local guide books is *A View of Plymouth-Dock, Plymouth and the adjacent Country* published by A Granville & Son of Plymouth in 1812 complete with a map by Neele. It was followed by *The Tourist's Companion* published by Longman, Hurst, Rees, Orme and Brown in 1823 which contained the same map. Not only does it contain one of the first directories in Devon but it also refers the reader to one of its primary sources

1. For example, John Speed's county maps were published with an adapted Camden text (**6**), and in 1617 the so-called *Miniature Speeds* were published with a Latin version of Camden (**4**).
2. John Vaughan; 1974; p. 81.
3. Devon (the second county map) appeared in 1809 (**74**).

The Beauties of England and Wales by John Britton and Edward Brayley, published in 1809 but seldom, if ever, accompanied by Cole and Roper's county map which had been especially produced for the issue in 1805 (**67**).

Among the first publishers to claim, or admit, the use of Ordnance Survey maps were the Croydons of Teignmouth. One of their earliest illustrated works[1] was *A Guide to the Watering Places between the Exe and the Dart.* This was published in 1817 with a map published 'by special Leave, from the Honourable Board of Ordnance'. Other guides, printed and published by the same company, approached the guide book format we would expect today. *The Teignmouth, Dawlish, And Torquay Guide with an account of the surrounding neighbourhood* was one of the earliest guides conforming to a modern-day format. Probably written by N T Carrington (whose name is on the cover) it was published by Edward Croydon. He and George Croydon were the owners of the news, reading, and billiard rooms at their public library in Teignmouth. The volume included lithographic views[2] and a map 'By Special Permission from the Right Hon'ble The Board Of Ordnance.'[3] It was published in 1830 as Volume I of the *Guide to The Watering Places*, and was followed by *The Torquay Guide* which reached a third edition in 1848.[4]

It is to Henry Besley that we must look for the first county 'guide' book of Devon. His *Route Book Of Devon* was first published *c.*1845. It, like the local guides, broke away from the old topographical dictionary style and, arranged by routes centred on Exeter, was designed for the 'tourist'. Few local printers published extensively but Henry Besley of Exeter could probably claim to have been the most prolific printer and publisher at the time. Although Tooley[5] lists only two cartographical works, one must presume that Besley was responsible for both the survey (if there was one) and the engraving of all those maps published with his imprint. The Besleys were important local publishers: Thomas Besley (1760-1834) and his son Thomas Junr (1790-1853) were registered in Exeter as printers as early as 1827. They published directories of the West country from 1828 and a map of Exeter in the same year. Henry Besley, the second son (1800-1886) was listed in both *White's* and *Martin's Directories* of 1851 and 1857 as printer, stationer and publisher of almanacks and guide books at 76 South Street. His son Robert Henry joined the firm in 1873, listed in *White's Directory* (1878/9) at 89 South Street as printers, booksellers and stationers, and *Besley's Post Office Directory of Exeter and Suburbs* was being produced from the same address in 1897. By 1898 the firm had become Besley and Dalgleish and Besley & Copp Ltd were still a well-known firm of printers in 1912. In addition to their popular illustrated guide books and directories which invariably included maps, they also produced other maps including special folding maps, such as *Dartmoor* (drawn to a scale of 6M to 100 mm). In 1854 they issued a *Devon, Cornwall, Somerset and Dorset Almanack*, in 1855 a *Route Book of Cornwall*, and in 1858 they acquired the *Exeter Journal* (**93**).

The Devon *Route Book* was reprinted many times until the 1870s. Reminiscent of Cary's itinerary with its roads out of London, *The Route Book* suggested fourteen roads out of Exeter and described each route in detail. With fine engravings by G Townsend (and later by S R Ridgway) and specially produced maps and plans it was the forerunner of the county guides to be followed later by the leading guide book producers. The engravings were often sold separately in booklets of twelve or more prints and are popular with print collectors. Ward & Lock proceeded to use Besley maps from *c.*1886 in their *Shilling Pictorial Guides.*

Two further early examples of local publishing were both produced by clerical writers;[6] *The Panorama of Plymouth; or, tourist's guide to the principal objects of interest, in the towns and vicinity of Plymouth, Dock, and Stonehouse* and *A Picture of Sidmouth.* The former was written, printed and published by Samuel Rowe (1793-1853) in 1821 and is a good example of a thorough local guide covering most topics succinctly and authoritatively. First apprenticed to a bookseller, Rowe's father bought a business for him and he was able to devote his time to writing. He became secretary of the Plymouth Institution in 1821. In the following year he was ordained, held livings in his home county and continued his topographical interests.

The other volume was by Edmund Butcher (1757-1822), a Unitarian minister. Poor health forced him to retire to Sidmouth, where he lived for many years producing sermons and the local guide. A fourth edition appeared in 1830 with the new title *A New Guide, Descriptive of the Beauties of Sidmouth.*

The first recorded mention of the term guide-book is to be found in Lord Byron's poem *Don Juan* in 1823. Earlier writers had used *survey*, *panorama* or *picture* but the word *guide* quickly caught on and Croydon and others proceeded to use it regularly as we have seen. Besley preferred *Route Book* and indeed many other writers used the idea of the route to plan the descriptions of those sights worth seeing. Another term that became popular

1. With very good engravings by W B Noble.
2. Drawn on stone by L E Reed or George Rowe and Printed by W Day, 17 Gate Street, London.
3. However, the map, although claiming to be *from the Honourable Board of Ordnance* was probably a copy of part of Cary's map of 1807.
4. Mr S A Croydon was managing director of The Teignmouth Printing and Publishing Co. Ltd in Station Road, Teignmouth in 1930: *Kelly's Directory.*
5. R V Tooley's *Dictionary of Mapmakers* lists the *Route Map of Roads*, and a *Plan of Ilfracombe* (1892).
6. John Vaughan; 1974; pp. 134-136.

from 1850 was *Handbook* [1], a term created by John Murray II and later adopted by, his great rival, Nelson's and then by Thomas Cook & Sons.

The two most famous names in the history of the genre of the guide book must be those of Murray and his continental rival, Karl Baedeker (1801-1859). Dissatisfaction with existing guides was Murray's motive for producing his new publications. In an article in *Murray's Magazine* for 1889 John Murray III explained how sixty years earlier he had brushed up his German and set out to tour Europe but found 'The only Guides deserving the name were: Ebel, for Switzerland; Boyce for Belgium; and Mrs Starke for Italy. Hers was a work of real utility, because ... it contained much practical information gathered on the spot'.[2]

Murray's (1778-1843)[3] famous series of red-bound guides began in 1836 with a general guide to European countries and by 1855 there were Murray guides to most places worth visiting. However, it was his son, also John III (1808-1892), who had the idea for the entire series and who wrote some of the first guides. Murray's guide to Devon and Cornwall was his first county guide. It appeared in 1851 with the Devon map by J & C Walker (**116**) and by 1899, when Warwickshire was published, there were some sixty volumes. An impressive tribute to their popularity was the recognition by Baedeker. His first *Handbook*, to Holland and Belgium, appeared in 1839; in the preface he wrote that Murray's *Handbook for Travellers on the Continent* had formed its basis and some of his later works also contained acknowledgements to Murray.[4] The Murray guides were so good that Professor Jack Simmons when writing about some twentieth century guide books wrote that 'Murray's Handbook of 1851 is still, from many points of view, the most profitable guide-book to use in Devon and Cornwall in 1951'.[5]

Murray's first three editions of Devon & Cornwall contained the two Walker county maps. The fourth edition, revised by T C Paris,[6] contained a new map of Devon by W & A K Johnston (**135**) which was included in the the fifth, sixth and eighth editions (there was probably no seventh edition). Murray liked to use local writers as much as possible and these later editions were revised by Richard John King (1818-79). King came from Plymouth and studied at Exeter College, Oxford. He not only carried out revisions to a large number of these guide books but he also contributed to other works including the *Encyclopaedia Britannica* published by A & C Black. A well-known historian, King became President of the Devon Association; he died at Crediton in 1879. Each edition was extensively revised and Murray employed other writers such as the Rev. H S Wilcocks, C Worthy and W O Goldschmidt. They brought the guides up to date and banished 'some of those minute details interesting only to the antiquary and those portions of the legendary lore which appeared so trivial' (Ninth Edition) or they introduced 'many facts which recent investigations or church restorations have brought to light' (Tenth Edition). Many editors of Murray's *Hand book* were local people: the Rev. Horace Stone Wilcocks, who contributed to the Devon issue of 1879, was born at Exeter (1835) became Vicar of St. James-the-Less church in Plymouth, resigned after a dispute with the Bishop of Exeter and died in Plymouth in 1912; Charles Worthy, paid £10 for corrections to the 9th edition and £52.10.0 to edit the 10th edition, was born at Exeter. Another contributor to the guide of 1879, and to the *Cornwall Handbook* of 1882, was Sir John C B Milton, a cousin of Anthony Trollope, who was awarded his knighthood for services during the Crimean War.[7]

The 9th, 10th and 11th *Murray* editions (1879, 1887 and 1895) separated Devon from Cornwall. Strangely Murray chose to use Walker maps again in the 9th and 10th Editions, possibly as they were more detailed than the W & A K Johnston map, but the Eleventh Edition had a larger, more detailed map by Bartholomew (**150**). When John Murray III died in 1892 the company was facing strong competition not only from the Blacks and Dulau but also from Baedeker. John Murray IV finally sold the series (excluding *Japan* and *India*) to Edward Stanford in 1901.[8]

1. The spelling seems to be optional: the third edition for example had Handbook (on the title page), Hand Book (spine title) and Handbook (cover).
2. John Vaughan; 1974; p. 42.
3. The firm was founded by John Murray (1737-1793). He printed his first guide in 1769: *A Description of Bath* by John Wood. John II is often erroneously credited with founding the company. John III introduced the famous red book series. See William Zachs; *The First John Murray*; OUP for The British Academy; 1998.
4. John Vaughan; 1974, p. 47.
5. J Simmons; *Parish and Empire: Studies and Sketches*; 1952; quoted in Vaughan, *op cit*.
6. This edition pays tribute to one of Murray's early contributors: in the passage concerning Heavitree, near Exeter, "the residence of the late Richard Ford, who here wrote his celebrated *Handbook for Spain*".
7. All biographies are taken from W B C Lister (1993).
8. Under the terms of the agreement, dated 30th April 1901, Stanford bought up the stock and copyright of all other guides for £2000. The accompanying list seemed to indicate that a total of 321 copies of Devon were available. (Our thanks to Mrs V Murray for providing us with a copy of the agreement.)

Murray's greatest rival was probably the Edinburgh firm of Adam & Charles Black. Adam Black (1784-1874) established the business in 1807 bringing his nephew Charles (1807-1854) into the firm in 1833[1]. *Black's Economical Tourist of Scotland* appeared in 1826 and the first of Black's popular series of guides (most of them written and revised by the family)[2] appeared in 1839 with the publication of guides to Edinburgh and Glasgow. The series progressed and in 1841 the first regional guide to the English Lakes appeared and two years later a guide to England and Wales. The series then expanded steadily, at first with city guides then with county guides from *c.*1855 including *Devonshire and Cornwall.*[3]

Unlike Murray, who strived to keep his guides up-to-date, the Blacks do not appear to have had such ambitions and the early editions of their guide to Devon remained virtually identical as far as the text is concerned. From the start they combined counties. Devon and Cornwall appeared in 1855 (**130**) and in 1862 Devon, Dorset and Cornwall. The later *Devonshire* volumes retained the pagination of the combined 3-county volume. Blacks' 1862 guide (the three county volume) contained a map of *Dorset, Devon and Cornwall* produced by Bartholomew from plates of a larger regional map (**142**). *Devonshire* appeared the same year and the map was taken from the same plate showing *Devon* alone. Extensive revisions to this map were carried out and editions after 1882 included a larger map, again drawn and engraved by Bartholomew; the same map that was issued in Murray's 11th Edition (**150.13**). The pages were renumbered and for the first time *Devonshire* starts on page 1 - twenty years after its first appearance! In 1892 text revisions were carried out, probably by Charles Worthy of Exeter. He had served in India and took up studying antiquities when invalided out of the army in 1864. He lived at Heavitree, contributed to the *Devon* issues of Murray's Handbooks of 1879 and 1887, and also wrote *Devonshire Parishes in the Archdeaconery of Totnes.*[4] A phenomenally successful series, Black's *Devonshire* was reprinted almost annually until the end of the century.

After Blacks' success in publishing a guide to the three south-western counties of Cornwall, Devon and Dorset followed by the guides to each of the individual counties it would have seemed reasonable to follow with specific guides for the tourist resorts, or at least areas. Yet it took them forty years to realise that there was a growing market for such guides and only in 1901 did *Black's Guide to Torquay and the South Hams, Paignton, Dartmouth, Totnes, Kingsbridge, Salcombe, Etc.* appear, priced sixpence and in paperback format.

Like the previous guides, the Blacks put little investment in the new series. Just as their original *Devonshire* text remained virtually unchanged from 1862 to 1881 the new Devon series remained but little changed until the end of the century, and the guide to Torquay was no exception. The Preface warns the reader that 'The bulk of these pages make part of our general Guide to Devon, which accounts for what may seem the irregular pagination of this section, in which pp. 9 to 65 come to be omitted'. The text is a virtual copy of that found in the *Devon* volume published in the same year. Four maps were included: The Map of *Torquay* that had appeared in the last two *Devon* books; sectional maps of *Torquay District* and *Dartmoor District* (also from the Devon guide) and the county map, minimally revised, was that used for the *Encyclopaedia Britannica* in 1877 (**152**).

In 1878 and 1879, possibly inspired by the successes of Murray and Black, Edward Stanford published two guide books covering north and south Devon (**154**). The editor, Richard Nicholls Worth (1837-1896), was a popular Devon journalist, geologist, author and editor. His books included a *History of Devonshire,* other local histories such as *Devonport, sometime Plymouth Dock, History of Plymouth from the earliest period to present time* and other county guides such as *Somerset* and *Dorset.* His *Tourist's Guides* were very popular and were reprinted until 1894 in six editions with the same maps by Edward Stanford.

One other book publisher of county guides was fairly successful. Dulau & Co. began to publish from 1882 their two companion *Thorough* guides to *North Devon and North Cornwall* and *South Devon and South Cornwall.* These also included sectional maps by Bartholomew, transfers from earlier Bartholomew plates which later appeared in the *Royal Atlas,* and were not maps of the whole county. However, like Blacks' the guides also had a key or index map printed on the inside front cover to help tourists. These guides were also very popular and ran to many editions through the last years of the century. The first volume of the series, *English Lake District,* is notable as here John George Bartholomew effectively introduced layer colouring into Britain (in the sectional maps).[5] Both Devon volumes were written and/or edited by M J B Baddeley and C S Ward (and included layer-coloured maps from *c.*1902).

1. Adam Black is known to have vistied Exeter in 1837 as part of a long selling trip. He arrived with just £5 in his pocket and had to borrow money for the return to Scotland. In Adam & Charles Black 1807-1957; A&C Black; 1957.
2. See Harold M Otness; *Index to 19th century city plans ...*; WAML; 1980; p. xv.
3. John Vaughan reports the progress of Black's series of guides as beginning with (1839) Edinburgh, Glasgow; 1841, the English Lakes; 1843, England and Wales; 1850, London to Edinburgh; 1851, Wales; 1853, Trossachs; 1854, Aberdeen, Belfast, Dublin, Ireland, Killarney; 1855, Derbyshire, Devonshire, Hampshire. J Vaughan; 1974; p. 50 footnote. However, Devon was published with Cornwall in one volume.
4. W B C Lister ((1993); p. 186.
5. L Gardiner; 1976; pp. 30-33.

Another company to exploit the growing leisure market was Ward, Lock and Co. Some time around 1886 they produced their first guides to Devon. These were issued for local areas or bound together as North Devon or South Devon illustrated guides. Originally these contained copious engravings and woodcuts, but these were replaced by photographs in the 1890s. The map of north or south Devon that accompanied these guides to the end of the century were those of Besley (**134**) first published in the *Route Book Of Devon.* The *Red Guide* series continued and guides were produced into the second half of the twentieth century with transfers of John Bartholomew's maps of 1895 (**174**) still being used to provide area maps.

There were other guides written throughout the period including many local guides with sectional maps, but few were as successful as those printed by the major publishing houses. Concurrently ephemeral works appeared such as *Ferny Combes* by Charlotte Chanter. Half guide book and half personal experiences, *Ferny Combes* was illustrated with ferns of the area and included a specially produced county map (**131**). Mackenzie Walcott's *Guide to the Coasts* was also published with a county map (**137**). These were never as popular: the former was printed only three times and the latter once as an omnibus edition (of the whole of the south coast) and once as a mini version (Devon and Cornwall). A local writer towards the end of the century, John Page, wrote three books on Devon: *The Coasts Of Devon* (**172**); *The Rivers Of Devon* (**170**)*; Dartmoor.* As far as is known none were reprinted but they are good descriptions of the countryside written in guide style and each contains a map. One guide book, frequently reprinted, was Arthur Norway's *Highways and Byways in Devon and Cornwall* (**178**). Appearing in 1897, it was reprinted in 1898 and 1900 and several times in the twentieth century, appearing as a pocket edition in 1923.

Two popular local guides that deserve mention are *The Hand-Book to South Devon and Dartmoor* and *The North Devon Handbook.* The author of the first was William Wood, a publisher with premises in Devonport. He published a great number of local guide books including ten editions of the *Hand-Book of Devonport, the Three Towns' Almanack,* from 1860 until 1896, *Rambles and Excursions* (Plymouth) and the *Handbook to Cornwall* (in 1880). All were illustrated with maps. The first two editions of the South Devon handbook included a reprint of the Ebden/Duncan county map first published in 1825 (**95**) but with a new title *Devonshire From A Recent Survey With The Railways*, and changes to the arrangement of hundreds, addition of railways and the imprint *Published by W. Wood No. 52 Fore St. Devonport* (CeOS). However, when the guide was reissued it only included a map of the southern part of the county.[1] Later copies of the almanack also contained a smaller map of South Devon, probably drawn and engraved by W G Cooper, who was an engraver and lithographer with premises in Union Street, Stonehouse.

Interestingly this would seem to be the first map of the county that showed the new Eastern boundaries. In an exchange of 1842 the parish of Thornecombe (the island in Dorset) was transferred to Dorset and the parishes of Stockland and Dalwood (the islands of Dorset in Devon) were transferred to Devon. In 1844 the parish of Maker, at the mouth of the Tamar was transferred to Cornwall. There was a second eastern exchange in 1896 when Churchstanton was transferred to Somerset and Chardstock and Hawkchurch transferred to Devon.[2]

The second publication *The North Devon Handbook .. edited by the Rev. George Tugwell* was published by both a London firm, Simpkin, Marshall & Co. and a local Ilfracombe firm, John Banfield. The original work contained a map of north Devon signed by W Gauci.

The Gauci family lived and worked in London but possibly had associations with the westcountry. M Gauci was a lithographer of costumes, portraits and topographical views. His two sons, Paul and William followed in his footsteps.[3] They both made many excursions to the west country and produced plates for J Banfield's *Scenery in the North of Devon* (*c.*1837). Paul produced plates for R Woodroffe's *Views in Bath* (*c.*1840). William produced many more plates including those for G H P White's *Four Views on the River Dart* (*c.*1830), J Baker's *Views of the Landslip at Axminster* (1840), C F Williams' *Six Views of Berry Pomeroy* (*c.*1840) and one for Rev. G Tugwell's *The North Devon Handbook* (*c.*1855), in which the other thirteen or fourteen illustrations were by or after W Willis. The Rev. Tugwell also produced *The North Devon Scenery Book* which was illustrated by H B Scougall with eight litho plates (2nd Ed. *c.*1863).

1. *SOUTH-DEVON* with imprint: *Engraved from the Ordnance Survey for the Hand-Book to South Devon.* (CeOS). The area covered was from Rame Head to Lyme Regis and north to Chulmleigh. Size: 215 x 270 mm. First appeared in: *The Hand-Book to South Devon and Dartmoor Third Edition;* Devonport. W Wood. (1859). Revised in *c.*1860, 1870, 1872 with changes to railways and date. In *c.*1872 with imprint erased and new imprint: *Engraved from the Ordnance Survey for the Hand-Book to South Devon. Pub. By W.Wood, Fore St. Devonport* (CeOS). Signature added: *W.G.Cooper, Sc* below the title (Ed).
2. W G Hoskins; *Devon;* Collins; London; 1959.
3. Ian Mackenzie; 1988; p. 127.

The first edition of his *The North Devon Handbook* contained a map of the north coast.[1] This had, across the top, a *'Section from the Foreland to Morthoe Hill'* showing the geological strata from Henry De la Beche's Report. Subsequent issues always included a new map originally published by Besley and son as *Besleys' Plan of Ilfracombe and Lynton.* In *c.*1865 the guide became *Stewart's Shilling Guide Book to North Devon. New Edition* (Ilfracombe. W Stewart);[2] and *c.*1872 as *Milligan & Co.'s (Late Stewart's) Shilling Guide Book to North Devon, New Edition* (Ilfracombe. Milligan & Co. until *c.*1879). Milligan's, established in 1854, had a thriving business as bookseller, printer and stationer at The Library, Ilfracombe, as well as selling anything to do with paper (and even offering pianofortes for hire).

A competitor to Henry Besley at the end of the nineteenth century was Thomas Doidge. A map of South Devon was issued by Doidge and Co., a local Plymouth firm.[3] *Doidge's Western Counties Annual* - a miscellany of useful, instructive and entertaining local and general information - was issued annually from *c.*1877 from their premises at 169-170 Union St. The Annual was produced in two editions with a special printing for the Army and Navy edition, which included county historical information in the calendar section. The annuals were packed with stories and contained an almanack for each month. Doidge advertised himself as *Great Book and Stationery Mart* and also as a *Discount Bookseller.* In 1879 Thomas Doidge was still registered at this address but the company was advertised as photographic studio, fancy goods dealers, booksellers, binders, stationers, &c. In 1892 Simpkin, Marshall, Hamilton, Kent & Co. became their London agents. Photos were added in 1893 and by 1894 Doidge was claiming sales of fifteen thousand copies. The printers at the end of the century were the Plymouth firm of Hoyten and Cole.[4] A map was included in 1899 - *Midland Railway Map of Great Britain.*

The map of Devon, *Doidge's New Large Scale Cycling and Touring Road Map,* shows the area from St Austell to Teignmouth and north as far as Stratton and Bradford by Black Torrington. Railways are shown to Holsworthy, opened in 1879, but not the line to Kingsbridge which opened in 1893 nor the extension to Bude (1898 though planned earlier).

There were a great number of local guides yet it was the major publishers and especially Murray and Blacks who made the guide book popular and, through their support and marketing, the guide book in its present form had arrived and was here to stay.

Although the majority of Victorian works containing maps of the county were dated many were still published with no date on the title page. Collectors must also be wary of the dates on the title page, especially during the latter half of the century. There are ways to date such works: the accompanying text may include helpful dates of recent events; advertising may be dated or contain a reference to another contemporary work; the evolution of the railway is a also a guide, though this may be misleading as lines optimistically included on the maps were not necessarily built.

From the 1860s guide book publishers often included advertisements at the back of their volumes and the discerning reader can spot dates embedded there, especially in those giving railway information. For example, Murray's Third Edition (1856) of the *Handbook to Devon & Cornwall* had no advertisements but in later copies of the Fourth Edition (first published 1859) there is a 24-page section with a separate title page *Murray's English Handbook Advertiser 1860*, showing that the guide was reprinted the year after the first publication but still with the original title page and date. *Black's Guide To Devonshire* 1868 has no adverts; that of 1869 a 48-page section with a separate title page *Black's Guide-Book Advertiser 1869.* The inside and back covers should also be inspected as these, too, can be dated. The inside back cover of the two Black's Guides mentioned above are dated and not dated respectively.[5] A copy of Murray's *Handbook to Devonshire* Tenth Edition, in the author's possession, is dated 1887 on the title page, dated on the inside covers as February 1892 yet has an advertising section for 1892-1893.

Although the guide books flourished in the second half of Victoria's reign, many of the books have lost their maps, and it is surprising how few guide books are available to enthusiasts and researchers in public libraries and museums, although they regularly appear at antiquarian book fairs. The authors have included many of their own guide books for the purpose of completeness and often these are the only copies known.

1. *A Map of the NORTH OF DEVON AND PART OF SOMERSET. Published by John Banfield, Ilfracombe, 1856.* Imprint: *W.Gauci, Litho. 72 Gt. Titchfield St. London.* below the title (Bb). The Key (Db) lists Picturesque Views and roads. The railway is shown to Bideford. Size: 330 x 743 mm. The map shows the coast from Harland to Watchet and inland to Torrington. The book was reprinted in 1862.
2. *COAST OF NORTH DEVON.. Published by W Stewart,* (later *Milligan*) *Ilfracombe,* usually dated. Imprint: *F J Halfyard, Litho, Exeter.* The railway is shown to Ilfracombe. Size: 310 x 390 mm. The map shows the coast from Harland to Minehead and there are town maps of Ilfracombe, and Lynton and Lynmouth.
3. Size: 598 x 742 mm with *Scale* (10 =128 mm). Imprint: *DOIDGE & Co. 169 & 170 UNION STREET, PLYMOUTH* (BeOS) and there is an inset map: *KEY MAP TO THE ROADS OF THE SOUTH OF ENGLAND.*
4. Hoyten and Cole, 39 Whimple Street, Plymouth, were still active in 1930 when they were listed in *Kelly's Directory.*
5. One wonders what was more important: Black's *Guide to Devonshire,* for example, had 232 pages of text and a further 128 pages of advertisements by the time of its 16th edition in 1898.

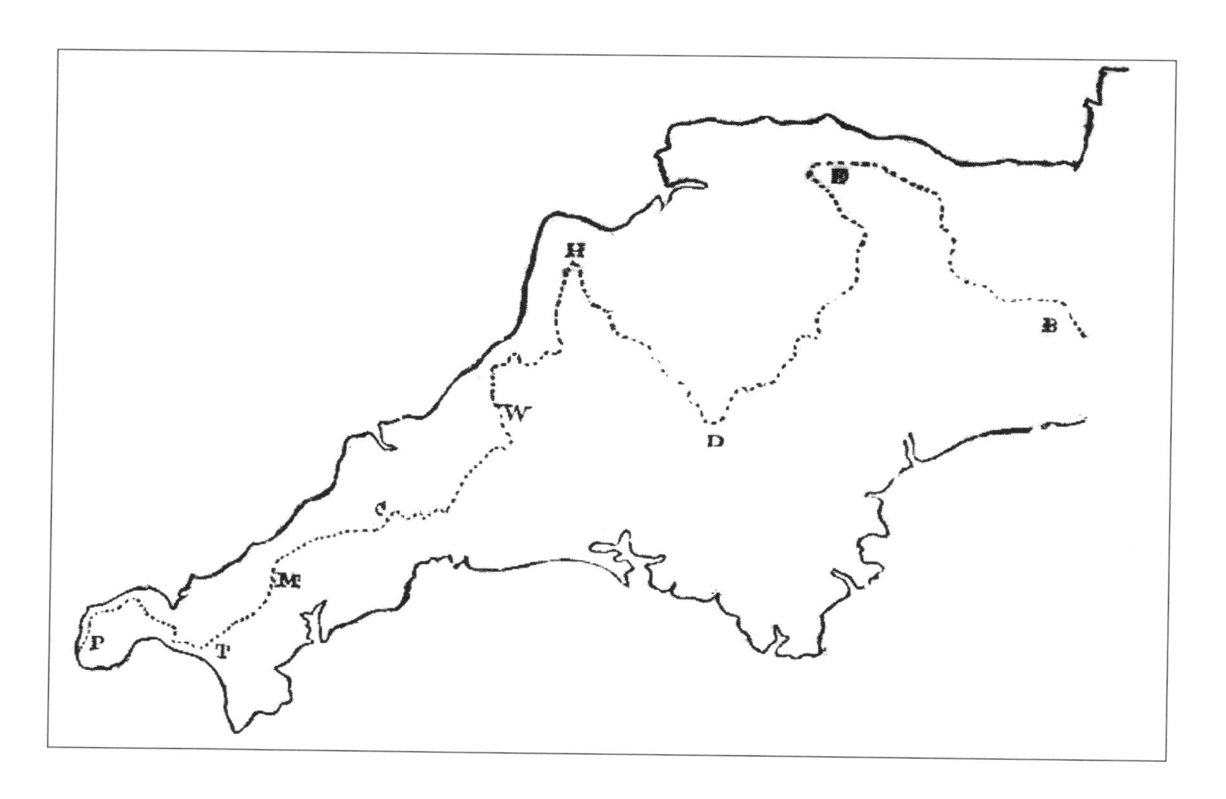

Fig 3. Henry de la Beche's hydrographical map of Devon

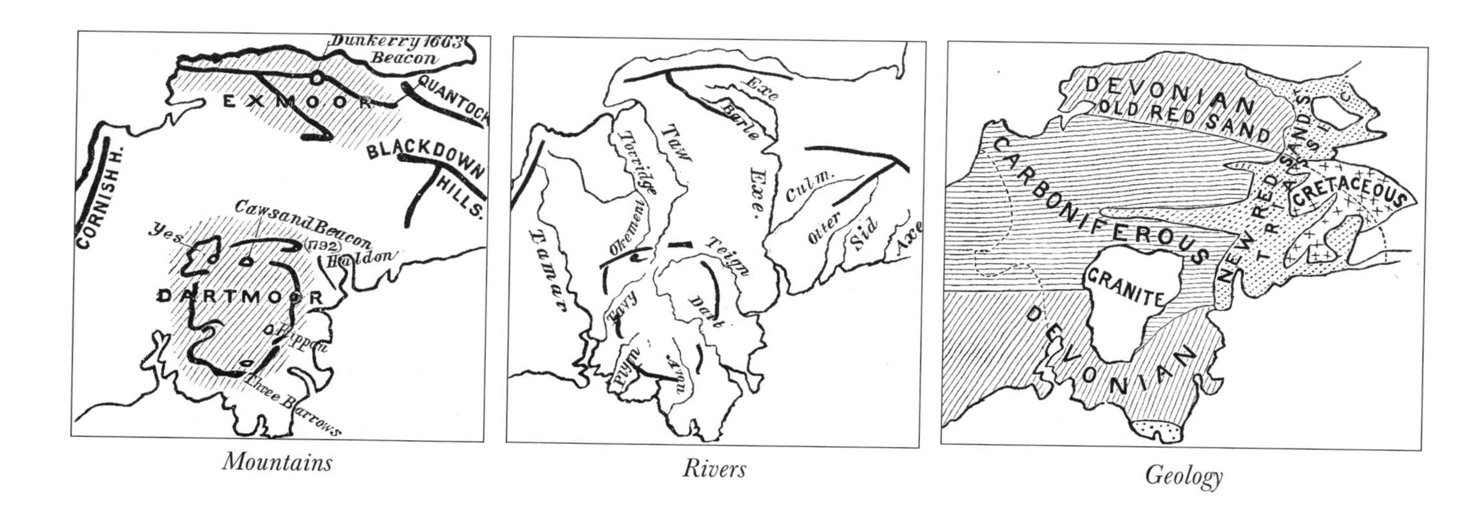

Mountains *Rivers* *Geology*

Fig. 4: Three small maps included in Faunthorpe's work.

EXPLANATION

Definition

For the purpose of this cartobibliography, a county map of Devon is a printed map published specifically to show the whole of Devonshire on one or two sheets, either individually, or with only one other complete county. Devon was not grouped with other counties in any of the atlases included in the current work; however, one geological work (**118** - two maps), three ecclesiastical works (**129**, **139** and **143** depicting the Diocese of the Bishop of Exeter), five guide books (**130**, **135**, **137**, **157** and **178**), *The Western Morning News* (**160**) and Firks and Son (**173**) show Devon together with Cornwall. These are all included.

There is one exception. Black's revised *Guide to South of England* (the spine title, **142**) originally had a map of Dorset, Devon and Cornwall, which was later cut into two maps *Dorset* and *Devon & Cornwall*, or the three separate counties. As all of these maps were taken from the same plate the first stage has been included. Cassell's produced a railway guide, and Gall & Inglis a cyclist's road book: both of these have maps of the county. Although not true county maps they show the majority of the county and have been included (**165** and **182**). The Gall & Inglis map is also of Devon and Cornwall.

Two key maps have also been included: the key map of Bartholomew/Black (1882, **156**) which depicts Devon; and the index map of Baddeley and Ward's *Thorough Guide Series* which depicts Devon and Cornwall (1882, **157**). Both of these maps were printed on the inside front covers. The earlier key map of Devon, with Cornwall & Dorset, is not included.

Other index maps have not been catalogued, as either too small to provide real information or as they are unlikely to be found loose:

> A label for Ordnance Survey sheets which was included only on boxed sets sold in 1819. (Fig.1 - p.xiii)
> Henry de la Beche included a rough sketch map of Devon in his work (**118**) (see previous page).
> A small index map by W J Sackett (80 x 60 mm) shown above the list of the Diocese (**143**).
> Three small maps in Faunthorpe's work (**149.1**) (see previous page).
> The inset map on Stanford's map of South Devon (**154**), (see p.109).

Maps of Devon which show only part of the county, coastal charts and road books are also excluded. It is hoped further volumes will list these.

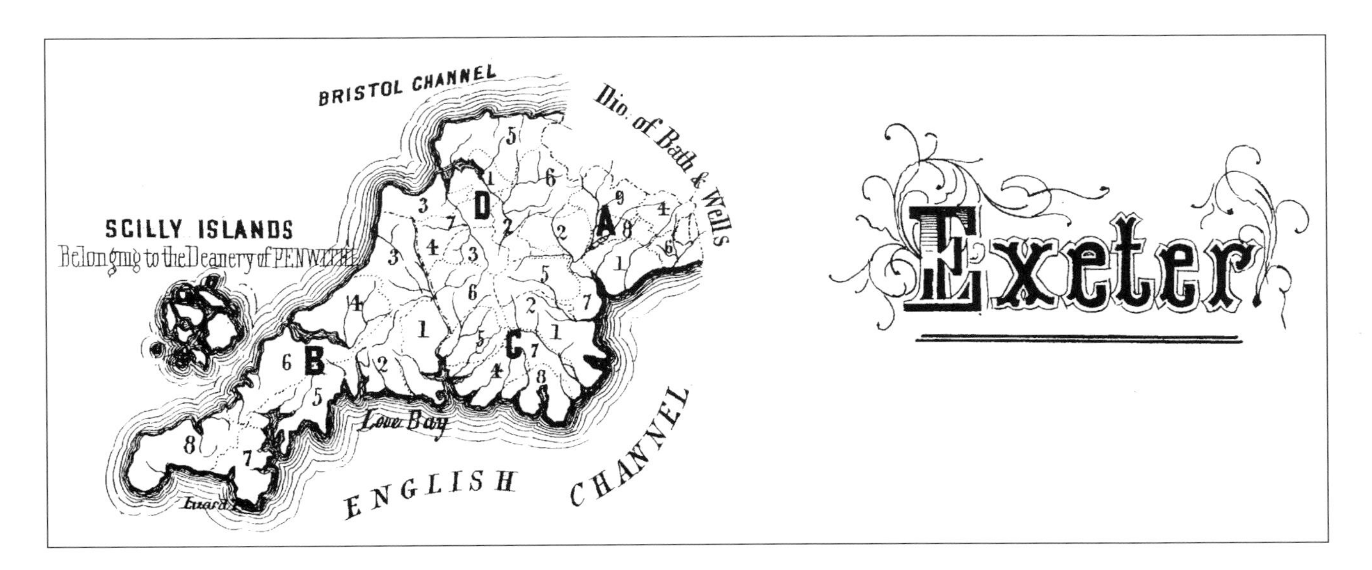

Fig. 5: Sackett's key map

Order

The maps are listed in the chronological order of the first appearance. Subsequent issues are dealt with as part of the main entry.

Numbering

Devon, being one of Britain's largest counties, poses problems for the carto-bibliographer. Devon was often published on more than one sheet, as it was often popular to show North and South Devon separately. When a county map is produced on a single sheet, a change to the map creates a new state. But when a map is on two (or more) sheets does a change to one of the sheets necessarily constitute a new state? We have resolved the problem as follows:

1. A single map is given a map number and each state a further number.
2. When the county was shown on two sheets in an atlas most sheets were changed for a new edition and we have given one map number and a subsequent state number when either (or both) sheets were changed.
3. When the county map is published on two sheets for the use in two guide books we have given both sheets the same map number but with a suffix N or S showing in which guide book, North or South, the half map occurs. This method is also used when the same map is used in two guide books as a separate index for north and south.

One map of Devon that was later changed to include Devon and Cornwall (and Dorset) was Bartholomew's map for A & C Black (1862, **142**). The subsequent states have been numbered according to system 1.

Numbers in brackets refer to other Devon maps; those up to and including **117** refer to maps contained in the previous cartobibliography of Devon; Numbers from **118** are in this volume. If maps are discovered later they will be numbered in their correct chronological place with the suffix **A** or **B** added (for example, the lately found Arrowsmith map of 1814 is **80A**, see **Appendix I**).

Heading

In labelling the maps the chosen order of preference throughout is surveyor, draughtsman, engraver, author, publisher. The map is given the name of a surveyor, if known, where the map is believed to be the result of an original survey. Otherwise the name of the draughtsman, the engraver, the author of the original work, or the first publisher in that order is used to help in identification. Names joined by a slash, eg Emslie/Reynolds, indicate that both (or all) were connected with the original issue and both (or all) names are found in works of reference and/or dealer's catalogues in connection with the map. A slash is also used to identify separate works by the same draughtsman or engraver, eg Bartholomew/Black and Bartholomew/Pattison. In some cases maps are commonly associated with another mapmaker or publisher, eg Weller's map (1858) for the *Weekly Dispatch* newspaper (**136**), or a later publisher, eg Bacon's publication of Cary's maps: in these cases the alternative name is added in brackets, eg Swiss & Co (Cary).

Date

The date listed is the one that appears on the first edition of the map or, if it is undated, the date of the publication in which it first appeared. (See also Publications below.)

Biographical Note

Wherever possible a short biographical sketch is given of the surveyor and/or publisher, together with a brief history and description of the map or atlas. There is also a note covering those copies that appeared after 1901.

Size

This gives the dimensions in millimetres, vertically then horizontally, between the outer frame lines. In the absence of frame lines, the dimensions are those of the printed map. Where the map has a panel of information or a title extension this is noted, together with an indication of the size of the complete engraving and of the map panel or area. Because of variations in paper shrinkage small differences can be expected. Lithographs may differ widely in size, being reduced or enlarged by photographic or other means.

Scale

The wording on the map is shown in bold type. The figure in brackets refers to the scale bar. The reference **British** (1+10=11 mm) **miles** means that the scale shows 1 mile to the left of zero and 10 miles to the right, the whole measuring 11 mm. **British** is written above or at front of the scale bar and **miles** after or below.

Inscriptions and Place-Names

In the map description the title, imprint and publisher's names etc. that appear on the map are given in bold print using the original spelling. Punctuation is omitted if unclear. In the text place names are usually given as they appear on the map in question, therefore either Okehampton or Oakhampton may be used.

Editions

The main entry shows the title and features that appeared on the first edition of the map. Subsequent states, with a note of the variations, are listed below. Information is complete up to the year 1901. Later issues have been noted in the previous biographical sketch.

Sources

Atlases and guide books seen by the authors have the source listed after the publication details, eg BL (see list page xxxiv). This also applies when folding maps were issued as separate publications and the copies seen are believed to be complete. Loose maps from atlases or other folding maps are in brackets, eg (Pl). Where atlases and other publications are mentioned but have not been seen by the authors the source is noted in square brackets, eg [EUL]. The authors have included items in their own collections where these are fairly scarce. Although many libraries have a selection of guide books, they rarely have complete runs.

Position of Features on Maps

Following the popular method, the position of each feature on the map is given, where appropriate, in brackets by reference to the following grid:

Aa	**Ba**	**Ca**	**Da**	**Ea**
Ab	**Bb**	**Cb**	**Db**	**Eb**
Ac	**Bc**	**Cc**	**Dc**	**Ec**
Ad	**Bd**	**Cd**	**Dd**	**Ed**
Ae	**Be**	**Ce**	**De**	**Ee**

Where a large feature extends over more than one square of the grid the position of the centre is given, except when the feature is in a corner when the corner reference is given. When the feature is outside the inner border the reference is followed by 'OS'. Position information is given where this feature changes during the lifetime of the map.

Publications

The first date is the date of issue of the atlas etc. or the date printed on the map, whichever is the earlier. Reissues with information on known changes follow. Facing each description is an illustration of one issue of the map.

After the description of an edition of each map there is a list of the titles of atlases, books or other publications in which the map appeared. The title (which may be shortened, omissions being shown by dots) is given in italics, using the original spelling but in lower case except for initial letters. This is followed by the place, the publisher and the year of publication. If the year is shown without brackets, eg 1839, this is the date given on the title page. If more than one date is shown, separated by commas, these are the dates of successive editions of the same publication. If the date is shown in the form 1882 (1892), this means that although 1882 is the date on the title page, there is other evidence to indicate a later (or earlier) publication. Dates in brackets alone, eg (1845), indicate that there is no date on the title page and the date is conjectured. Some maps, due to their size, are illustrated only in part.

CATALOGUE REFERENCES

The dates of issue are followed by the locations of atlases and maps. Locations in brackets imply loose sheets: the exact source of these is not certain, they could come from any edition of that state. Locations in square brackets identify those atlases not inspected by the authors. We would like to thank the curators and staff of the following institutions without whose help this book could not have been written.

B	Bodleian Library, Oxford.
BCL	Birmingham Central Library.
BL	British Library (and British Museum), London.
BPR	Bartholomew & Co. Listing at NLS.
BRL	Bristol Central Library, Reference Library.
C	Cambridge University Library.
CB	Collection of Clive Burden.
DEI	Devon and Exeter Institute, Exeter.
E	Exeter Westcountry Studies Library.
EB	Collection of Eugene Burden.
EUL	Edinburgh University Library.
FB	Collection of Francis Bennett.
GCL	Gloucester Central Library.
GL	Guildhall Library, London.
GPL	Grimsby Public Library.
GUL	Glasgow University Library.
KB	Collection of Kit Batten.
Hull	Kingston upon Hull Central Library.
Leeds	Leeds Reference Library.
Leics	Leicester Central Library.
Leics UL	Leicester University Library.
Liv	Liverpool Central Library.
MCL	Manchester Central Library.
Midd	Middlesborough Central Library.
MW	Collection of Malcolm Woodward.
NDL	North Devon Library, Barnstaple.
NLS	National Library of Scotland, Edinburgh.
Norwich	Norwich Central Library.
Notts	Nottingham University Library.
NMM	National Maritime Museum, Greenwich.
OrU	University of Oregon.
P	Private collection.
Pl	Plymouth Public Library.
PRO	Public Record Office, Kew, London.
RGS	Royal Geographical Society.
SGL	Spalding Gentleman's Library, Spalding Gentleman's Society.
T	Torquay Public Library.
TB	Collection of Tony Burgess.
TN	Collection of Tim Nicholson.
TQ	Torquay Natural History Society (Torquay Museum).
Truro	Truro Public Library.
ULL	University Library London.
W	Whitaker Collection, University of Leeds.
WM	Wisbech and Fenland Museum.
WSL	William Salt Library, Stafford.

VICTORIAN MAPS OF DEVON

A List of the Maps

The maps are numbered in chronological sequence of first publication and the title refers to that publication. The date given is that which is known or surmised. The chosen order of preference throughout the catalogue is surveyor, draughtsman, engraver, author, publisher. Names joined by a slash indicate that both/all were connected with the original issue. A slash is also used to identify separate works by the same draughtsman/ engraver. Where maps are commonly associated with another mapmaker or printer the alternative name is added in brackets.

118.	1839	De la Beche - index map	*Report on the Geology of Cornwall, Devon ...*
119.	1842	Archer/Dugdale	*Curiosities of Great Britain*
120.	1845	Becker/Fisher	*Fisher's County Atlas*
121.	1845	Becker/Besley - Route map	*The Route Book Of Devon - Route Map*
122.	1846	Becker/Besley - Devon I	*The Route Book Of Devon - Devonshire I*
123.	1848	Emslie/Reynolds	*Reynold's Travelling Atlas of England*
124.	1850	T Rowlandson	*Illustrated London News*
125.	1851	Rock & Co.	*Devonshire*
126.	1852	Walker/Knight	*The Imperial Cyclopaedia. ...*
127.	1852	Archer/Collins	*Collins Pocket Ordnance Railway Atlas*
128.	1854	Becker/Besley - Devon II	*The Route Book of Devon - Devonshire II*
129.	1854	G Oliver	*Supplement to Monasticon Diocesis Exoniensis*
130.	1855	Schenk-McFarlene/Black	*Black's Tourist's Guide to Devonshire & Cornwall*
131.	1856	V Brooks	*Ferny Combes*
132.	1856	Becker/Kelly	*The Post Office Directory...*
133.	1857	M Billing	*Martin Billing's Map of Devonshire*
134.	1857	Becker/Besley - Devon III	*The Route Book Of Devon*
135.	1858	W & A K Johnston	*A Handbook for Travellers in Devon & Cornwall I*
136.	1858	Weekly Dispatch I	*Weekly Dispatch*
		Weekly Dispatch II	*Cassell's Universal Atlas*
		Weekly Dispatch III	*Bacon's County Atlas*
137.	1859	Walker/Stanford	*A Guide to the South Coast of England*
138.	1859	G F Cruchley	*Cruchley's Map of Devonshire*
139.	1860	L Becker	*An Ecclesiastical Map of the Diocese of Exeter*
140.	1861	W McLeod	*Physical Atlas of Great Britain*
141.	1862	G Philip & Son	*Philips' Atlas of the Counties of England*
142.	1862	Bartholomew/Black	*Black's Guide To Devonshire*
143.	1864	W J Sackett	*A New Set of Diocesan Maps*
144.	1864	W Hughes	*The National Gazetteer*
145.	1865	T Spargo	*The Mines of Cornwall and Devon*
146.	1868	H James	*Report of the Boundary Commissioners*
147.	1868	Jackson & Partridge	*The Children's Friend*
148.	1870	Barnes & Howe	*A Parochial Boundary Map of the County of Devon*
149.	1872	G Philip & Son	*Philips' Handy Atlas*
150.	1872	Bartholomew/Heydon	*Plymouth, Devonport & Stonehouse*
151.	1874	T Murby	*Murby's County Geographies*
152.	1875	W Collins	*Collins County Geographies*
153.	1877	Bartholomew/Black	*The Encyclopaedia Britannica*
154.	1878	Stanford/Worth	*Tourist's Guide to Devon*
155.	1881	E Stanford	*The London Geographical Series - Stanford's Handy Atlas*
156.	1882	Bartholomew/Black	*Black's Guide To Devonshire - key map*
157.	1882	Bartholomew/Dulau	*Thorough Guide - index map*
158.	1883	Anon/JP	*Early Days*
159.	1883	Bryer/Kelly	*Kelly's Directory II*
160.	1885	Western Morning News	*Devon & Cornwall Electoral Districts*
161.	1885	E Stanford	*Parliamentary County Atlas*

162.	1885	Owen Jones (OS)	*Report Boundary Commission - 1885*
163.	1885	G W Bacon	*Hand Book Of England And Wales*
164.	1888	Ordnance Survey	*Index to the Ordnance Survey 4M:1"*
165.	1888	Cassell & Co./L&SWR	*Official Guide to the L&SWR*
166.	1890	Swiss & Co. (Cary)	*Hunting Map*
167.	1890	J Jaques	*Skits - A Game of the Shires*
168.	1891	W J Southwood	*The Birds of Devonshire*
169.	1892	Ordnance Survey	*Index to the Ordnance Survey 6M:1"*
170.	1893	J L W Page	*The Rivers Of Devon*
171.	1894	F S Weller	*The Comprehensive Gazetteer of England & Wales*
172.	1895	J L W Page	*The Coasts Of Devon*
173.	1895	Firks/Everson	*Westward Ho! Map of Devon & Cornwall*
174.	1895	Bartholomew	*New Reduced Ordnance Survey*
175.	1897	Bartholomew / Pattisons	*Pattison's Cyclist's Road Map of Devon*
176.	1897	Cassell & Co.	*The Rivers of Great Britain*
177.	1897	J Jervis	Torrington & Okehampton Railway map
178.	1897	Clark/Norway	*Highways and Byways in Devon and Cornwall*
179.	1899	E G Ravenstein	*Philips' Series Of District Maps ... Devon*
180.	1901	Anon - Gazetteer	*Gazetteer Of Devon For Tourists ...*
181.	1901	Ordnance Survey	*Devonshire*
182.	1901	H Inglis	*The Contour Road Book*

The

VICTORIAN MAPS OF DEVON

PRINTED MAPS 1838-1901

118

HENRY THOMAS DE LA BECHE

1839

Thomas Colby was Director, then Superintendent and later Director-General, of the Ordnance Survey from 1820 to 1846 under the Board of Ordnance before it was taken over by the War Office. Colby was a man of many interests who saw the Ordnance Survey as not just a survey team but as a centre for studies into the history, geology and natural history of the country. He was very interested in the geological nature of the land and even encouraged his officers to acquire geological information in areas surveyed.

In the early 1830s Henry Thomas De la Beche offered to add geological information to the Ordnance Survey sheets of the southwest for a fee of £300. He was given the go ahead and during the period 1832-1840 the various Ordnance Survey sheets of William Mudge's 1809 survey (**74**) were improved with the addition of geological information.[1] Although it took longer than he anticipated his results were impressive. *Mr De la Beche, ... has produced a geological map of the county of Devon, which, for extent and minuteness of information and beauty of execution, has a very high claim to regard. Let us rejoice in the complete success which has attended this first attempt of that honourable Board to exalt the character of English topography by rendering it at once more scientific and very much more useful to the country at large.*[2] The Geological Society subsequently campaigned for the establishment of a national geographical survey and, in 1835, the Ordnance Geological Survey was founded. Henry De la Beche became its first Director. From this date all of the Ordnance Survey's topographical sheets were adapted to convey geological information.

Henry De la Beche was born in London in 1796 but his family moved to Charmouth when he was young. They later lived in Lyme Regis (1817-21) and Henry began his coastal surveys in that area. He remained Director of the Geological Survey until his death in 1855.

His major work on the west country, *Report On The Geology Of Cornwall, Devon, And West Somerset* appeared in 1839 complete with index map. This detailed report, printed by W Clowes and Sons, had 648 pages and numerous maps and plans besides the index map. Some of the maps are signed *H De la Beche del* and *J Peake sculpt*, some are *Printed from Stone by Standidge & Co., London.* There were cross-sections of the county, scenes including Wicca Pool at Zennor and the interior of Fowey Consols copper mine, and plans of the Dolcoath and Fowey mines. The final chapters were devoted to the Action of the Sea on the Coast and Economic Geology.

A second simple hydrographical map of Devon and Cornwall was included which showed, dotted, the water-shed line of the district (see Fig.5 p.xxx).

Henry De la Beche (later Sir Henry) also wrote *A Geological Manual* (1831), *Researches in Theoretical Geology* and *Geology* (1834 and 1835 which were both published by Charles Knight), *Report on the State of Bristol and other Large Towns* (1845), *The Geological Observer* (1851), as well as translations of others' works and numerous reports and articles including one of his earliest on the *Present Condition of the Negroes in Jamaica* (published by Thomas Cadell in 1825).

No scale bar.

Size: 305 x 410 mm [Scale 1M=2.5 mm]

INDEX to the ORDNANCE GEOLOGICAL MAPS OF Cornwall, Devon, AND WEST SOMERSET.
Imprint: **Standidge & Co. Lith. London**. below title.

1. 1839 *Report On The Geology Of Cornwall, Devon, And West Somerset. By Henry T. De La Beche*
London. Longman, Orme, Brown, Green, and Longmans. 1839.

BL, RGS, GUL, E, T, TQ.

1. However, unlike the original series, the maps were never sold as a county set. The individual sheets all have full piano-key borders, numbers (Ea) and the title: *Ordnance GEOLOGICAL MAP of DEVON, WITH PARTS OF CORNWALL & SOMERSET, BY Henry T. De la Beche FRS &c.* and some have the mapseller's imprint: *Sold by J Gardner Agent for the sale of the Ordnance Surveys, 163 Regent St. London.* In addition sheets have *Index of Colours and Explanation of Signs.* All maps have signature *Engraved at the Drawing Room at the Tower by Benj n. Baker & Assistants* and also *Scale of Statute Miles* (CeOS). Plymouth Library has Sheet XXIII (Start Bay), Sheet XXIV (Whitsand Bay to Prawle Point) and Sheet XXV (Dartmoor).

2. Quoted by David Smith 1985; p. 60.

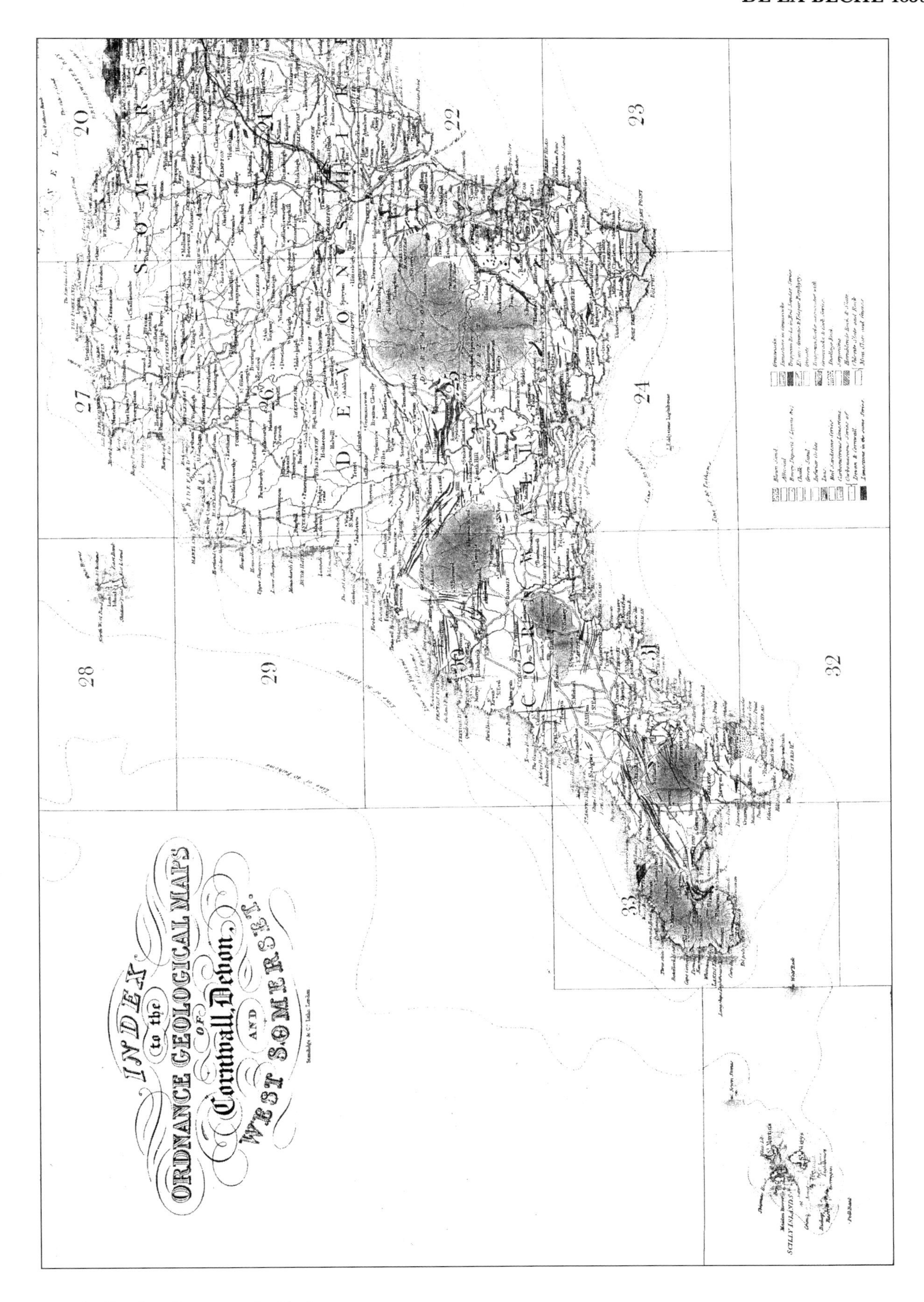

118.1 De la Beche *Report on the Geology*

119

ARCHER/DUGDALE

1842

Joshua Archer (*fl.*1841-1861) was a prolific engraver of maps. He had produced a series of maps for William Pinnock (**108**), a part series of Diocese maps for the *British Magazine* (1841-1844), some of the maps for Henry Fisher (see **120**) and engraved a set for Thomas Dugdale's *Curiosities of Great Britain.* These Dugdale maps were engraved by Archer to replace those by Cole and Roper (**67**) found in the early copies of Dugdale's (*fl.*1835-60) historical and topographical, gazetteer-style work. This was first published in 1835 using maps produced 30 years earlier. These were gradually replaced but the Tallis firm used up copies of the Cole and Roper maps and some copies of the *Curiosities* contain maps by both engravers.[1]

The maps continued to be used in works ostensibly by Dugdale, but edited by the Rev. J Barclay, William Goldsmith or E L Blanchard and published by John (*d.*1842) or Lucinda Tallis and company which often contained the same text. E L Blanchard (1820-89) was responsible for staging an annual pantomime at the Drury Lane Theatre in London.[2]

The sequence of railway improvements made to the maps is confusing: projected lines appear and are later deleted, eg the Crediton to Bideford line, added in two stages (states 4 and 5), was removed in 1858 (state 6) due to the delays in building the line (see p. xviii).

Size: 180 x 230 mm. **SCALE** (20=44 mm) **Miles**.

DEVONSHIRE. Plate no.**11** (EaOS). Signature: **Drawn and Engraved by J. Archer Pentonville London** (EeOS).

1. 1842 *Curiosities of Great Britain. England and Wales Delineated ... by Thomas Dugdale Antiquarian, ... Assisted by William Burnett ... Vol.I .*
London. John Tallis[3]. (1842). B.

2. 1842 Railway to Exeter. Imprint: ***Engraved for* Dugdales England and Wales *Delineated*** (CeOS). (E).

Curiosities of Great Britain. England and Wales Delineated ... by Thomas Dugdale
London. L Tallis. (1842). BCL, C[4], W.

3. 1842 *Dugdale* imprint deleted.

Curiosities of Great Britain. England and Wales Delineated ... Vol.II
London. L Tallis. (1842). CB, BL.

The Universal English Dictionary ... by The Rev. J Barclay
London and Glasgow. J & F Tallis. (1846). MCL.

1. David Smith; 1985; p. 123.
2. R Carroll; 1996; p. 287.
3. Many of the volumes containing Devon are without publisher's name. Other volumes in one and the same series have either J Tallis or L Tallis, ie both names are found in the same series. The dates of all volumes are speculative and the leading county carto-bibliographers often differ as to dating. This is more difficult for multi-volume series such as the Dugdale works offered in sets of up to 9 volumes.
4. Collection of maps without title or text.

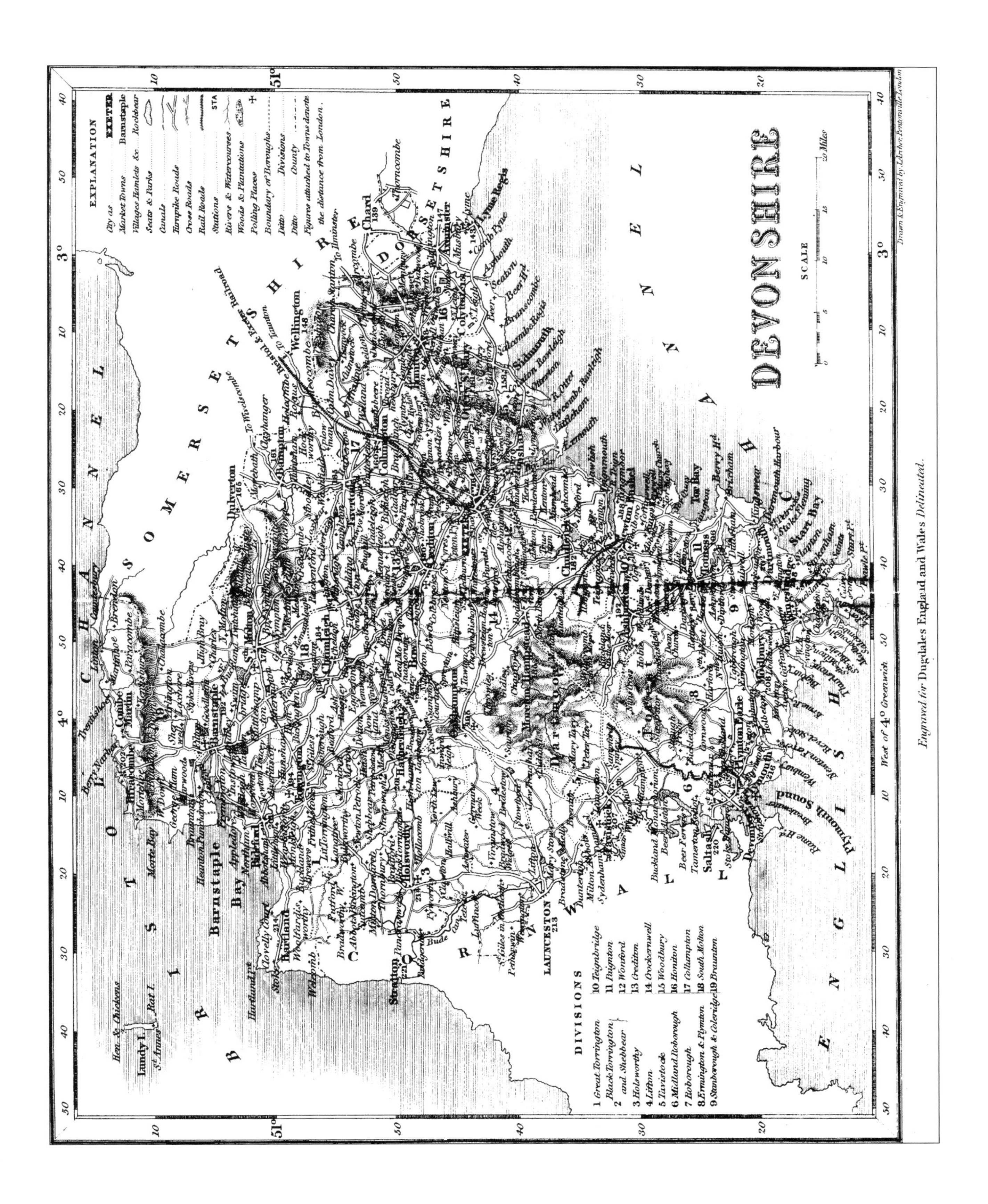

119.2 **Archer/Dugdale** *Curiosities of Great Britain*

4. 1850 Railways to Barnstaple (Taw Vale Ry), Exmouth and line complete to Kingswear.

Curiosities of Great Britain. England and Wales Delineated ... Vol.II
London. L Tallis. (1850). W, B, BL.

The Universal English Dictionary ... by The Rev. J Barclay
London and New York. John Tallis & Co. (1850). CB.

2. Barclay's Universal English Dictionary Improved
London and New York. John Tallis & Co. (1850). WSL.

5. 1850 Railway: Bideford, South Molton (from Umberleigh Bridge), Berry Head, Ashburton (from Newton Bushel).

Curiosities of Great Britain
London. L Tallis. (1850). W.

Modern & Popular Geography. Vol. I ... *By William Goldsmith*
London, Edinburgh & Dublin. J & F Tallis. (1850). CB.

6. 1858 Railways erased: Crediton to Bideford (V & R still just visible), South Molton and Torquay to Kingswear. Roads revised: Torquay, Brixham and Lympstone (Exmouth).

Dugdale's England and Wales Delineated ... edited by E L Blanchard
London. L Tallis. (1858). BL.

7. 1860 The Crediton-Bideford railway is reinstated with slightly different route, ie passes through last two letters of Eggesford (the V of Vale just visible); Ashburton route erased. (E).

The Topographical Dictionary of England & Wales Vol. I
London. L Tallis. (1860). C.

8. 1860 Railways: Plymouth-Tavistock, line into Cornwall, Torquay-Paignton reappears.
(The V now barely visible)

Tallis's Topographical Dictionary of England & Wales
London. L Tallis. (1860). CB.

Topographical Atlas of England & Wales
London. L Tallis. (1860). CB.

Curiosities of Great Britain Vol. II (of IX)
London. L Tallis. (1860). CB.

The Topographical Dictionary of England & Wales Vol. I (of VI)
London. L Tallis. (1860). CB.

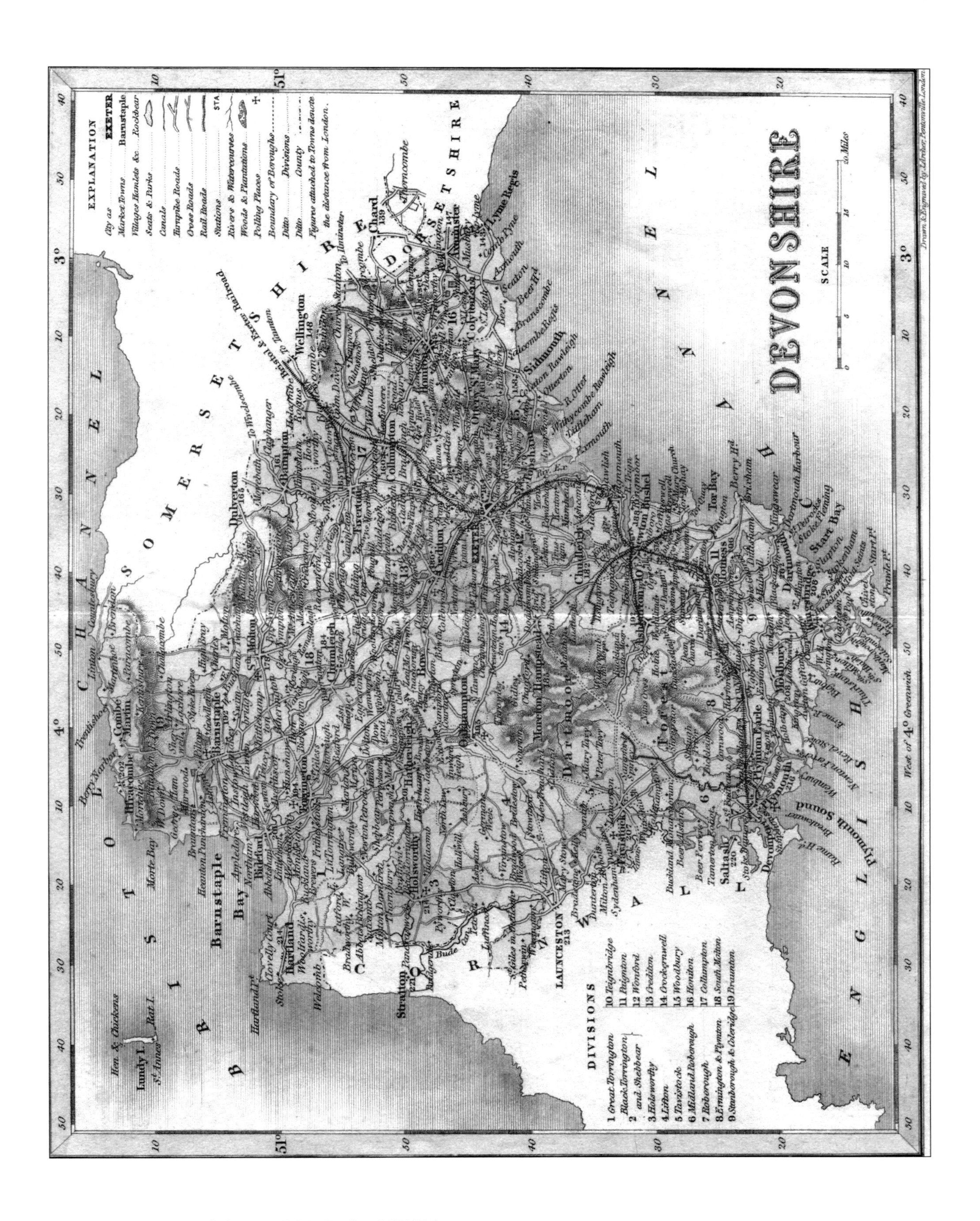

119.6 Archer/Dugdale *Dugdale's England & Wales*

120

BECKER/FISHER

1845

Henry Fisher (*fl.*post-1816) was a publisher with premises at Caxton Press, Angel St., St. Martin-le-Grand, London. Although Henry died in 1837 the firm continued to operate as Fisher, Son & Co. The company had already printed and published an attractive map of Devonshire engraved by J & C Walker for *Devonshire Illustrated c.*1831 (**102**).

The preparation of *Fisher's County Atlas of England and Wales* was started by James Gilbert and the earliest maps[1] bear both his name and imprint: *London, Published April 1. 1842 for the proprieter M. Alleis, by James Gilbert, 49 Paternoster Row* or *Published for the proprieters by Grattan & Gilbert, Map agents by appointment to the Hon. Board of Ordnance, 49, (removed from 51) Paternoster Row.* These maps were *Drawn and Engraved by J. Archer, Pentonville, London. for GILBERT'S COUNTY ATLAS.* Apart from those noted above, only the map of England and Wales is dated - 1845.

The project seems to have been taken over by Fisher, Son & Co. at a very early stage and his imprint was substituted on the maps of Gloucester and Oxford although the reference to Gilbert's atlas was retained. The remainder of the maps all bore Fisher's imprint and usually the draughtsman's and engraver's signatures but the maps were sometimes trimmed losing these. Although the first ten maps were engraved by Joshua Archer, the later maps were prepared by Francis Paul Becker who went on to produce a large number of maps including others of Devon (see **121**, **128** and **132**).

Most of the maps were based on J & C Walker's maps of 1836 (**116**) as is Devon which is a lithograph map with hand colouring with emphasis placed on the eight boroughs. The railway is shown as far as the outskirts of Exeter, which it reached in 1844, and some canals are shown, eg Bude to Launceston (1825-1891), but neither the Tavistock (1817-1832) nor Stover canals. The Haytor granite tramway is shown which was in operation from 1820 to 1858.

Size: 355 x 505 mm **SCALE OF MILES** (20=90 mm).

DEVONSHIRE. Signatures: **Engraved on Steel by the Omnigraph, F. P. Becker & Co. Patentees**. (EeOS) and **Drawn by F. P. Becker & Co. 12, Paternoster Row**. (AeOS). Imprint: **FISHER SON & CO. LONDON & PARIS** (CeOS)[2]. Railway to Exeter with (erroneous) line from Exeter north to Tiverton.

1. 1845 *Fisher's County Atlas Of England And Wales*
London, Liverpool & Manchester. Fisher, Son & Co. (1845), (1845[3]).
CB, BL, B, Leeds; C, NLS.

1. The original Gilbert maps were of Berks, Derby, Glocs, Leics with Rutland, Lincs, Northants, Notts, Oxford and Warwick.
2. Most maps have been trimmed close to the edge and the imprint is missing.
3. The Gilbert maps now bear altered imprints.

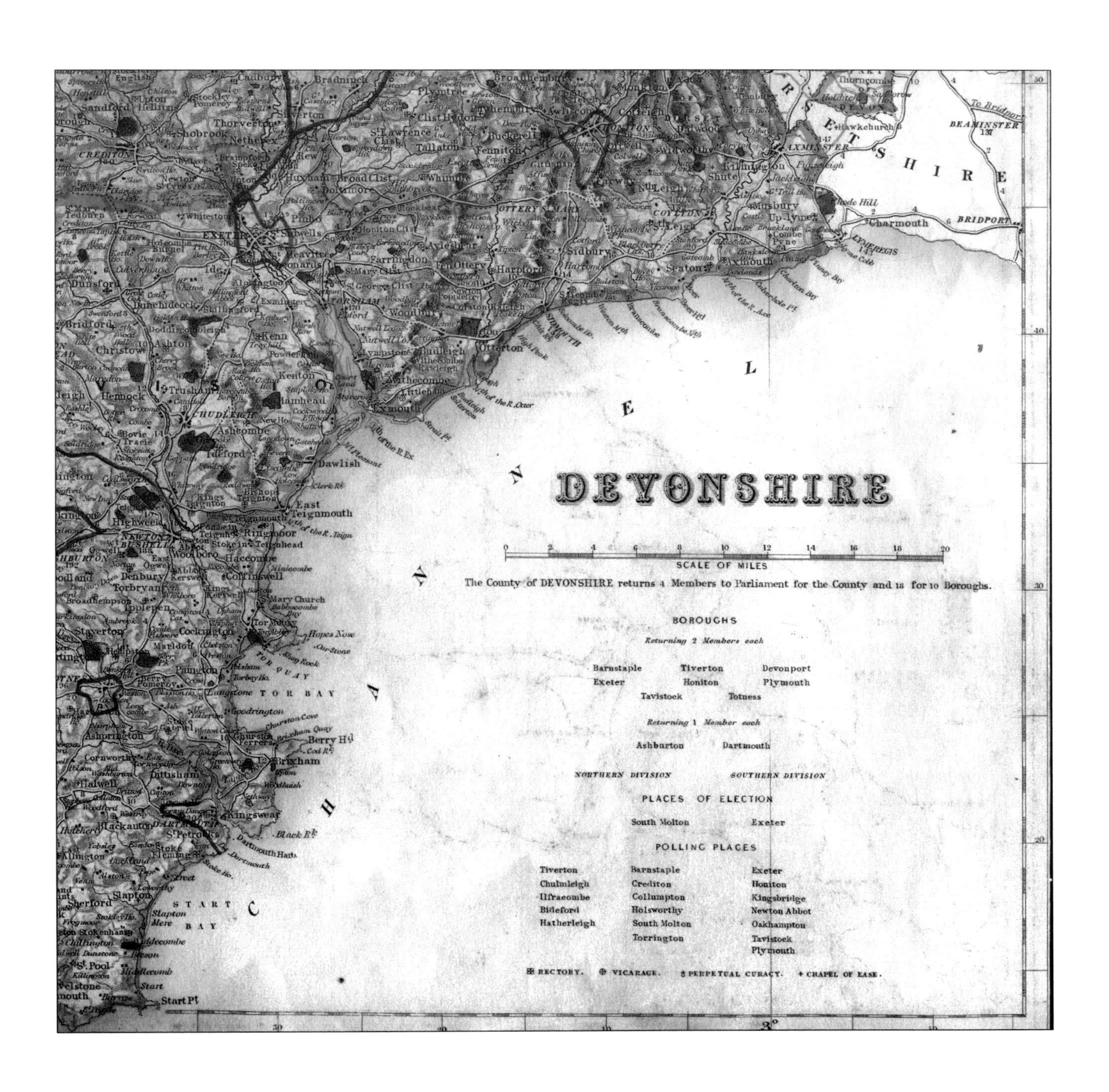

120.1 Becker/Fisher *Fisher's County Atlas of England And Wales*

121

BECKER/BESLEY – Route Map

1845

Very few county maps of Devonshire were produced by local publishers, but the maps of Henry Besley are among the few which were produced in Exeter. The Besleys were important local printers, stationers and publishers of almanacks, guide books and directories which often included maps.

Besley produced various maps for his series of Guide Books; both county maps on one or more sheets (see **122**, **128** and **134**) and a *Route Map.* His guide, *The Route Book Of Devon,* was first published *c.*1845 with a *Second Edition* appearing approximately a year later. The next issue was the *New Edition* published *c.*1850. There were (at least) three separate issues of the *New Edition* on the evidence of dates contained in the text, engravings and adverts. The issues of 1850 and 1854 have the same text.

The *Route Map* was a folding map with suggested touring routes: fourteen routes all emanating from Exeter, beginning in a clockwise direction starting with Exeter to Crediton. The map appeared in early issues of the *Route Book* upto *c.*1856.

Size: 235 x 285 mm. No scalebar. [Scale 1M=3 mm]

ROUTE MAP of the ROADS OF DEVON. TO ACCOMPANY ROUTE BOOK OF DEVON. Protected by Provisions of Act of Parliament 5 and 6 Victoria. Cap 45. PRINTED AND PUBLISHED BY HENRY BESLEY, DIRECTORY OFFICE, SOUTH ST. EXETER. (Ee) Signature: **Engraved by the Omnigraph F P Becker & Co. Patentees.** (AeOS) **12, Paternoster Row.** (EeOS). The Exeter & Bristol Railway is dotted through Somerset, then drawn full to Exeter. From there it is dotted to Plymouth, presumably planned. There is no scale or compass and the map extends into the border north and south.

1. 1845 *The Route Book Of Devon: A Guide For The Stranger And Tourist*
Exeter. H Besley. (1845[1]). BL, E.

The Route Book Of Devon ... Second Edition
Exeter. H Besley. (1846). BL, RGS, E.

The Route Book Of Devon ... A New Edition
Exeter. H Besley. (1850[2]), (1854[3]), (1856[4]). E, T[5]; E; E, RGS.

1. Advertised as *Shortly will be published* in Henry Besley's *West of England, Or, Improved Almanack, 1845.*
2. Last date in text relates to expected completion of a basin and lock at Devonport in 1851 (p. 313).
3. Last dates in text as above but engravings are dated 1853 and 1854.
4. Last dates in text as above but an advert relates to Queen Victoria's visit to the county in 1856.
5. Lacking the county map (see Besley **122**).

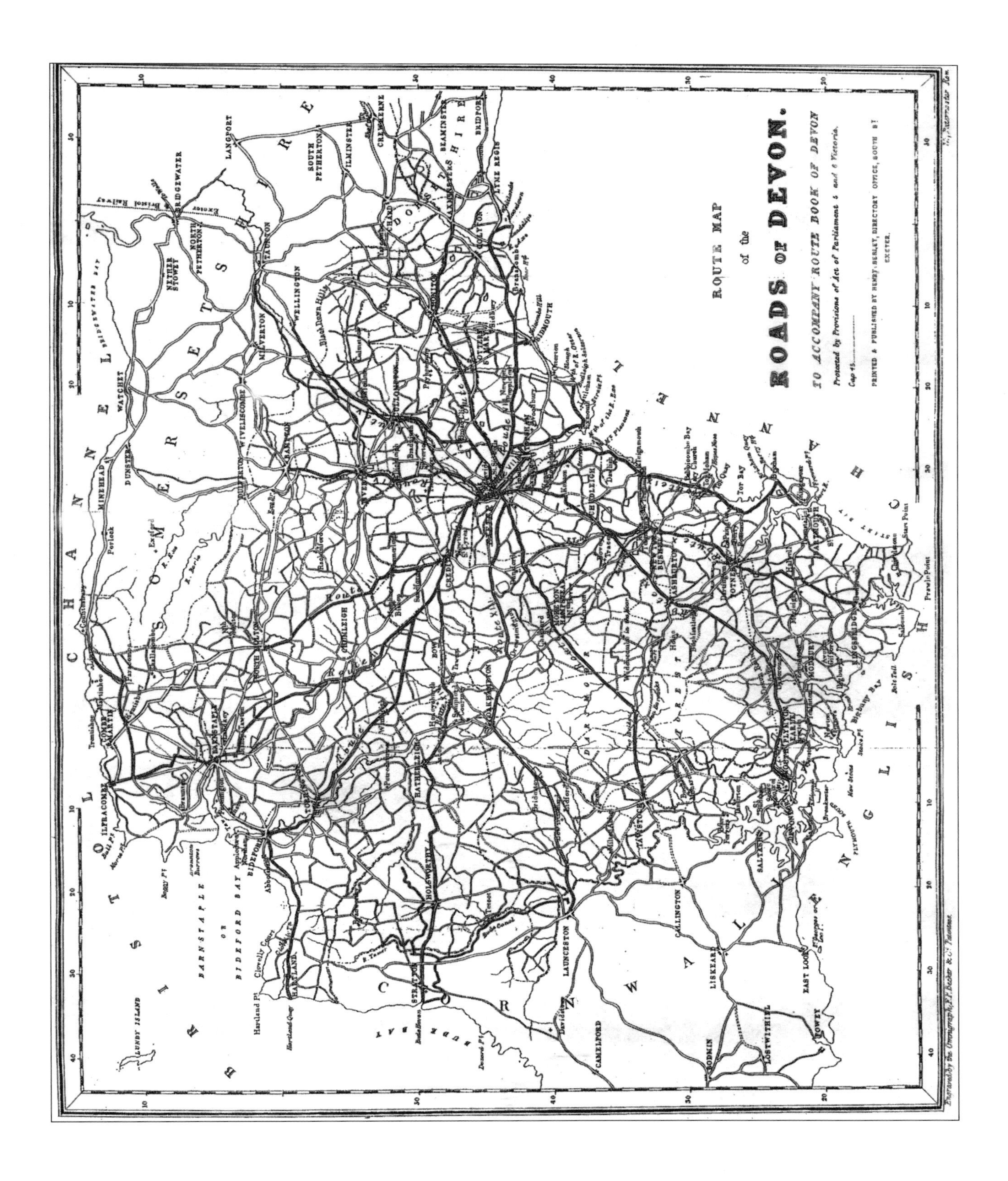

121.1 Becker/Besley *The Route Book Of Devon* – Route Map

122

BECKER/BESLEY - Devon I

1846

The editions of the *Route Book Of Devon* (see **121**) included a *Route Map*. For the second edition Henry Besley also included a general county map, based on the larger J & C Walker map of 1836 (**116**). At first glance it seems similar in style and layout, however the Walker map has *NORTHERN DIVISION* and *SOUTHERN DIVISION* written across the map whereas Besley simply has the two letters *N* and *S*. The Walker is also larger (320 x 385 mm).

The text of the first and second editions refers to the railway terminating at Teignmouth; this was opened on 31st May 1846. The county population is given as 495,186 and Parliamentary Representation is 18 members for 10 Boroughs. Tables show Places of Election, Divisions and Hundreds.

This county map only appeared in two editions of the *Route Book*. A second *New Edition* appeared *c.*1854 and a new, larger map was included (see **128**).

Size: 238 x 288 mm. **English Miles 69.1 - 1 Degree** (8F+15=48 mm).

DEVONSHIRE (Ed). Imprint: **Printed & Published by Henry Besley, Directory Office, South Street, Exeter.** (CeOS). The proposed railway to Exeter is shown as a pecked line.

1. 1846 *The Route Book Of Devon ... Second Edition*
Exeter. H Besley. (1846). BL, RGS, E.

2. 1850 The railway is now shown as solid line to Plymouth.

The Route Book Of Devon ... New Edition
Exeter. H Besley. (1850[1]). E.

1. Last date in text relates to expected completion of a basin and lock at Devonport in 1851 (p. 313).

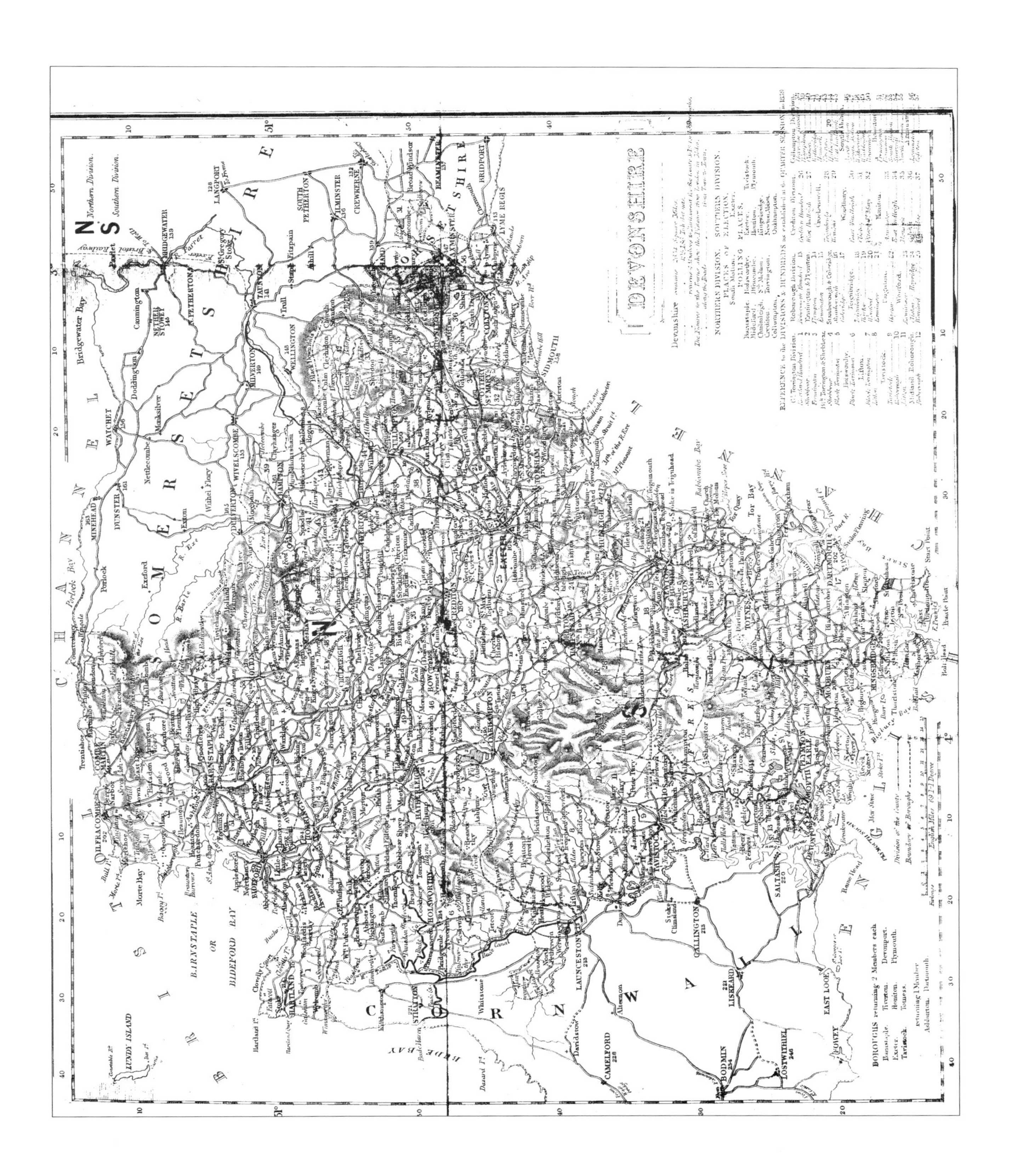

122.# Becker/Besley *The Route Book Of Devon* – County Map

123

EMSLIE/REYNOLDS

1848

The publisher of only one important county cartographic work, *Reynolds's Travelling Atlas*, James Reynolds had premises at 174 Strand, London, from 1836 until his death in 1876. In addition to the successful series of these maps he also produced Booth's *Plan of London* 1845, a map of *Suffolk* 1850, and *Astronomical Diagrams* 1851. His business continued as James Reynolds & Sons.

John Emslie (1813-75) was in business from 1843 and was elected to the Royal Geographical Society in 1863 (eleven years before Reynolds).[1] The maps he engraved were based on the maps of Sidney Hall (**101**).

The original maps had no geological information. This was added in 1860 together with a detailed text. The plate of this map was used for intaglio printing from 1848-1860. From *c.*1860 lithographic transfers were used and continued in use until at least 1927, in successive editions of *Reynolds's Travelling* or *Portable Atlas* and *Reynolds's Geological Atlas* before they were incorporated into Stanford's *Geological Atlas of Great Britain* by H B Woodward. When that atlas was issued in 1903 the map of Devon had a second table of New Series Ordnance maps (Ad) superimposed on an index map of Devon (22 x 20 mm) as well as the table of the Old Series (Da, but retitled) and Stanford's imprint: *London. Edward Stanford. 12, 13 & 14, Long Acre, W.C.*

Size: 175 x 235 mm. **ENGLISH MILES** (16= 40 mm).

DEVONSHIRE. Signature: **Drawn & Engraved by John Emslie**. (AeOS). Imprint: **Published by J Reynolds 174 Strand**. (CeOS). Plate **10** (EeOS). Railways to Plymouth Earl, Crediton, and Tiverton from Tiverton Junc. Note: Ringmore occurs twice, and South Huish is mistaken for Thurleston (sic).

1. 1848 *Reynolds's Travelling Atlas Of England*
London. James Reynolds. 1848. CB.

Reynolds's Travelling Atlas Of England
London. Simpkin, Marshall & Co. and James Reynolds. 1848. CB[2], NLS, B, EB.

2. 1848 Proposed railways to Barnstaple, Saltash, Exmouth and Newton Abbot to Ashburton.

Reynolds's Travelling Atlas Of England
London. Simpkin, Marshall & Co. and James Reynolds. 1848, (1849), (1854). BL; C; W.

3. 1854 Railway completed to Barnstaple.

Reynolds's Travelling Atlas of England
London. Simpkin, Marshall & Co. and James Reynolds. (1854), (1857). C; C.

4. 1860 **a**) Maps are engraved, have geological contours and reference numbers and are geologically coloured. Railways to Tavistock, Kingswear, Bideford and Saltash with L&SWR to Exeter.

Reynolds's Geological Atlas Of Great Britain
London. James Reynolds. 1860. BL, W.

b) Maps in same state but produced lithographically.

Reynolds's Geological Atlas Of Great Britain
London. James Reynolds. (1860). EB, KB[3].

1. R Carroll; 1996; p. 291.
2. Date of atlas has been erased by hand!
3. Undated: variant copy with pages set incorrectly (11-16 in back to front).

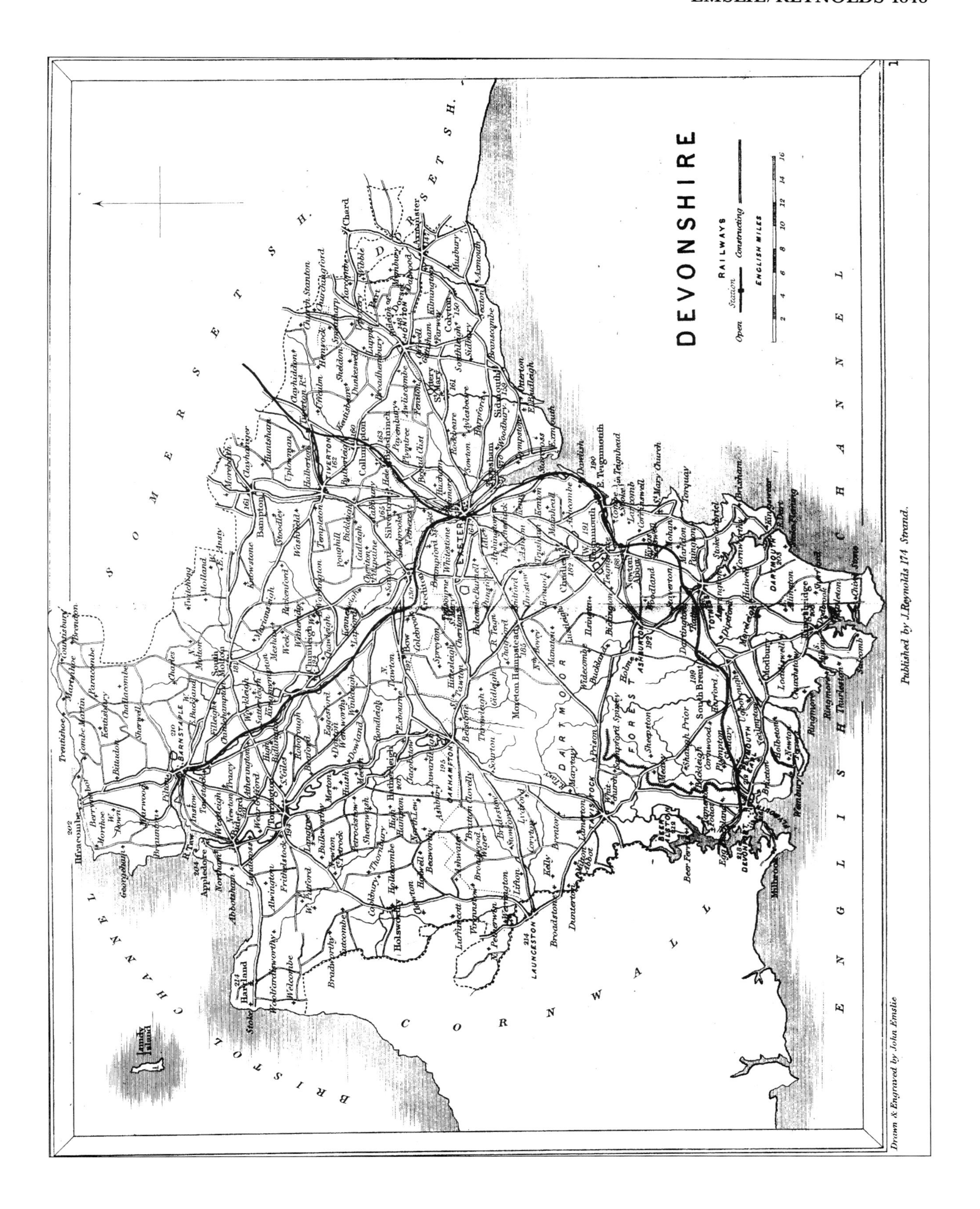

123.3 Emslie/Reynolds *Reynold's Travelling Atlas*

5. 1864 Outer frame, signature and plate number removed leaving inner border line, the border is now broken for Foreland and Prawle Point. Plate number **10** is raised to suit. Landslip note at Axminster, Mt Edgcumbe and Plymouth Breakwater added.

Portable Travelling Atlas Of England And Wales
London. James Reynolds. (1864). C[1].

6. 1864 Planned railways: Dartmoor loop to Launceston, Newton Abbot to Ashburton, Moreton Hampstead and a line southeast from Axminster. Exmouth line added alongside previous double line. Uncoloured but geological information still present.

Portable Atlas of England & Wales; With Tourist's Guide
London. James Reynolds. (1864). BL.

Portable Atlas of England And Wales; With Tourist's Guide
London. James Reynolds and George Musgrove. (1864). CB, NLS.

7. 1864 Number **10** is now inside the border (Ee). Planned railway to Okehampton and Lydford is shown. Linton added. The second Ringmore is deleted.

Reynolds's Geological Atlas Of Great Britain New Edition
London. James Reynolds. (1864). BL, W, C[2].

8. 1868 Line to Moreton Hampstead and Dartmoor Loop completed.

Reynolds's Geological Atlas Of Great Britain New Edition
London. James Reynolds. (1868). EB.

9. 1875 South Huish replaces the (old) Thurleston (sic). Railways now shown from Barnstaple to Ilfracombe and to Taunton.

Reynolds's Geological Atlas Of Great Britain New Edition
London. James Reynolds. (1875). C.

10. 1889 Imprint: **London. Published by James Reynolds & Sons 174 Strand.** Inset table; *Index to Sheets of Geological Ordnance Map* added (Ad). Railways shown to Seaton, Holsworthy, Sidmouth, Ilfracombe, Ashton, Hemyock and from Launceston to Halwell. Comprehensive additions.

Reynolds's Geological Atlas of Great Britain Second Edition
London. James Reynolds & Sons. 1889. BL, KB.

1. Cambridge copy has hand-written '[1864]'.
2. The lower (false) Ringmore is erased on the sheet of the Cambridge edition.

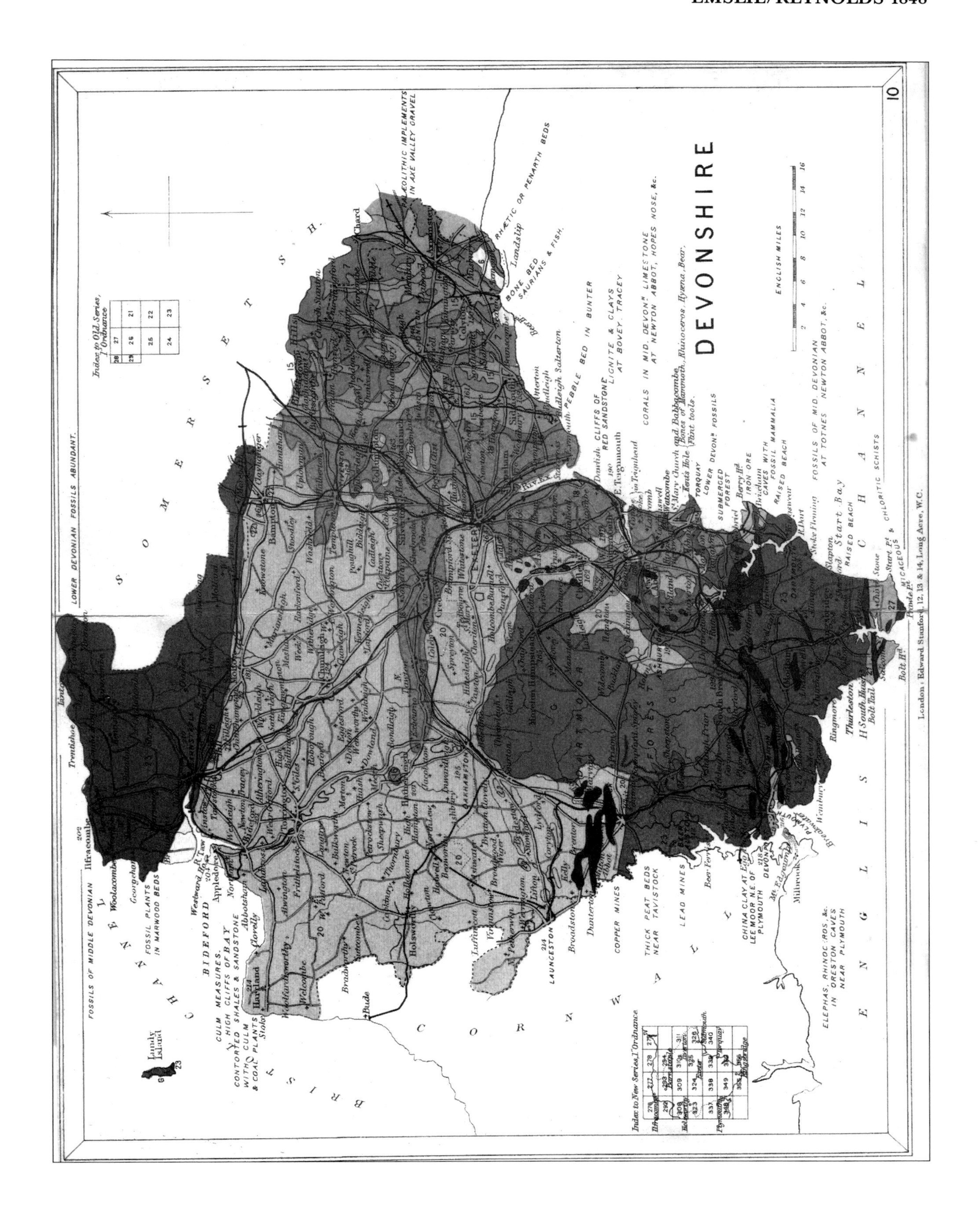

123.11 Emslie/Reynolds Stanford's 1904 edition with added inset map

124

THOMAS ROWLANDSON

1850

In 1850 three reports on the agriculture of the westcountry appeared in the *Illustrated London News.*[1] The first in the series, Cornwall, appeared on August 17th on page 151 which was possibly part of a supplement.[2] A week later Part II was published with a report on Devon in the edition of August 24th. The report appeared on page 178 which was not a supplementary page. Part III, covering Somerset, was published on another supplementary page (p. 218) on September 7th.

The first report on Cornwall has a short introductory note written by Thomas Rowlandson[3]: ... *it is intended to give a series of papers on British Agriculture.* However, it would appear that only these three articles were published. Devon was signed Thomas Rowlandson and Somerset merely T.R. suggesting that Rowlandson wrote all three and intended to write more.

The maps themselves are small untitled maps of each county showing only the principal towns by number with a reference below. Somerset and Devon have the main railway lines, eg Devon with the railway from Exeter to Plymouth, but the line to Exeter is erroneously drawn from the east, ie from Yeovil (which was only completed by the L&SWR in 1860) and not the line from Bristol completed in 1844. The sea is shaded and there is a North point (Bd on Devon). Cornwall and Somerset also have coats of arms.

Size of engraved area: 65 x 60 mm. No scale.

No title, but text begins **PART II. - DEVONSHIRE**

1. 1850 *British Agriculture. By Thomas Rowlandson. Part II - Devonshire.*
London. The Illustrated London News. 1850. BL, (E).

1. Kit Batten; Illustrated London News; *IMCoS Journal 75*; Winter 1998; pp. 45-48.
2. The page reverse bears the word SUPPLEMENT top right. Somerset has this on the same page as the article.
3. Somers Cocks (1977) has one entry for a Thomas Rowlandson (S.50); *Sketches from Nature*, publ. T Rowlandson, London, 1818. This contains one aquatint *A View in Devonshire.*

BRITISH AGRICULTURE.

BY THOMAS ROWLANDSON.

PART II.—DEVONSHIRE.

DEVONSHIRE is famed for its cider and clouted cream. Every writer on the agriculture of this county has remarked the little attention paid by its orchard cultivators to the fermentation of the expressed juice of the apple, on a due attention to which the strength and flavour of cider depends. Whilst in Devonshire during the late meeting of the Royal Agricultural Society, we took every opportunity of ascertaining what was the general character of this celebrated local tipple, we found, however, the qualities so various, as to render a detailed description impossible. From information we had previously received, we were prepared to meet with an article different from that exported to the various large cities and towns of the empire — the exported being designated as "sweet,' in contradistinction to that locally consumed, which is termed "dry" cider. In our imagination we anticipated to find that, with one exception, the dry cider was something similar to the dry wines of the Rhine; but we regret to state that all which fell in our way was either vapid, or, if astringent, it was due to the alcohol having been converted into vinegar. Both these faults might be remedied by a careful attention to the process of fermentation—for we tasted some of excellent "*bouquet*," but of deficient strength.

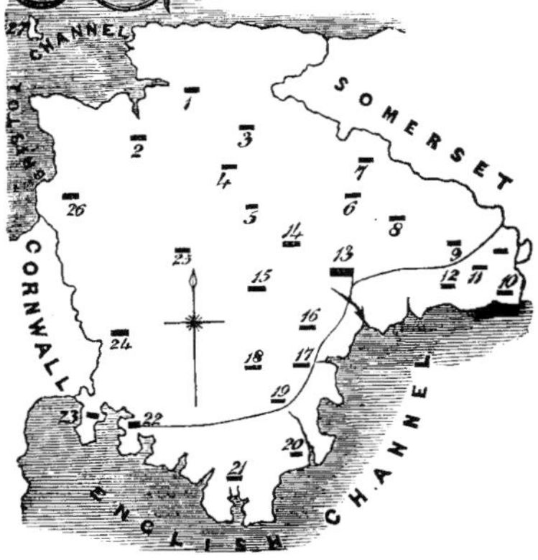

1 Barnstaple
2 Bideford
3 South Molton
4 Dorrington
5 Chumleigh
6 Tiverton
7 Bampton
8 Chilhampton
9 Honiton
10 Lyme Regis
11 Axminster
12 Colyton
13 Exeter
14 Crediton
15 Moreton Hampstead
16 Chudleigh
1 Newton Abbott
18 Ashburton
19 Totness
20 Dartmouth
21 Kingsbridge
22 Plymouth
23 Devonport
24 Tavistock
25 Oakhampton
26 Holsworthy

From these remarks it will be perceived that we are no admirers of the bulk of the home-made "undoctored" ciders, so emphatically praised by native Devonians. The fact is, fermentation is a most delicate process, requiring constant attention and care—so much so, that the necessary superintendence can only be remunerative when carried out on a large scale, and cannot be given in the management of the produce of the majority of orchards. For this reason we are Goths enough to give a preference to the cider manufactured by the dealers, even at the risk of having it "doctored"—which, we believe, is nothing more than using a little sugar occasionally, to give the cider strength—when, as sometimes happens in cold summers, the apples are deficient in the saccharine material.

We must confess to enjoying a glass of cider (we presume, home-made), free from the defects alluded to, at the hospitable *fête champêtre* of Sir T. Dyke Acland, Bart., M.P., at Killerton. It ought, however, to be remarked, that at Killerton and its vicinity there are irruptions of volcanic ash amongst the pervading red sandstone rock—a circumstance which, we believe, is invariably found to increase the quantity and quality of orchard produce.

Good general farming land appears to hold a proportional value for growing apples. It is worth noticing that some of the most celebrated orchards for producing cider in Devonshire are situated on a similar geological formation with those occupied by the best orchards in the counties of Hereford and Worcester, viz. on the cornstones and marls of the old red sandstone.

The value of cider varies from £2 to £5 per hogshead; but the greatest portion does not fetch above £2 10s. The average produce is said to be ten hogsheads per acre.

124.1 Rowlandson *Illustrated London News*

125

ROCK & CO.

1851

Although vignettes had appeared on a number of earlier maps, only very few vignettes were engraved on county maps during Victoria's reign. Martin Billing's large map had two full engravings of Plymouth buildings and Pattison's cyclists' map, published towards the end of the century, had advertising vignettes around the Devon map.

One unusual map with vignettes illustrating towns was that published by Rock & Co. of London. Two brothers, William[1] and Henry Rock, formed a London company in 1834 to produce line engravings, topographical bookplates depicting places in England and Christmas cards which continued very successfully until the end of the century.[2] Towards the middle of the century they produced a series of small specialised maps. To date only five county maps, including Devon, have been discovered. While the other copies have been folded and inserted into hard covers[3], the only known copy of the Devon map is a plain single sheet suggesting that it may have been designed for an atlas or work on the county rather than as a folding map. The firm is known to have produced other cartographic work, eg a map of London for James Reynolds in 1860.[4]

The Devon map has a total of 19 circular mini-vignettes (*c.*8 mm) engraved within the map detail at important Devon towns. If the loose copy so far discovered was from a Devon guide then the towns depicted could have been described in detail in the text. The signature is difficult to read but may be *Stevenson Sc.*

The railway to Exeter and Plymouth with the Tiverton branch line is shown. This was completed *c.*1848. The line to Torre (Torquay) was completed about the same time but is not shown. However, the line from Exeter as far as Crediton is illustrated; this was opened in 1851 with the extension to Barnstaple not being completed until 1854.

William Rock was born in Barnstaple in 1802 but moved to London where he later established his successful publishing business. He kept his attachment to north Devon as evidenced by endowments, the establishment of the North Devon Athenaeum and, not least, by the large number of views his company produced of the west country. Between 1848 and 1876 Rock & Co. published at least 260 views of Devonshire (out of some 7000 produced by the firm)[5]. They also produced engravings for the *Hand-Book To South Devon* by W Wood. The 1st and 2nd[6] editions have a number of delightful engravings (engraved area approx. 7 x 10 cm) some of which are dated, eg the engraving of Sidmouth executed by E M Underdown which is dated *Sept 20th 1849* or that of Exmouth dated *Feby 10th 1850.*

Size: 185 x 245 mm. **Scale of Miles** (20=50 mm).

DEVONSHIRE (Ed). Signature: **Rock & Co. London.** (AeOS). Signature: **Stevenson Sc** (?). Railways to Exeter, Plymouth, Tiverton and Crediton. Graticuled border broken for north coast and Prawle Point.

1. 1851 Loose sheet, source unknown. (DEI).

1. William Rock was recorded at 5 Cumberland Row, Walworth, Newington as printer in 1844. W B Todd; 1972.
2. B Adams; *London Illustrated*; London; 1983.
3. The other counties are Derby, Kent, Lincoln and Sussex. The map of Lincolnshire is known in two states: the first is dated 1850 and was issued in a cover; the second state is found in a book on agriculture, *A Farming Tour, Or Hand Book On The Farming of Lincolnshire,* by A Lindsey Yeoman and sold by Simpkin, Marshall And Co. 1854. See R A Carroll; 1996; p. 298.
4. R Hyde; *Printed Maps of Victorian London*; Folkestone; 1975.
5. Somers Cocks; 1977; p. 9.
6. This is identical to the 1st edition but has a label pasted on cover.

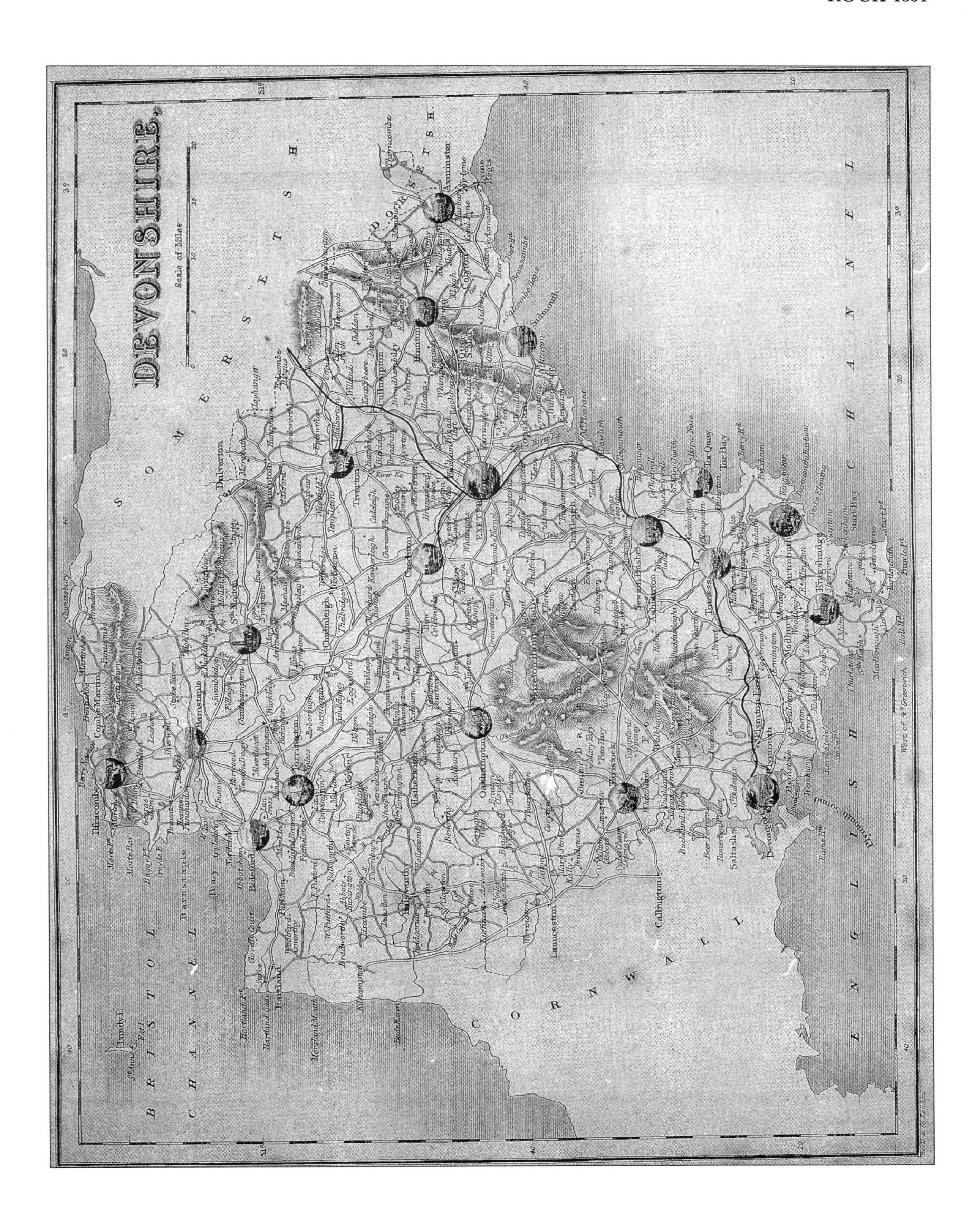

125.1 Rock Loose sheet

126

WALKER/KNIGHT

1852

Charles Knight (1791-1873), author and publisher, was the son of Charles Knight, bookseller and Mayor of Windsor. In 1812, having been a local reporter, he joined his father as proprietor of the *Windsor & Eton Express.* By 1814 he had become interested in popular instruction to bring all kinds of knowledge, mixed with lighter matter, within the reach of the poorest; this led him in 1828 to agree to superintend the publications of the Society for the Diffusion of Useful Knowledge (S.D.U.K.), which had just been formed by Henry Brougham, Rowland Hill and 'the leading statesmen, lawyers and philanthropists of the day'. In the next year he became the Society's publisher, a post he held until the Society closed in 1846.

He, with other publishers, served on the committee of the Association for the Abolition of Duty on Paper. In 1836 the tax of 3d per pound was halved and in 1860 it was abolished.[1] In 1840 he produced the first English colour-printed atlas using the relief process he had patented in 1838.[2] A difficult process it was soon superceded by chromo-lithography. He was a noted publisher of a number of magazines designed to impart knowledge: *Plain Englishman* 1820-22, *Knight's Quarterly* 1823-4, the *Penny Magazine* 1832-45, which in its first year sold over 200,000 copies, and the *Penny Cyclopaedia* 1833-44. After the Society ended he continued in the same vein with, amongst many others, the *English Cyclopaedia* 1853-61.

As a map publisher Knight, like many others, borrowed from the Ordnance Survey. In 1830 Baldwin and Cradock published maps for the SDUK[3], by *J & C Walker.* These were acquired by Knight and used in his *Imperial Cyclopaedia,* either issued in weekly parts or in 16 parts, completed in 1853.

Knight also published some of Henry De la Beche's works (**118**) as early as 1834. He compiled a *Popular History of England* which became, in abridged form, a best-selling school book. He was also an inventor and dreamer: his proposal to collect the newpaper duty by means of a stamped wrapper is said to have given Rowland Hill the idea of the penny post.

Size: 210 x 160 mm. **Statute Miles 69.1 = One Degree** (5+25=64 mm).

DEVONSHIRE (Ee). Imprint: **London, Charles Knight. 90, Fleet Street**. (CeOS). Small inset map of **Lundy Island** (Aa) with double line border. Railways to Tiverton, Crediton, Tor Quay and Devonport.

1. 1852 *The Imperial Cyclopaedia. Cyclopaedia of Geography.*
London. Charles Knight. (1852[4]). BL, (DEI), (KB).

2. 1852 Imprint deleted. The map has been realigned: the right hand border moved slightly with Axminster inside the border; Wambrook and Membury in Somerset are shown and St.(Keyne) is clear on the left hand border. The Lundy inset is slightly repositioned and now has a single line border. The top border is lower and most of Wales is omitted. Title and scale bar are redrawn; the title now aligns with Start and the scale bar is lengthened at both ends (10+30=85).

The Imperial Cyclopaedia. Cyclopaedia of Geography.
London. Charles Knight. (1852). B.

1. David Smith; 1985; p. 30.
2. Others were experimenting in colour about the same time. William Hughes, in his *Journey-Book of England,* printed only Berks, Derby and Hants in colour (Kent is sometimes found hand coloured). Burden; 1988 (1991); p. 166.
3. Sheet **ENGLAND IV** (260 x 380 mm) depicts everything west of Salisbury and south of Bridgend. It has the imprint: *Published by Baldwin and Cradock 47 Paternoster Row June 15th 1830* and the Walkers' signature (EeOS). Surprisingly the map was copied and reprinted. Later versions show everything west of St Albans Head and as far north as Caermarthen (315 x 385 mm). Copies of this are imprinted: *Published by Charles Knight & Co. Novr, 1845.* Sheet **ENGLAND No. 6.** Both maps were produced *Under the Superintendence of the Society for the Diffusion of Useful Knowledge.*
4. Both maps could have been printed earlier. Although both prefaces are dated (June 23) 1852 this is possibly when parts were collected and bound. The title pages have no date - the BL copy has a second title page: *The Cyclopaedia of The British Empire.* It has not been clearly established which copy is the earlier. Two additional states of Berkshire have been noted; one with the imprint revised by deleting *London.*

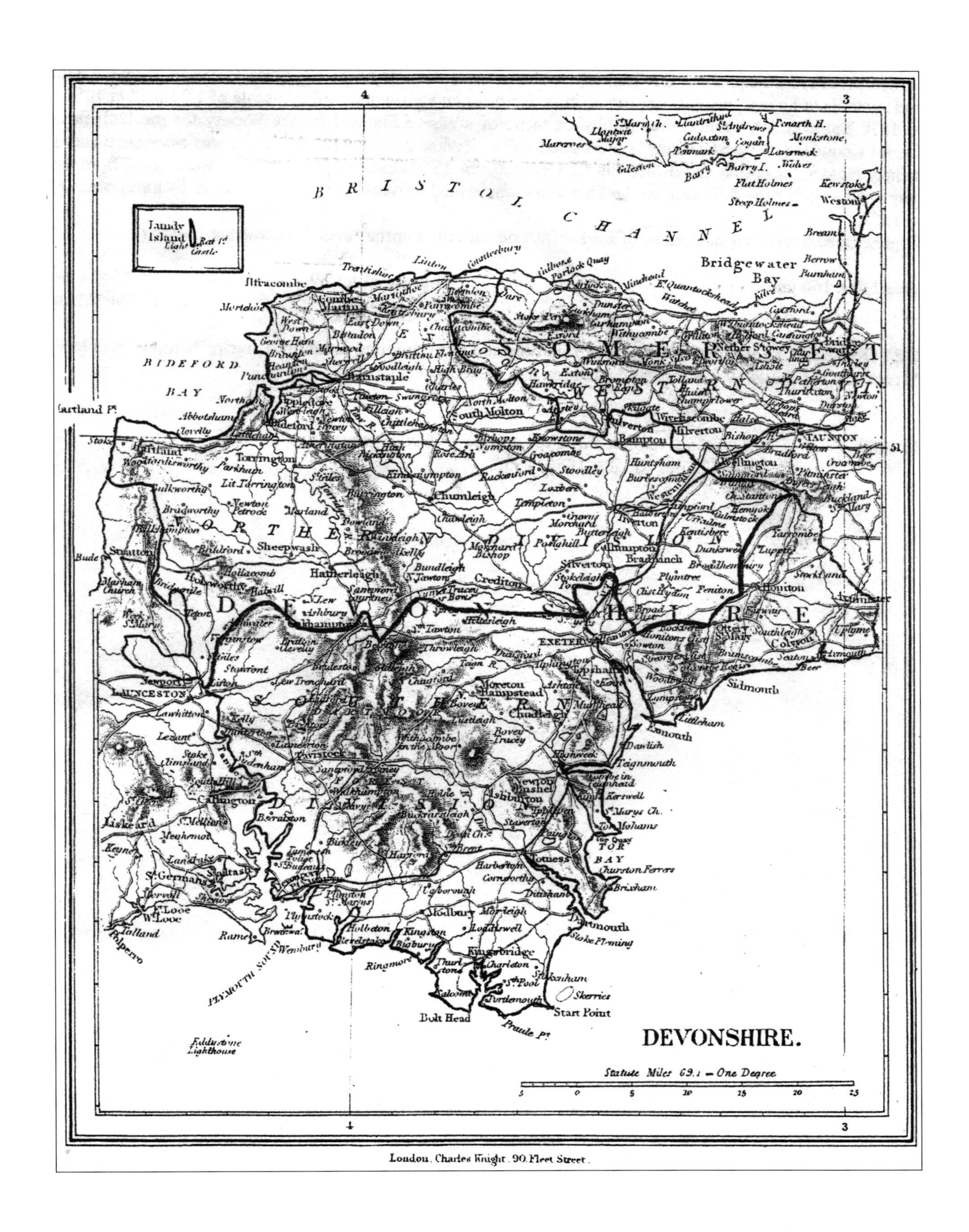

126.1 **Walker/Knight** *The Imperial Cyclopaedia of Geography*

127

ARCHER/COLLINS

1852

Considering the number of atlases published by Henry George Collins (fl.1849-59) surprisingly little is known about him. He had premises at 22 Paternoster Row in London, a popular address for publishers, and seems to have specialised in buying old plates and producing reissues. Probably between 1850 and 1852[1] Collins produced his *Pocket Ordnance Railway Atlas of Great Britain.* As with Collins' previous atlases and folding maps, the maps in the *Pocket Ordnance* were based on earlier engraved maps. Collins had already exploited maps by Cole and Roper (**67**), Robert Rowe (**81**), William Ebden (**95**), and Henry Teesdale (**99**) and the new series was no exception.

Although it is not sure that he produced the county maps specially for the *Railway Atlas*, Joshua Archer, a well-known engraver of county maps (see **67** and **108**), had produced a map of England and Wales for *Collins Indestructible Atlas* and *Collins One Shilling Atlas.* It was transfers of these plates which were then taken and used to produce the Railway atlas. The maps were on stiff card which was glued together in a leather wallet to be carried in the pocket.

The small maps have no scale but the text explains what scale was used and the Devon text reports that the width of the map equals 70 miles. No roads are shown, only hachured hills and some towns with the railways completed by 1848.

Size: 76 x 53 mm. Scale is irrelevant.

DEVON. above plate number - **7** - (CaOS). Railway to Plymouth and branches to Torquay (Torre) and Tiverton.

1. 1852 *Collins Pocket Ordnance Railway Atlas of Great Britain*
London. H G Collins. (1852). BL, (FB).

1. Chubb (1972) suggested 1852, but Raymond Carroll (1996; p. 299) suggests 1850 as a possibility.

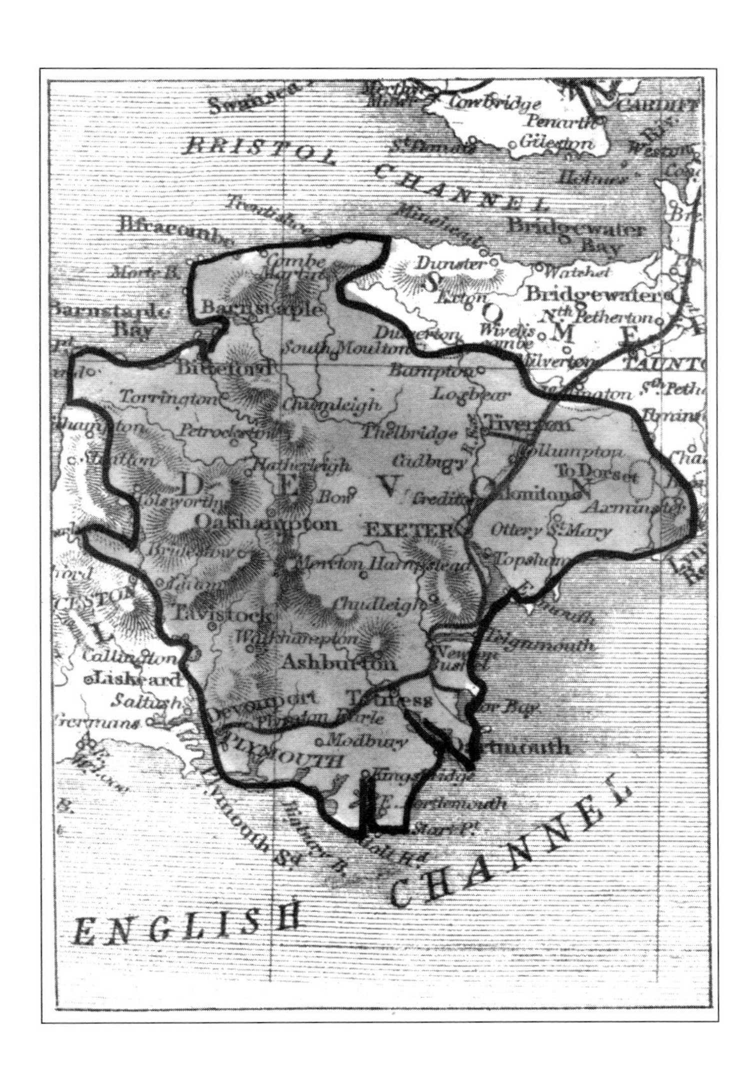

127.1 Archer/Collins *Collin's Pocket Ordnance Railway Atlas Of Great Britain*

128

BECKER/BESLEY - Devon II

1854

Henry Besley's *Route Book Of Devon* was first published in 1845 complete with a route map (**121**). A county map was added for the second edition in 1846 (**122**). A new, larger map was introduced for a second reprint of the 1850 *New Edition* similar to, but replacing, the previous county map which had been based on Walker's map of 1836. In 1871 and subsequent editions further maps appeared and the county was also shown in two parts (**134**).

The text was written before 1851 on the evidence of dates included. The first issues had no illustrations, but later Besley introduced a series of views drawn and engraved by G Townsend; the south coast in 1853 and the north in 1854. These first appeared in the *New Edition* and provide an accurate 'earliest date'. George Townsend (1813-1894), Artist, Lithographer and Drawing Master of 59 High Street, Exeter went on to produce some 117 illustrations of the county until *c.*1870 and was then succeeded by S R Ridgway who completed the series with another 40 engravings.[1]

As for the accompanying Route Map (included in the 1854 and 1856 editions), Besley again employed the company of F P Becker & Co. Omnigraph (litho) Engravers. This firm also produced maps for a variety of different publishers: eg Adlard's *Ireland*, Fisher's *County Atlas of England and Wales*, Frederic Kelly's *Post Office Directories*, the map of England and Wales in *Barclay's Dictionary*, and Scotland in Reynolds' *Geological Atlas.* Becker was renowned for the *Omnigraph* (a versatile ruling machine see p. xi & xiv).

This map also appeared in *The West of England and Trewman's Exeter Pocket Journal* in 1861 but though the journal continued until 1869 there are no Devon maps in copies extant for 1862, 1864 or 1869. In 1870 the journal became *Besley's West of England and Exeter Pocket Journal.* This journal was published from 1870 through to 1877 but the map has only been found in issues up to 1873.[2]

William Tunnicliffe in his *Topographical Survey* of 1791 (**58**) lists a Robert Trewman as *Printer* in Exeter who subscribed to 100 copies of his book. *The Exeter Pocket Journal* of 1816 was published by Trewman & Son of Exeter complete with a map by E A Ezekiel (**83**), and continued to be published by the same company from 1822 to 1856 but with a new map by Smith and Davies (**93**).

Size: 315 x 380 mm. **ENGLISH STATUTE MILES** (15=60 mm).

DEVONSHIRE (Ed). Imprint: **Printed & Published by Henry Besley, Directory Office, South Street, Exeter.** (CeOS) and signature: **Engraved by BECKER'S Patent Process on Steel, 11, Stationer's Court, London.** (EeOS). Note: **N** Northern Division **S** Southern Division (Ea). A topographical note, election and polling places and Poor Law Unions tables are below the title. A list of Boroughs (Ae) and Key to Divisions are shown (Be). Railways: into Cornwall, to Tiverton, Torquay and Bideford.

1. 1854 *The Route Book Of Devon ... A New Edition*
Exeter. H Besley. (1854[3]). E.

2. 1856 Railways added: Exmouth, Kingswear and Tavistock.

The Route Book Of Devon ... New Edition
Exeter. H Besley. (1856[4]). RGS, E.

The West of England and Trewman's Exeter Pocket Journal 1861
Exeter. H Besley. 1861. E.

3. 1863 Proposed railways to Moreton Hampstead, Launceston and Oakhampton.

The West of England and Trewman's Exeter Pocket Journal 1863
Exeter. H Besley. 1863. E.

1. J V Somers Cocks; 1977; pp. 10 and 306.
2. Copies mentioned are held by the Exeter Westcountry Studies Library.
3. Engravings are dated 1853 and 1854.
4. An advert relates to Queen Victoria's visit to the county in 1856.

4. 1865 Proposed railway Tavistock-Oakhampton added.

The West of England and Trewman's Exeter Pocket Journal 1865
Exeter. H Besley. 1865. E.

5. 1866 Railways: Launceston and Oakhampton. Projected: Oakhampton-Bude, Oakhampton-Bideford, Sidmouth, Budleigh, Ashburton, Ilfracombe and Barnstaple-Taunton.

The West of England and Trewman's Exeter Pocket Journal 1866, 67, 68
Exeter. H Besley. 1866, 1867, 1868. E; E; E.

6. 1869 All notes removed. Railway to Seaton.

The Route Book Of Devon ... New Edition
Exeter. H Besley. (1869[1]). E.

Besley's West of England and Exeter Pocket Journal. 1870
Exeter. H Besley. 1870. E.

7. 1871 Key to divisions and boundaries restored (Be).

Besley's West of England and Exeter Pocket Journal. 1871
Exeter. H Besley. 1871. E.

The Route Book Of Devon ... New Edition
Exeter. H Besley. (1871). BL, NLS, E, Pl.

8. 1872 Railways to Sidmouth, Ilfracombe (via Braunton) and Barnstaple-Taunton added.

Loose sheet possibly from *West of England and Exeter Pocket Journal. 1872* (DEI).

9. 1873 Railways Bideford-Torrington, Oakhampton-Tavistock and Ilfracombe now west but still incorrect.

Besley's West of England and Exeter Pocket Journal. 1873
Exeter. H Besley. 1873. E.

The Route Book Of Devon ... New Edition
Exeter. H Besley and Sons. (1877). BL.

1. Contains advertising section in which two pages have *(69)* in the bottom right corner.

BRISTOL CH
BARNSTAPLE OR BIDEFORD BAY
LUNDY ISLAND
Hartland Point
ILFRACOMBE
COMBE MARTIN
BARNSTAPLE
BIDEFORD
TORRINGTON
HARTLAND
CORNWALL
BUDE BAY
STRATTON
HOLSWORTHY
HATHERLEIGH
OKEHAMPTON
LAUNCESTON
CAMELFORD
BODMIN
LISKEARD
LOSTWITHIEL
FOWEY
EAST LOOE
SALTASH
CALLINGTON
TAVISTOCK
DARTMOOR FOREST
PLYMOUTH
DEVONPORT
PLYMPTON EARLE
MODBURY
KINGSBRIDGE
ASHBURTON
CREDITON
MORETON HAMPSTEAD
CHULMLEIGH
SOUTH MOLTON
DEVON
CORNWALL RAILWAY
ENGLISH CH
Bolt Head
Prawle Point
BOROUGHS returning 2 Members each
Barnstaple Tiverton Devonport
Exeter Honiton Plymouth
Tavistock Totnes
returning 1 Member
Ashburton Dartmouth
Division of the County
Boundary of Boroughs
Boundary of Poor Law Unions
Printed & Published by Henry Besley, Directory Office,

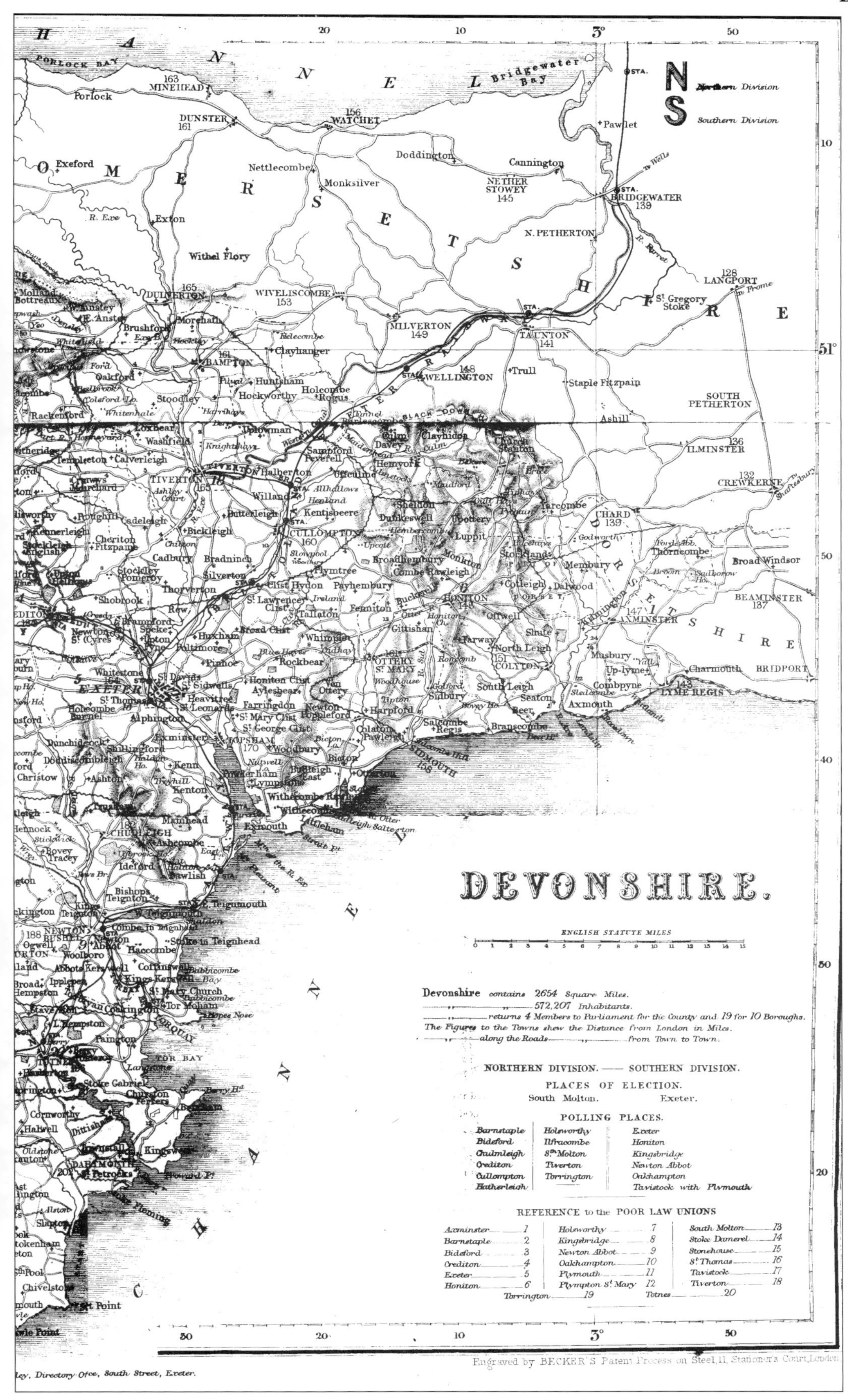

128.1 **Becker/Besley** *The Route Book Of Devon* – new county map

129

GEORGE OLIVER
1854

Aaron Arrowsmith engraved a map of Devon and Cornwall in 1814.[1] This map, DIOC' EXON' was copied in 1854 by the Reverend George Oliver. Slightly smaller than the original this map shows the deaneries, principal towns, rivers and sites of religious houses, including much of the earlier information but with fewer towns. The two panels in the original map have been omitted but an inset map of the Scillies, *Insule de Sully*, has been added. The title remains the same. It was drawn to accompany a supplement to the *Monasticon Diocesis Exoniensis,* a work on the state of the church in the Bishopric of Exeter which included the four arch-deaconries of Barnstaple, Exeter and Totnes and Cornwall.

George Oliver, Catholic divine and Exeter historian, was born at Newington, Surrey in 1781. He was promoted to holy orders in 1806 and moved to Exeter, to the mission of the Society of Jesus at St. Nicholas, in 1807. He served the mission for 44 years and continued to live in the priory until his death in 1861. He wrote numerous works concerning the history of Devon, especially with regard to the church and the catholic faith in the west. His most important works included *Collections Illustrating the History of the Catholic Religion in the Counties of Cornwall, Devon, Dorset, Somerset, Wilts, and Gloucestershire* (1847), *History of the City of Exeter* published in Exeter in 1821, *Ecclesiastical Antiquities in Devon* issued in three volumes 1839-42, and the *Lives of the Bishops of Exeter* and a *History of the Cathedral* 1861. He also wrote many articles for the *Exeter Flying Post,* the *Catholic Herald* and the *Catholic Magazine* as well as editing other publications such as Thomas Westcote's *View of Devonshire.*

His *Monasticon Diocesis Exoniensis, being a Collection of Records and Instruments Illustrating the Ancient Conventual, Collegiate, and Elemosynary Foundations in the Counties of Cornwall and Devon* was first published by P A Hanaford in Exeter and Longman, Brown of London in 1846. This contained illustrations of initials printed in gold and tinted plates of monastic seals but no maps.

Size: 340 x 477 mm.

No scale bar.
[Scale 1M=2.5 mm]

DIOC' EXON'. Graticulated border with degree lines. Explanation (Ee). Inset map of **INSULE DE SULLY**.

1. 1854 *Additional Supplement to the Monasticon Diocesis Exoniensis*
Exeter: A Holden. London: Natali and Bond. 1854. BL, GUL, E.

1. See Appendix **I**.

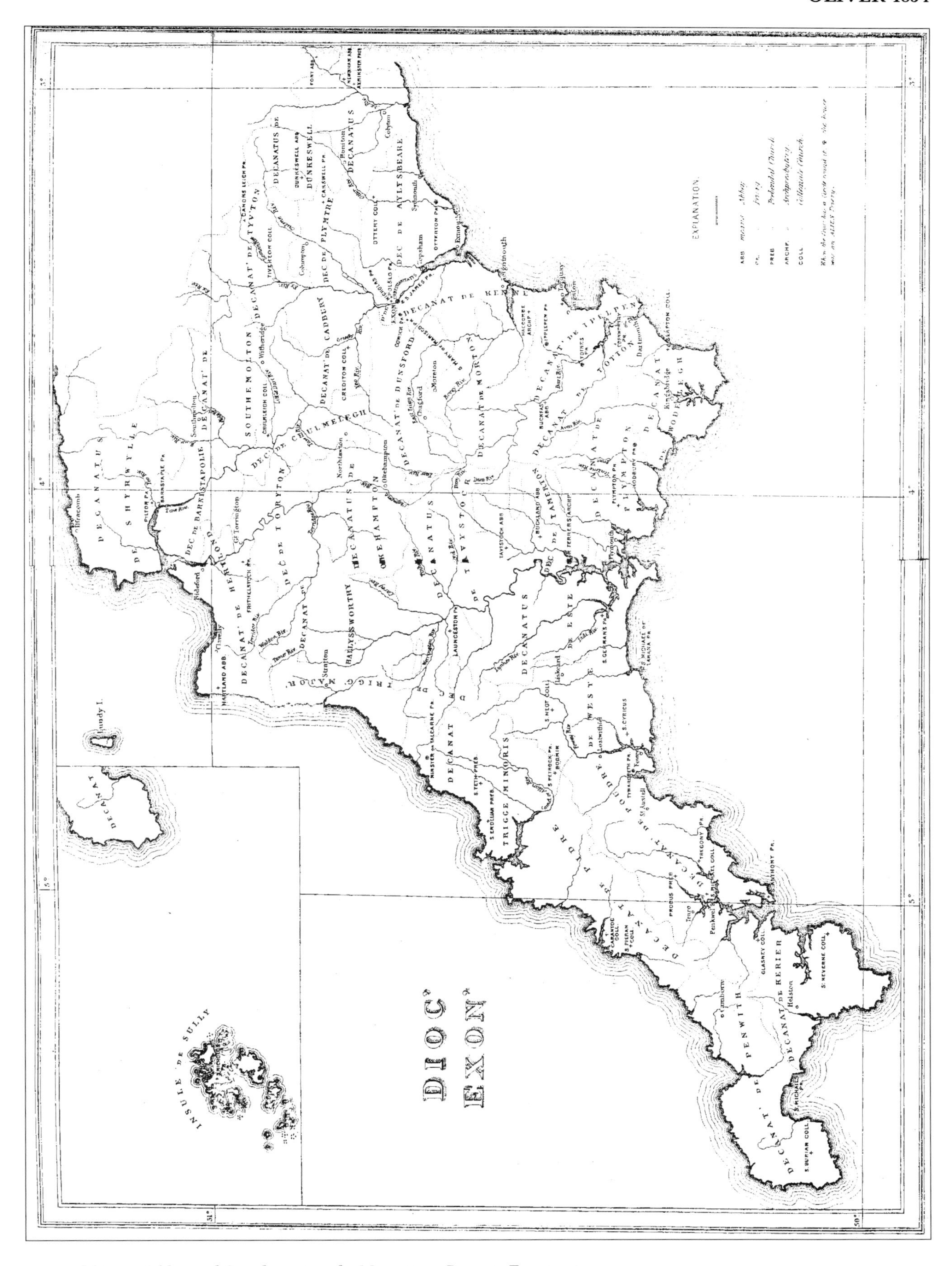

129.1 Oliver *Additional Supplement to the Monasticon Diocesis Exoniensis*

130

SCHENK & MCFARLANE/BLACK
1855

Besley's *Route Book to Devon* appeared in 1845 and five years later John Murray published the first of his long series of county handbooks. Originally combining Devon with Cornwall he later split them into two separate volumes. The Blacks, who had been publishing city and area guide books since 1839, were not slow to follow, their guide to Devon and Cornwall being one of their first county guides.

Established in Edinburgh in 1807 by Adam Black, the company became famous for its familiar series of green bound guides. They began shortly after Adam's nephew, Charles, joined the company to issue guides, firstly for Scottish cities and northern England, then for Wales and Ireland and from 1855and from 1855 for the English Counties. The first county guides were of Derbyshire and Hampshire and a guide of Devon with Cornwall and the Scilly Isles. The maps were taken from Black's *Map of England and Wales* which was published as a folding map *Black's Road & Railway Travelling Map of England.* The map was engraved by S Hall, Bury St. Bloomsbury. This was Selina Hall who took over the Hall's engraving business after the death of Sidney in 1831. The map can be provisonally dated to before 1853 when Edward Weller acquired the business on the death of Selina.

From the beginning a map of the area was included, and in the 1855 issue it was a map of Devon and Cornwall printed by Schenk and McFarlane. Although the title page is dated 1855, the copy of the guide we have seen was issued as is shown by the advertisements, one of which is dated May 1859. The railways are somewhat anticipatory, Exmouth was opened in 1861 and Truro was not reached until 1865. How often the guide was reissued is not known; however, it must have seemed promising as the Black's published a new edition in 1862. This was issued as a combined volume including Dorset, Devon and Cornwall, the counties were also published separately (with the complete volume pagination). For this issue a new Bartholomew map was produced (**142**). The series was so successful that editions were reprinted almost every year until the end of the century and a new Bartholomew map was included when the text was extensively revised in 1882 (**150**).

Size: 225 x 220 mm. No scale bar.
[Scale 1M=2.5 mm]

To Accompany BLACK'S GUIDE to DEVONSHIRE and CORNWALL. Signature: **Schenk & McFarlane Edin^{r}.** Graticulated border with degree lines.The border broken for Lands End. Inset map of **SCILLY ISLES(Ae).** Railways are shown to Plymouth and Torquay, and projected to Truro, Falmouth, Barnstaple and Exmouth.

1. 1855 *Black's Tourist's Guide To Devonshire & Cornwall including the Scilly Isles*
Edinburgh. A & C Black. (1855). KB.

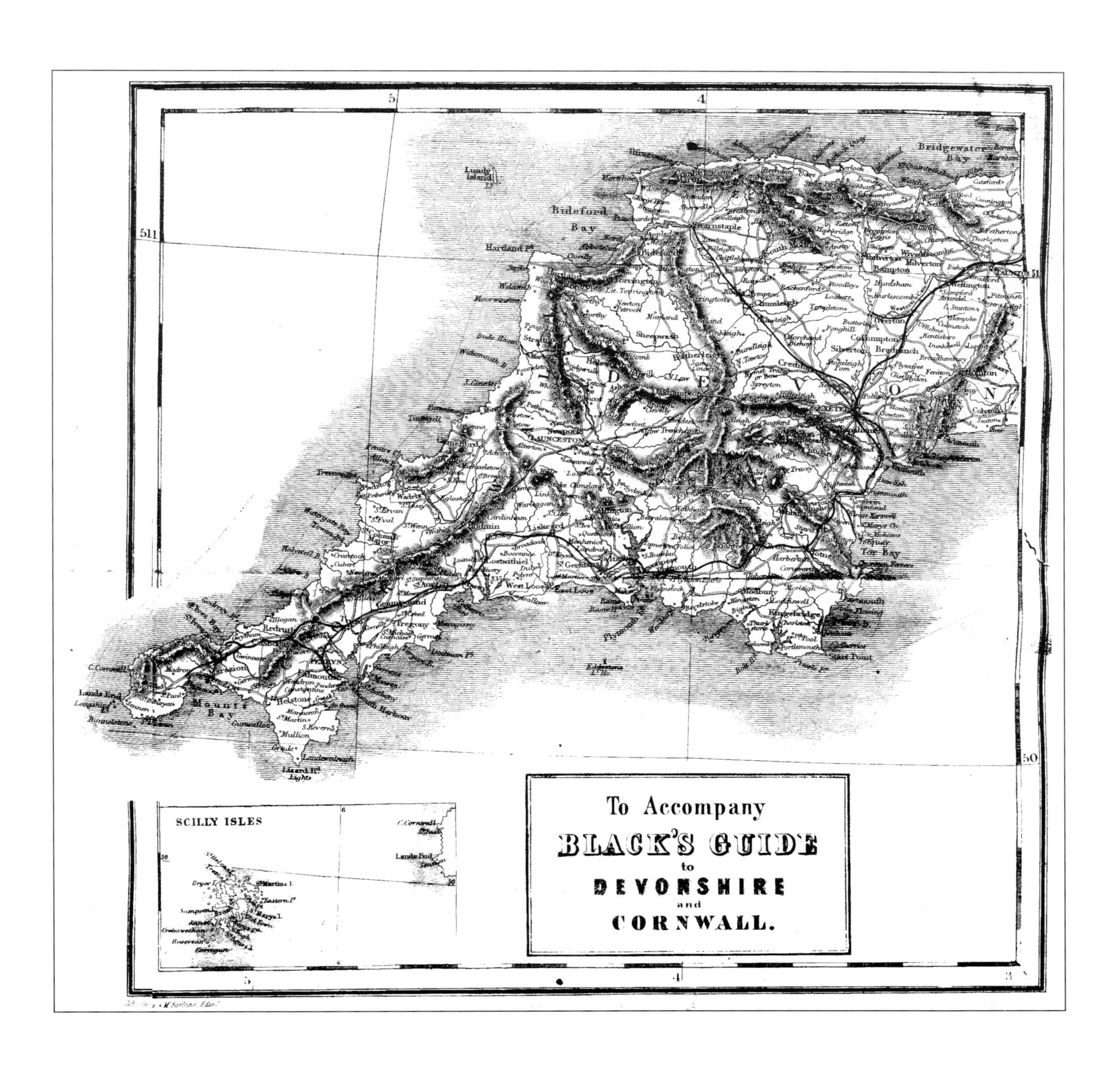

130.1 Schenk & McFarlane/Black *Black's Guide to Devonshire and Cornwall*

131

VINCENT BROOKS

1856

Charlotte Chanter (1828-1882) in 1856 wrote a short 'guide' book to the area around Ilfracombe and dedicated it to her parents, the Rev. Charles and Mrs Kingsley. Charlotte illustrated her book with her own drawings of ferns and (her own) map of Devonshire. The map details the areas of her own excursions, largely along the north coast and Dartmoor. Although the first edition did not include a map, one was added to the second and third editions.

Charlotte married John Chanter, the vicar of Ilfracombe, in 1849. The Chanter family wrote a number of topographical works on North Devon, largely concerning Barnstable. Charlotte's sister Gratiana wrote and illustrated *Wanderings in North Devon* in 1887, a book about Ilfracombe, the Combes and her father's life.

Vincent Brooks (1814-1885) was a lithographer well-known for his topographical views and naval subjects, often copying artists of the day. After working for Day & Son[1] he set up his own printing and publishing company, buying his former employer's business after William Day (1797-1845), the co-founder, died.[2] Day had started up the business together with Louis Haghe (1806-1895), a Belgian from Tournai. Haghe moved to England in 1823 and worked with Day; he later gave up lithography and turned to watercolours.

In about 1872 the company Vincent Brooks, Day & Son Lith. are known to have produced a few maps of other counties. If it was intended to be a series it was never completed. The maps may have been printed for an atlas, *Cassell's County Geographies*, to be published in parts at 1d or 2d per county,[3] but Devon was not one of the series. The size of the maps is similar to that in *Ferny Combes* and were also signed by Brooks. They showed only county divisions, main roads, railways and rivers.

There are eight delightful plates in *Ferny Combes*.[4] The first is a view of the Entrance to Clovelly, the others are varieties of fern. The book was printed by Lovell Reeve[5] who had premises at 55 Henrietta Street in Covent Garden.

Size: 149 x 174 mm.

No scale bar.
[Scale 1M=2.25 mm]

MAP of DEVONSHIRE. Signature: **Vincent Brooks Lith.** (Ee). Rivers and principal towns.

1. 1856 *Ferny Combes. A Ramble After Ferns In The Glens And Valleys Of Devonshire. Second Edition*
London. Lovell Reeve. 1856. NLS, E.

Ferny Combes ... Third Edition
London. Lovell Reeve. 1857. BL, E.

1. Day & Son *Lithographers to the Queen* produced some of the *Weekly Dispatch* maps (**135**). A local guide book, *The Teignmouth, Dawlish, And Torquay Guide* (by Edward Croydon, 1830) included views: Printed by W Day, 17, Gate Street. The map included was in this guide a small section copied from John Cary's 1807 map of the county (**71**).
2. Ian Mackenzie (1988); pp. 65, 95 & 145.
3. David Smith (1985); p. 135.
4. These are signed *Vincent Brooks Lith.* in the second edition. In the third edition the signature reads *Vincent Brooks. Day & Son. Lith.*
5. According to Carroll (1996), Lovell Reeve was publishing *The Naturalist: A Monthly Journal of Natural History for the North of England* in 1895 when they included a copy of the Lincolnshire map from *Philips' Handy Atlas* complete with new soil information.

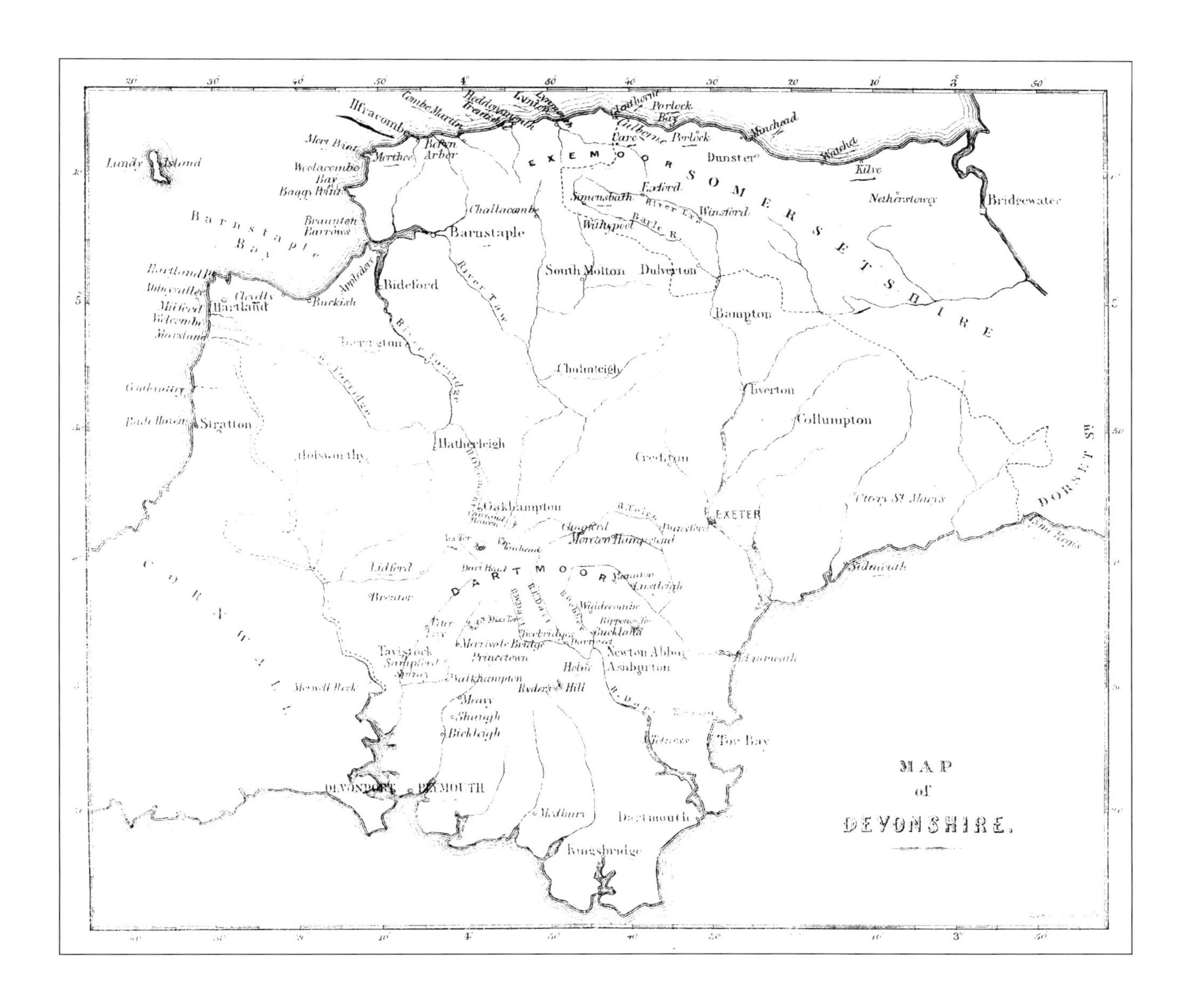

131.1 Brooks *Ferny Combes*

132

BECKER/KELLY

1856

Frederic Kelly (*fl.*1845-1900) founded Kelly & Co., the London-based publishers who specialised in directories. The first, London, was produced in 1843 from his Post Office Directory offices at Old Boswell Court (later from Temple Bar, 1845-68). The first County Directories appeared *c.*1845 and they included maps of the relevant county/counties. Devon appeared in 1856 and was re-issued at approximately five year intervals well into the next century with updated maps (easily identifiable having the title *Post Office Map Of* ...). In *c.*1860 a complete work was issued, the *Post Office Directory Atlas.* Francis Becker, (noted for his work for Besley, **121**, **134**) and Benjamin Rees Davies (*Dispatch Atlas,* **136**) engraved most of the 45 maps. In 1883 the directory was issued with a new map (see **159**).

Cheffins & Sons carried out work for Kelly and produced zincographed railway maps and one county map.[1] C F Cheffins was a well-known lithographer of topographical views, railway and shipping subjects and even westcountry views (eg *Scenery in the North of Devon, c.*1837). He worked for Day and Haghe (the forerunners of Day & Son, later Vincent Brooks, Day & Son) before setting up on his own.[2]

Size: 220 x 270 mm. **Scale of Miles** (16 = 48mm).

POST OFFICE MAP OF DEVONSHIRE 1856. Imprints: **Kelly & Co. Post Office Directory Offices 19, 20, & 21, Old Boswell Court , Clements Inn, Strand.** Signature: **Drawn & Engraved by F.P.Becker & Co. 11 Stationers Court, City** (EeOS). Note below title (Ed) of Polling Places and Post Office Money Order Towns. Shows incorrect routes of railways to Torquay, Barnstaple, Tiverton and projected (pecked) to Exmouth, (incorrect) Newton Abbot to Ashburton and to Stolford in Somerset (never completed).
Further signature (assumed): **Printed from Stone by C F Cheffins & Son London.** (AeOS).

1. 1856 *The Post Office Directory Of Devonshire*
London. Kelly & Co. 1856. E[3].

2. 1860 Date removed. L&SWR Exeter to Axminster.

The Post Office Directory Atlas Of England And Wales
London. Kelly & Co. (1860). BL[4], C, W, CB.

3. 1866 Signatures erased. Date **1866** added. Railways to Plymouth, Exmouth and Plymouth-Lydford-Launceston; projected to Okehampton and Bude, Lydford, Torrington, Taunton to Barnstaple as well as into Cornwall from Plymouth. Population revised to 584,531.

The Post Office Directory Of Devonshire
London. Kelly & Co. 1866. E[5].

4. 1873 Imprint: **Kelly & Co. Post Office Directory Offices 51 Gt. Queen St. London W. C.** and new signature: **J M JOHNSON & SONS, LITHO. 56, HATTON GARDEN, LONDON.** (AeOS). Date **1873**. Lines to Kingswear, Brixham, Moreton Hampstead and Totnes-Ashburton (correct); projected to Ilfracombe, Kingsbridge, Hemyock, Ashton, Seaton and Watchet to Dulverton (never built). Polling Places and Money Order Towns note removed.

The Post Office Directory Of Devonshire
London. Kelly & Co. 1873. E.

1. *Hertfordshire*, appeared *c.*1857 by *Cheffins & Sons, Lith, Southampton Bgs. London*, at a scale of 7 miles to 70 mm.
2. Ian Mackenzie; 1988; p. 78.
3. The Exeter map is partly destroyed and has right half only.
4. Includes an advertisement for an 1861 P.O. Directory.
5. Very torn.

132.4 Becker/Kelly *The Post Office Directory*

133

MARTIN BILLING

1857

Martin Billing was an active publisher and printer in Birmingham who produced a number of county directories. It is believed that the first directory was that for the *Counties of Berks and Oxon* published in 1854 in which there is a reference to a *Sheet Map* although this is missing from the editions known.[1] In 1857 he published *M Billing's Directory and Gazetteer of the County of Devon* from the Steam Printing Offices, Birmingham. The edition was compiled by John Hughes, who in his forward apologises for the delay that had unavoidably occurred, suggesting a start some few years earlier. The text includes the Exeter diocese officers up to 1856 (including Thomas Pickard, Dog Whipper) and there are advertisements with testimonials dated as late as May 30th 1857.

The publication includes, as an appendix, some 192 pages of advertisements. Although the last twenty pages are devoted to Birmingham advertisers, Billing or a representative must have visited Devon as the first 175 pages are all from local Devon firms. One company who purchased a half page advert was John Heydon, who published a map of Devon in 1872 (**150**). In 1857 Heydon was advertising himself as *Bookseller, Stationer, Printer, Music Seller, News Agent, &c.* at 104 Fore Street Devonport and 47 Treville Street in Plymouth. He was also a major seller of second-hand and new books *for sale at very low prices.*

Although there is no actual mention of a map it is possible that one was sold separately to accompany the gazetteer. William White's *History, Gazetteer and Directory of the County* had first appeared in 1850 (without a map) and Billing possibly thought the time was ripe for a new edition. However, Billing never reprinted his work while White went on to produce a new edition 1878/79 with a map by Walker (**116**). Two copies of the Devon map by Billing have been discovered but both, sadly, are in extremely poor condition.
The British Library has some engravings of furniture for grocer's shops, dated 1855, by Billing.

Size: 630 x 448 mm. **SCALE OF MILES** (20=86 mm).

MARTIN BILLING'S MAP OF DEVONSHIRE Engraved and Printed at his Offices. LIVERY STREET, BIRMINGHAM. (Da). Line border with elaborate corners. The Plymouth arms and those of Exeter are both present (Be and De). There are two vignettes: Exeter Cathedral (Ba) and the Royal Hotel Theatre and Athenaeum Plymouth (Ce). The railway proceeds as far as Torre and the projected route to Dartmouth is shown. Folded and hard-backed.

1. 1857 *Martin Billing's Map of Devonshire ...*
Birmingham. M Billing. (1857). (E), (Pl) [2].

1. See Eugene Burden; 1988 (1991); p. 179.
2. These are in extremely poor condition and cannot be seen to have come from the gazetteer.

133.1 Billing *Martin Billing's Map of Devonshire*

134

BECKER/BESLEY - Devon III

1857

Henry Besley's *Route Book Of Devon* first appeared with a route map in 1845 (**121**) and a map of the county in 1846 (**122**). In 1857/1858 he published new maps of North and South Devon for his hand books of *North Devon* and *South Of Devon and Dartmoor* (each *extracted from The Route Book*). The two maps were used in both the *Route Book* and the *North* and *South* handbooks until *c.*1880, and subsequently by Ward and Lock in their Devon guides published from *c.*1886.

The two new maps do not cover the complete county (a small portion is missing, north and east of Broad Clyst) but with the discarding of the single county map, it is apparent that these two maps were intended to map the whole county. In the 1871 *Route Book* the small missing portion was almost covered by the introduction of a *South East Devon* map[1], but this was produced in London to a different scale and does not complement the others. The map of *North Devon* was published *c.*1857 and *South Devon and Dartmoor c.*1858.

Besley's maps were also used in *Ward Lock's Pictorial Guides*. Ward and Lock started their own firm on Midsummer's Day 1854. The Locks were an influential Dorchester family and George Lock's father had married Eliza Galpin making George Lock (1832-1891) first cousin to Thomas Dixon Galpin, later to make his name with Cassell, Petter and Galpin. Ebenezer Ward (1819-1902) was a manager for the *Illustrated London News* publishers' book business. The two were introduced by Thomas Galpin and Ward Lock was launched with a £1000 loan from George Lock senior, the articles of partnership being signed June 23rd 1854. After initial friendly[2] relations and support, contact between the two cousins was broken when Galpin and his partner, George William Petter, took over Cassell's to found a rival business partnership. From 1891 until *c.*1897 James Bowden was a partner and the company traded under the name Ward, Lock and Bowden Ltd.

Ward and Lock guides often included smaller maps (eg Exeter or Lynmouth) printed by G Philip and Son. Another map of Devon was also engraved by Bartholomew and published by Ward, Lock and Co. This was prepared for *A Pictorial and Descriptive Guide to Plymouth, Stonehouse and Devonport,* and shows the area from Seaton in Whitsand Bay to Dunscombe by Sidmouth and north as far as Beaworthy.[3]

134 - NORTH

Size: 310 x 400 mm. **Scale of English Miles** (10=78 mm).

NORTH DEVON. Imprint: **Printed and Published by Henry Besley, Directory Office, South Street, Exeter** (CeOS). Signatures: **Engraved by Becker's Patent Process on Steel, 11, Stationer's Hall Court, London.** (AeOS) and **Drawn by Becker & Sons** (EeOS). Inset map of **Lundy I^{d}**. Railways: GWR and L&SWR to Exeter, lines to Tiverton, Exmouth and Bideford.

1. 1857 *The Hand Book Of North Devon*
Exeter. H Besley. (1857). E.

The Hand Book Of North Devon with a trip on the Crediton and North Devon railways
Exeter. H Besley. (1862). E.

1. *SOUTH EAST DEVON*: 280 x 382 mm. *Scale of English Miles* (7 = 88 mm). Imprint: *Printed & Published by Henry Besley, Directory Office, South Street, Exeter* (CeOS): and signature *Engraved by W E Trott* (AeOS) *2 Tysoe St London EC* (EeOS). Graticuled border. Area from Dawlish to Lyme and inland to Cullompton. Exeter Westcountry Studies Library has a copy of a Besley guide to *South East Devon* *c.*1872 but it lacks the map.
2. Edward Liveing; *Adventure in Publishing - The House of Ward Lock 1854-1954*; Ward, Lock & Co. Ltd; 1954.
3. Also with title: *SOUTH DEVON.* Imprint: *WARD LOCK & CO. LTD. Warwick House, Salisbury Square London* (CeOS). Signature: *John Bartholomew & Co. Edinburgh* (EeOS). Size: 260 x 330 mm. *English Miles* (8 =51 mm). The Exeter Westcountry Studies Library has copies of the Second Edition (lacking its map), the Fifth Edition and (possibly) Sixth Edition. The volumes were probably published *c.*1893, *c.* 1898 and *c.*1898 respectively; there is a note dated 1892 (on p. xiv) but no reference to the Yealmpton railway (opened 1898) which is shown on the map contained in the *Fifth Edition.* The *Sixth Edition* is assumed as the title page is missing. The signature has been changed: *John Bartholomew & Son Lt^{d}. Edin^{r}* (EeOS).

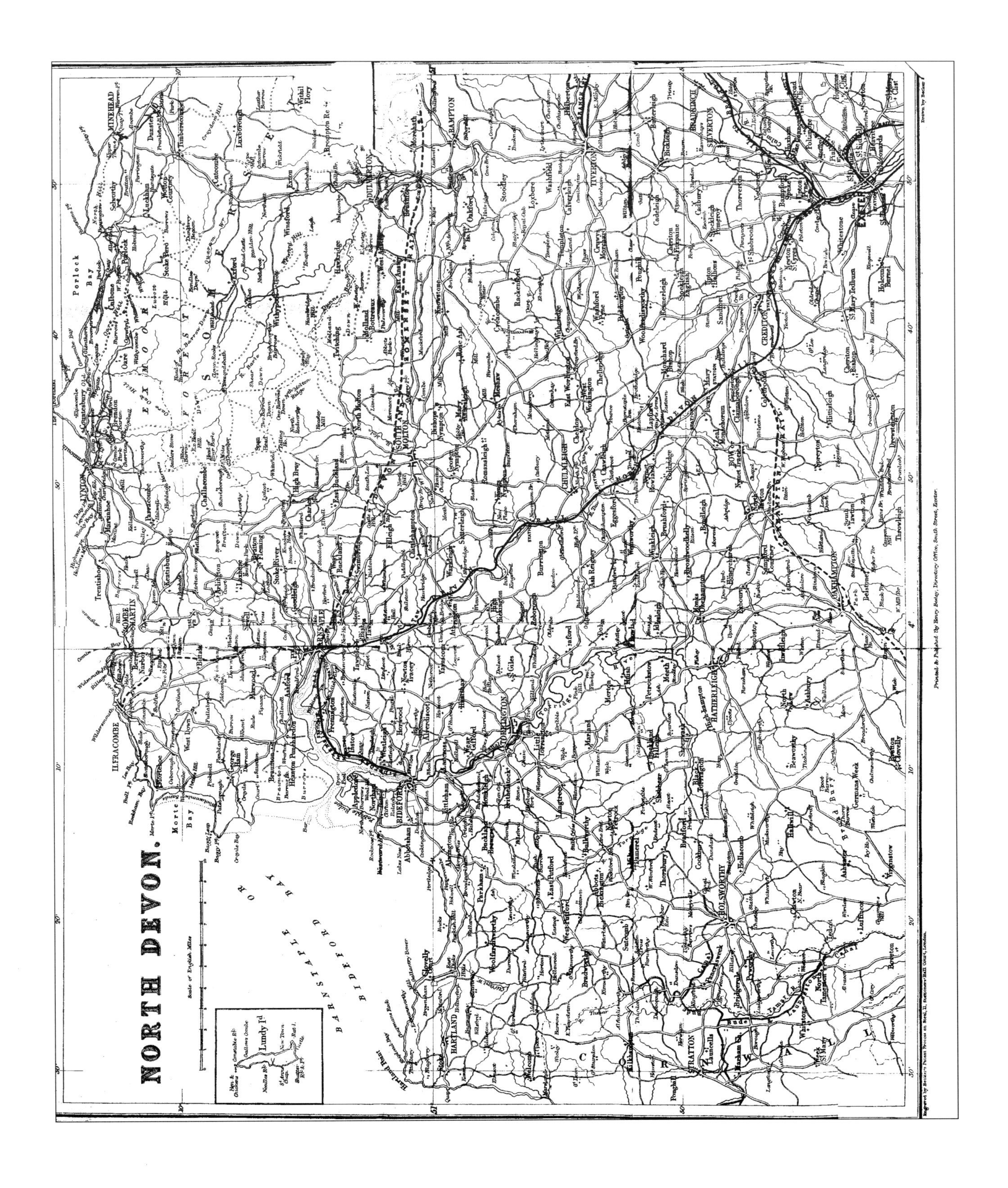

134.N.2 Becker/Besley *The Route Book Of Devon – North Devon*

2. 1871 Projected railways: Ilfracombe (via Bittadon), Colebrooke (Crediton) to Okehampton, Taunton to Barnstaple.

The Route Book Of Devon ... A New Edition
Exeter. H Besley. (1871). BL, E, NLS, P1.

The Hand Book Of North Devon
Exeter. H Besley. (1871). NDL.

3. 1873 Railways to Ilfracombe (passing east of Morthoe), Okehampton, Torrington and Taunton-Barnstaple now complete. Both signatures removed leaving imprint.

Besley's West of England and Exeter Pocket Journal
Exeter. H Besley. 1873. E.

4. 1874 Besley imprint deleted.

The Hand Book Of North Devon, Bude, Launceston, Tavistock, &c. with a trip on the Crediton and North Devon Railways
Exeter. H Besley. (1874). BL, E.

5. 1877 Imprint reinstated. The graticules at the full degrees have been deleted from the border.

The Route Book Of Devon
Exeter. H Besley and Son. (1877). BL.

The Hand Book Of North Devon, Bude, Launceston, Tavistock, &c. with trips on the North Devon, Devon and Cornwall Central, and Ilfracombe Railways
Exeter. H Besley and Son[1]. (1880) [2]. KB.

6. 1887 Imprint **H. B. & S., E.** added (EeOS). Key now includes Coaches and Steamers (routes). Steamer route Clovelly-Ilfracombe-Lynton. Coach routes include Holsworthy-Bude.

Ward Lock's Pictorial And Historical Guide To North Devon
London and New York. Ward, Lock, And Co. (1887). [3] KB.

7. 1894 Railway Holsworthy to Bude added.

Ward Lock's Pictorial Guide To Lynton, Lynmouth, Minehead &c.
London. Ward, Lock, And Co. (1894). E.

1. Besley was using the company of Adams and Francis (W J Adams & Sons) to distribute his books and views in London (p.17 of adverts).
2. This edition has text dated to Sept. 1877 (p. 32) and a reference to Bideford Grammar School expected to be completed Michaelmas 1878 but the adverts section appears to be dated 1880, ie **80** printed on first advertising page.
3. Spine title. Title page: *Ward Lock's Pictorial And Historical Guide to North Devon.* This work contains four volumes in one. Contents include Exeter and its Cathedral; Ilfracombe, Barnstaple, &c; Bideford, Clovelly, &c; and Lynton, Lynmouth, Exmoor, Minehead, &c. Each section is paginated separately and each has its own index. There are probably many Ward & Lock guides still to be discovered.

134 - SOUTH

Size 305 x 405mm. **Scale of StatuteMiles** (10=78 mm).

SOUTH DEVON AND DARTMOOR. Signatures: **Engraved by BECKER'S Patent Process on Steel, 11, Stationer's Hall Court, London.** (AeOS) and **Drawn by Becker & Sons** (EeOS). Imprint: **Printed & Published by Henry Besley, Directory Office, South Street, Exeter** (CeOS). Graduated border from Whitesand Bay to Beer and inland to Exeter. Railways: to Torre, Plymouth and into Cornwall.

1. 1858 *The Hand Book Of South Of Devon And Dartmoor - Extracted from the Route Book Of Devon*
Exeter. H Besley. 1858. E.

2. 1862 Railway to Tavistock and to Churston with remainder still projected to Kingswear.

Single sheet linen backed and bound into small booklet (100 x 70 mm).
Exeter. H Besley. (1862). E.

3. 1865 Railways completed to Kingswear, Exmouth and projected Tavistock-Oakhampton and northward with branch to Lifton and west.

The (handbook) *south of Devon and Dartmoor*
Exeter. H Besley. (1865) FB

4. 1870 Railways completed to Lidford and projected from Buckfastleigh to Ashburton, Chudleigh and Exeter.

The Route Book Of Devon ... A New Edition
Exeter. H Besley. (1870). BL, NLS, E, Pl.

5. 1870 Railways Moreton Hampstead, Tavistock-Lifton and still projected to Oakhampton.

The Hand Book Of South Of Devon And Dartmoor
Exeter. H Besley. (1870). E.

6. 1873 Becker signature removed. Railways: Ashburton from Totnes, Brixham, Lidford to Oakhampton and Launceston line.

The Hand Book Of South Of Devon And Dartmoor
Exeter. H Besley. (1873), (1875). KB; E.

The Route Book Of Devon
Exeter. H Besley & Son. (1877). BL.

7. 1886 Besley imprint erased and new imprint **H. B. & S., E.** (EeOS) added. A new key below the title showing Railways, Coaches and Steamers. Railways: Ashton, Princetown and Sidmouth with projected from Sidmouth to Exmouth and Ashton to Exeter (dotted). Steamer routes: Plymouth to Kingsbridge and Kingswear to Totnes. Coach routes: in the coastal area of the South Hams and Exmouth to Budleigh Salterton.

Ward Lock's Pictorial & Historical Guide To Torquay, Teignmouth, Dawlish, Dartmouth, Totnes And Other South Devon Watering Places
London and New York. Ward, Lock & Co. (1886[1]). MW.

Ward Lock's Pictorial Guide to South Devon
London. Ward, Lock & Co. (1888[2]). E.

1. Two pages of the Guide Book Advertisement at the back are dated 1886.
2. The Guide Book Advertisement at the back is dated 1888.

8. 1893 Railways: L&SWR Tavistock-Plymouth, Launceston to Halwill via Ashwater and Plymstock,. Station names in the west added in thick script. Projected line Ashton to Exeter (dashed) and Ashburton to Exeter (Teign Valley Ry) and coach route Plymstock to Yealmpton, Modbury and Kingsbridge. Steamer route Kingswear to Totnes deleted.

Ward and Lock's Pictorial and Historical Guide To Dartmoor
London, New York, Melbourne and Sydney. Ward, Lock & Bowden, Ltd. (1893). KB.

9. 1895 Railway and coach route to Kingsbridge.

Ward, Lock & Bowden's Plymouth, Tavistock &c
London, New York and Melbourne. Ward, Lock & Bowden, Ltd. (1895). KB.

Ward Lock's Pictorial Guide To Plymouth, Tavistock &c
London. Ward, Lock & Co. (1896). E.

10. 1896 **Steamers** erased from key and route Plymouth to Kingsbridge deleted. Railways: Budleigh Salterton, Ashton-Exeter, Yealmpton, Turnchapel, L&SWR Tavistock-Lidford. Projected line Yealmpton-Modbury, but the Ashburton-Chudleigh-Exeter is erased together with Sidmouth to Exmouth line. Coach routes are modified, erased or added, eg Venn Cross to Kingsbridge deleted.

Ward Lock's Pictorial Guide To Dartmoor
London. Ward, Lock & Co. (1896). E.

A New Pictorial And Descriptive Guide To Dartmoor
London, New York and Melbourne. Ward, Lock & Co. Ltd. (1897)[1]. KB.

A New Pictorial And Descriptive Guide To Torquay, Paignton, Dartmouth & Totnes
Torquay. Andrew Iredale[2]. (1897). KB.

1. An edition dated 1899-1900 is in the author's collection. Although a map of South Devon And Dartmoor is called for this is missing.
2. The spine and cover still have Ward & Lock's names. Andrew Iredale's address was The Library, 13 The Strand, Torquay. He also issued the 7th edition (*c.*1910) with his name on the title page as publisher.

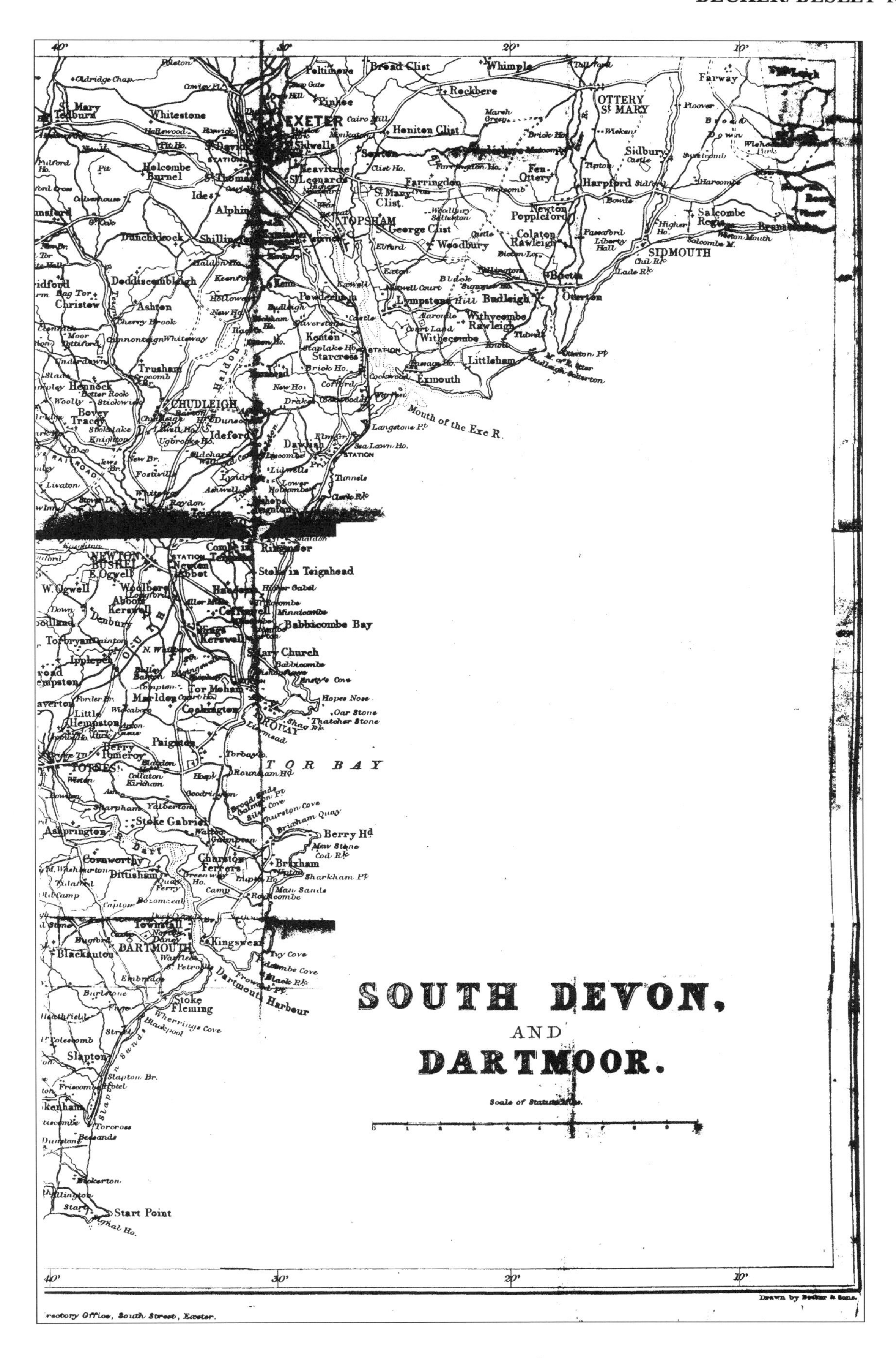

134.S.1 **Becker/Besley** *The Route Book Of Devon – South Devon And Dartmoor*

135

WILLIAM AND ALEXANDER KEITH JOHNSTON

1858

William Johnston (1802-1888) opened his printing business in 1825 after training with two well-respected companies including W & D Lizars (a company he and his brother would acquire in 1862). The following year his brother, Alexander Keith Johnston (1804-1871), joined the flourishing company. They produced their first maps in 1830. Its present publishing activities are carried out in partnership with G W Bacon & Co.

The Johnstons produced a map of the whole country, *Modern Map of England and Wales constructed by W & A K Johnston* (BL 1175 (99)) and transfers of Devon with Cornwall were used for *Murray's Handbook Devon, Cornwall* 4th, 5th, 6th and 8th editions; a folding map of Devon and Cornwall being slipped into a pocket in the inside back cover.[1] The handbooks were all printed by W Clowes and Sons.

Sectional maps covering the county were taken from the same (virtually unchanged) plates of the whole country and used in *Cassell's Gazetteer of Great Britain and Ireland.* This was published by Cassell and Co. in parts from 1893 - 1898 and Devon is found mainly on sheets XIX, XX and XXV in Volumes II and III. The Johnston brothers produced a series of maps as advertising for Dexter's Weatherproof Clothing: this contained one map of the complete southwest peninsula with imprints *Dexter Proofing is proof against time as well as tempest* and *As British As The Weather - But Reliable.*

In 1889 W & A K Johnston published their county atlas *The Modern County Atlas of England & Wales.* Each county map was a transfer taken from their earlier map of the whole country. These maps show the county in detail, the surrounding areas are left plain.

In 1892 the transfers were also used to provide a set of maps for *Deacon's Devon and Cornwall Court Guide.* These contained four maps of Devon and Cornwall: Political; Geological showing rock strata etc.; Climatological, the weather; Hydrographical, depicting rainfall and run-off areas. This last was probably the first hydrographical map of the county excluding De la Beche's small map (**118**).

Size: 335 x 485 mm. **SCALE 7 Miles to one INCH** (20=75 mm).

HANDBOOK MAP OF DEVON AND CORNWALL (CaOS) (length 100 mm). Signature: **Engraved by W. & A. K. Johnston. Edinburgh**. (EeOS). Imprint: **Published by John Murray, Albemarle Street, London, 1858.** (CeOS). Inset map of the **SCILLY ISLANDS** (Ad) 12 mm from border. Railways to Bideford, Tor Mohan, Tiverton and Penzance.

1. 1858 *A Handbook for Travellers in Devon and Cornwall. Revised by T C Paris Fourth Edition*
London. John Murray. 1859, 1859 (1860). C, E; KB.

2. 1863 Date changed to **1861**. Wolf Rock deleted. Railways to Kingswear, Tavistock and Exmouth.

A Handbook for Travellers in Devon and Cornwall. Fifth Edition
London. John Murray. 1863. C.

3. 1865 The date is removed from the imprint, Scilly inset moved nearer to the border.

A Handbook for Travellers in Devon and Cornwall. Sixth Edition
London. John Murray. 1865. E.

1. The first edition appeared in 1850 (reissued 1851, 1856) and contained copies of the J & C Walker map (**116**); the 9th and 10th Editions also had the Walker map; the 11th Ed. has a Bartholomew map (150.14 - 1895).

4. 1867 Date **1866** is added to the imprint. Title re-written (125 mm). Bristol Channel re-written (ST lower than B and L). Note on authorised railways added. Railway to point on Torquay-Brixham road marked Sandridge and then to Brixham then projected to Kingswear. Planned railway Taunton-Barnstaple added. L&SWR as far as North Tawton only and projected lines from there south and west. Railway to Bideford and Watchet and proposed lines Ilfracombe via Bittedon, to Torrington etc.

A Handbook for Travellers in Devon and Cornwall. Sixth Edition ....
London. John Murray. 1865 (1867), 1865 (1871). TQ; E[1].

5. 1872 The imprint date is changed to **1872**. Railways are now shown to Ashburton, Launceston, Seaton and Exmouth and the L&SWR line to Okehampton and Bodmin-Wadebridge.

A Handbook for Travellers in Devon and Cornwall ... Eighth Edition revised[2]
London. John Murray. 1872. BL, KB.

6. 1875 The imprint date is changed to **1874**.

A Handbook for Travellers in Devon and Cornwall ... Eighth Edition revised
London. John Murray. 1872 (1875). KB.

7. 1889 Map is reduced to show Devon alone and has new title **DEVON** (Dd). Imprint removed, new signature: **W. & A. K Johnston. Edinburgh & London**. (EeOS). Scale **English Miles** (15=55 mm). Detail in surrounding counties reduced. Size 245 x 340 mm. Grid ABCD and abcd added. Note: 'Plymouth to Southampton 150 miles'. Foreland, Prawle & Lundy break the border, Eddystone is outside. Railways shown to Berry Head, Torrington, Holsworthy, Ashton, and Exeter-Morebath via Tiverton. Scilly Isle map deleted.

The Modern County Atlas Of England & Wales
Edinburgh and London. W & A K Johnston. 1889. BL, RGS, B, CB, GUL, (DEI[3]).

8. 1894 Railways added to Kingsbridge, Launceston-Halwill and L&SWR line Plymouth-Tavistock.

Loose sheet possibly from revised copy of *The Modern County Atlas* (DEI).

9. 1896 **a**) Map of Devon and Cornwall with new imprints and signatures: **Deacon's Devon and Cornwall, Political** (CaOS); **Charles William Deacon & Co., London**.(EeOS). Johnston signature as above. The Explanation key, Lundy Island and Scillies inset map are all removed. Scale bar now (CeOS). Wolf Rock shown (Ae). Railways extend to Sidmouth and Budleigh Salterton, Ashton (and pecked to Exeter), Moreton Hampstead and Okehampton-Tavistock.

The Devon and Cornwall Court Guide and Country Blue Book ... with coloured maps ...
London. Charles William Deacon & Co. 1896. E, TQ.

b) Map as above. **Deacon's Devon and Cornwall, Geological.** Coloured to show strata with Explanation Key (Bb). Sources as last.

c) Map as above. **Deacon's Devon and Cornwall, Hydrographical.** Coloured to show rainfall and boundaries of river basins with Explanation Key (Ab). Sources as last.

d) Map as above. **Deacon's Devon and Cornwall, Climatological.** Coloured and lined to show temperature and rainfall with Explanation Key (Ab). Sources as last.

1. Both maps are badly torn.
2. According to Lister (p. 69) there was no Seventh Edition. This is likely as the Sixth Edition was revised as late as 1871. The Exeter Westcountry Studies Library copy is torn and the central section is missing.
3. Has number 15 on reverse.

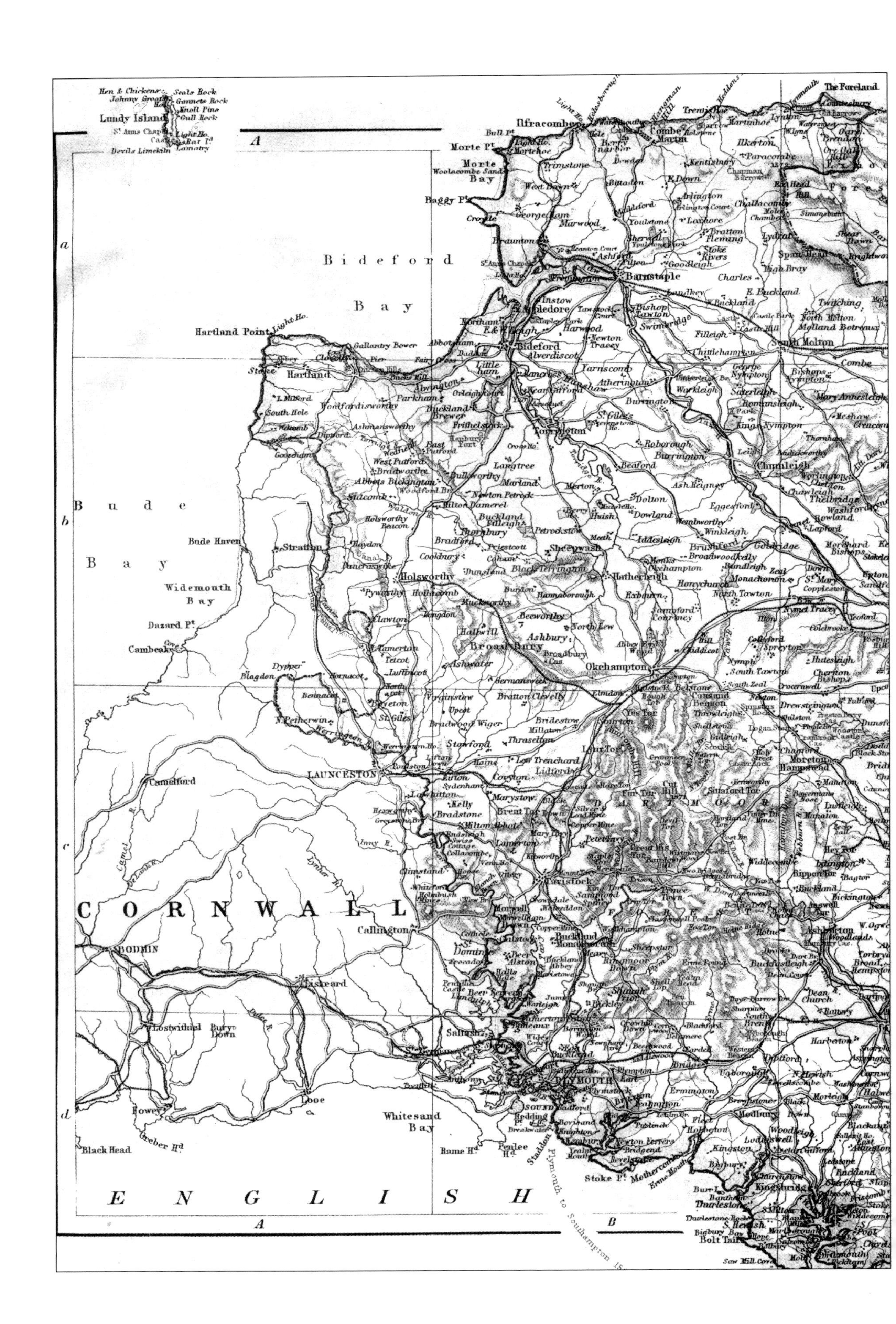
Lundy Island
Ilfracombe
Morte Pt.
Morte Bay
Baggy Pt.
Braunton
Barnstaple
Bideford Bay
Hartland Point
Hartland
Bideford
Appledore
Instow
Torrington
South Molton
Chumleigh
Holsworthy
Stratton
Bude Bay
Bude Haven
Widemouth Bay
Dazard Pt.
Cambeak
Okehampton
Hatherleigh
Launceston
Camelford
Tavistock
CORNWALL
BODMIN
Callington
Liskeard
Lostwithiel
Saltash
Looe
Fowey
Plymouth
Whitesand Bay
Black Head
Ashburton
Kingsbridge
Bolt Tail
ENGLISH
A
B
a
b
c
d

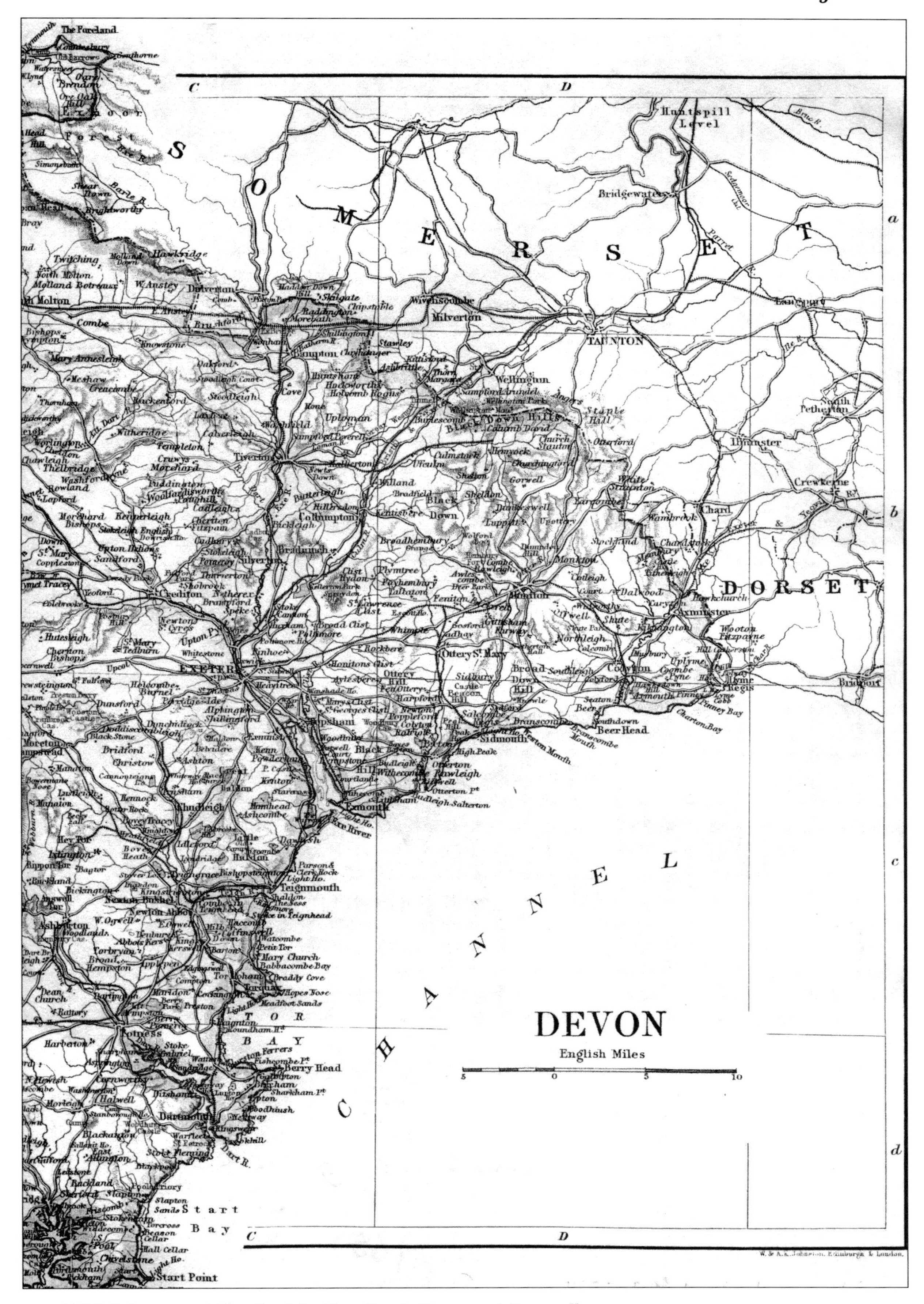

135.7 Johnston *A Handbook for Travellers in Devon and Cornwall*

136

EDWARD WELLER - Weekly Dispatch I

1858-1864

Between 1858 and 1862 maps of the counties were printed in the *Weekly Dispatch* newspaper. These were engraved by Edward Weller, Benjamin Rees Davies, T Ettling, J W Lowry and John James Dower (whose father engraved a Devon map for Moule, **111**) and lithographed by Weller or Day & Son *Lithographer to the Queen.* The plates were taken over first by Cassell, Petter and Galpin (p. 40) and later by G W Bacon (p. 54).

Devon was printed on two sheets, appearing in November 1858 and April 1859, the date being embossed: these were removed when the atlas was reissued. However, some atlases contain maps with and without the embossed emblem implying that some of the first atlases used up sheets prepared for distribution with the newspaper. Sometimes the maps had page numbers vertically (EeOS): a copy in the author's collection has no. 22 on the South sheet; and the 1863 edition in the Whitaker collection has a blind stamp - 16 - on reverse of North sheet. The South Division map is incorrect: Thornecombe had been exchanged for Stockland and Dalwood in 1842.

Devon was engraved by Edward Weller, a very successful engraver who produced work for McLeod (**140**); for *Collins County Geographies*; *Philips' Atlas of the Counties*, and he also produced maps for John Murray, eg the town plans of Plymouth and Exeter. His son, Francis Sidney Weller, also produced a large number of maps including some in the *Comprehensive Gazetteer* (1894) and for natural histories (eg Lincs).

Size: 300 x 425 mm. **British Statute Miles** (15=95 mm).

DEVONSHIRE (SOUTH DIVISION) BY EDW^D WELLER. F.R.G.S. (Ee) surmounted by small engraving of Mercury flying over a 'half' globe, which has a ribbon with **THE DISPATCH ATLAS**. Imprint: **Weekly Dispatch Atlas 139 Fleet Street** (AeOS), Signatures: **E Weller Lithogr** (CeOS) and **Engraved by Edwd Weller, Duke Street, Bloomsbury** (EeOS). Map has embossed emblem in top left corner outside map - **SUPPLEMENT TO THE WEEKLY DISPATCH OF SUNDAY NOVR 21ST 1858**. Bristol and Exeter Railway to Exeter. South Devon Railway to Tor Mohan, Plymouth, and Taw Valley Railway to map border.

and

DEVONSHIRE (NORTH DIVISION) BY EDW^D WELLER. F.R.G.S. (Ea) surmounted by Mercury with **THE DISPATCH ATLAS**. Imprint (AeOS) and signatures (CeOS and EeOS) as above. Map has embossed emblem in top left corner outside map - **SUPPLEMENT TO THE WEEKLY DISPATCH OF SUNDAY APRIL 17TH 1859.** Bristol and Exeter Railway to Exeter including Tiverton branch. South Devon Railway to border and North Devon Railway to Bideford.

1. 1858/9 *The Weekly Dispatch*
London. *Weekly Dispatch* Office. 1858 and 1859. (KB).

2. 1863 Without the embossed emblem of Weekly Dispatch.

The Dispatch Atlas
London. *Weekly Dispatch* Office. 1863. BL, C, RGS, BCL, CB, W.

Cassell's Complete Atlas
London. Cassell, Petter and Galpin. (1864). C, TB.

Folded and mounted on card with marbled end cover as part of set. (RGS).

CASSELL, PETTER & GALPIN – Weekly Dispatch II

1864-1867

Publishers at La Belle Sauvage Yard, Ludgate Hill, Cassell, Petter & Galpin purchased the plates and stock of the *Weekly Dispatch Atlas* in 1864 and immediately advertised their atlas. They used old stock plus sheets with amendments and new imprints side by side (see below) until all plates had been revised. The maps were also issued weekly as loose sheets to the readers of *Cassell's Illustrated Family Newspaper.*[1] The maps were used in other geographical works: Cassell's Folio Atlases, Cassell's Pocket Maps and Cassell's Tourist Guides. From 1869 the maps were being offered by Bacon (see p.54).

John Cassell (1817-1865) was a keen supporter of the teetotal movement and lectured against drinking and is known to have visited the westcountry in 1840/41.[2] He founded a tea and coffee business before going into publishing and printed a number of works aimed at the working man such as almanacks, tracts, a temperance monthly and a weekly radical newspaper. He ventured into book publishing in 1850 and in 1851 was advertising guides to the Great Exhibition in English and French. Thomas Dixon Galpin (*b.*1828) and the five years senior George William Petter ran a printing business with which Cassell worked frequently. In 1855 Cassell ran into financial problems and Petter and Galpin bought his business, retaining the Cassell name and moving into his printing premises. John Cassell had little more to do with the running of the firm and died in 1865. Thomas Galpin was related to George Lock of Ward, Lock & Co.[3]

Petter and Galpin both had westcountry connections: a Petter of Barnstaple was High Bailiff of North Devon at the beginning of the century and his grandson, George William, began as an apprentice draper in Barnstaple before moving to London; Galpin (*b.*1828) moved to London *c.*1851 from Dorset.

3. 1864 North sheet is as State 2. South sheet has the designs of Mercury, scroll and hemisphere deleted and new imprints: **LONDON, PUBLISHED BY CASSELL; PETTER & GALPIN, LA BELLE SAUVAGE YARD, LUDGATE HILL, E C.** (EeOS) and **MACLURE, MACDONALD & MACGREGOR'S STEAM LITHO: MACHINES, LONDON.** (AeOS).[4] Railways: L&SWR through Honiton to Exeter, lines to Chard, Tavistock, Exmouth, Kingswear and a line from Tor Mohan to Torquay town centre.

(KB).

Cassell's County Atlas
London. Cassell, Petter and Galpin. (1864). NMM.

4. 1864 The two sheets are numbered **36** and **37** (EeOS). South sheet has title, signature, scale and list of symbols now raised to make room for table of statistics for 1861 (Ee). Reference to **MACLURE** etc. removed. North sheet has L&SWR through Honiton to Exeter and the designs of Mercury, scroll and hemisphere have been deleted. [5]

(KB), (FB).

Cassell's *Illustrated Family Newspaper*
London. Cassell, Petter and Galpin. 1864. BL.

Cassell's Complete Atlas
London. Cassell, Petter and Galpin. (1866). RGS.

Cassell's British Atlas (issued in monthly parts)[6]
London. Cassell, Petter and Galpin. (1867). BL, W, C, Leeds.

1. One of Cassell's more successful ventures, this paper began in 1853 and, under different titles, survived until 1932.
2. S Nowell-Smith; 1958; pp. 8, 29, 42, 51ff and D Smith; Cassell and Company; in *The IMCoS Journal*; Issue 70; 1997.
3. Edward Liveing in *Adventure in Publishing* (Ward Lock & Co. Ltd; 1954; p.29) recounts that Cassell, Petter and Galpin sold quantities of maps from the *Dispatch Atlas* to Ward and Lock in 1863.
4. The printers Maclure, Macdonald and Macgregor were soon changed. Maclure & Macdonald also printed a late version of Dix and Darton's (**82**) Devon map for Simpkin, Marshall & Co. in *Devonshire The Official County Map and Guide* published *c.*1877. The much revised map had a new title: *Official Map of Devon* and the deletion of the vignette.
5. Other counties are known in this state either as folding maps in a cover entitled *Cassell's County Maps For Road And Rail*
6. Issued initially in 19 monthly parts of six maps each from February 1864 to august 1865 with five supplementary parts published September 188 to January 1867. The BL copy has a second title page: *Cassell's Universal Atlas*.

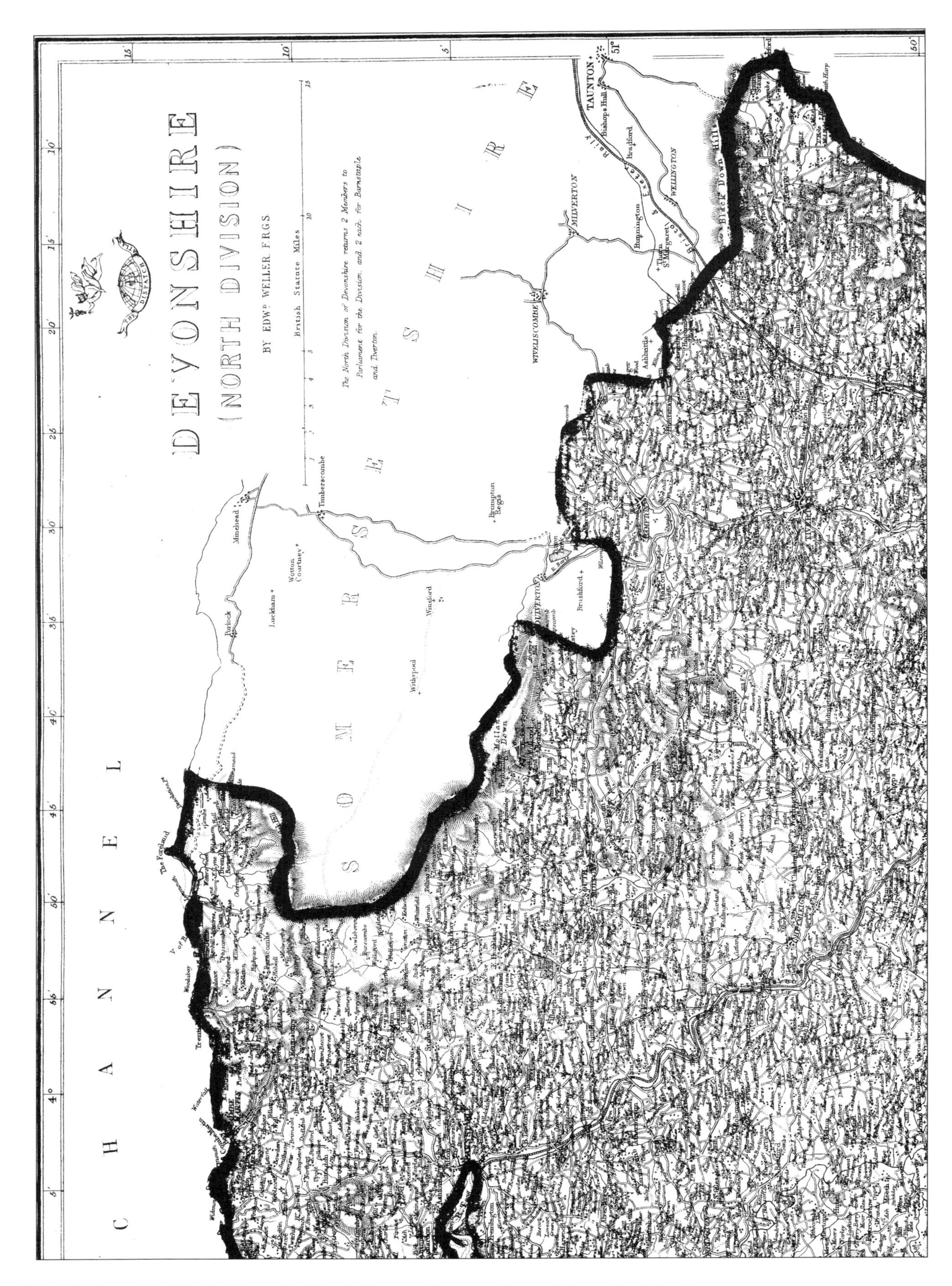

136.1 Weller *The Weekly Dispatch* – North Sheet without embossed date of publication

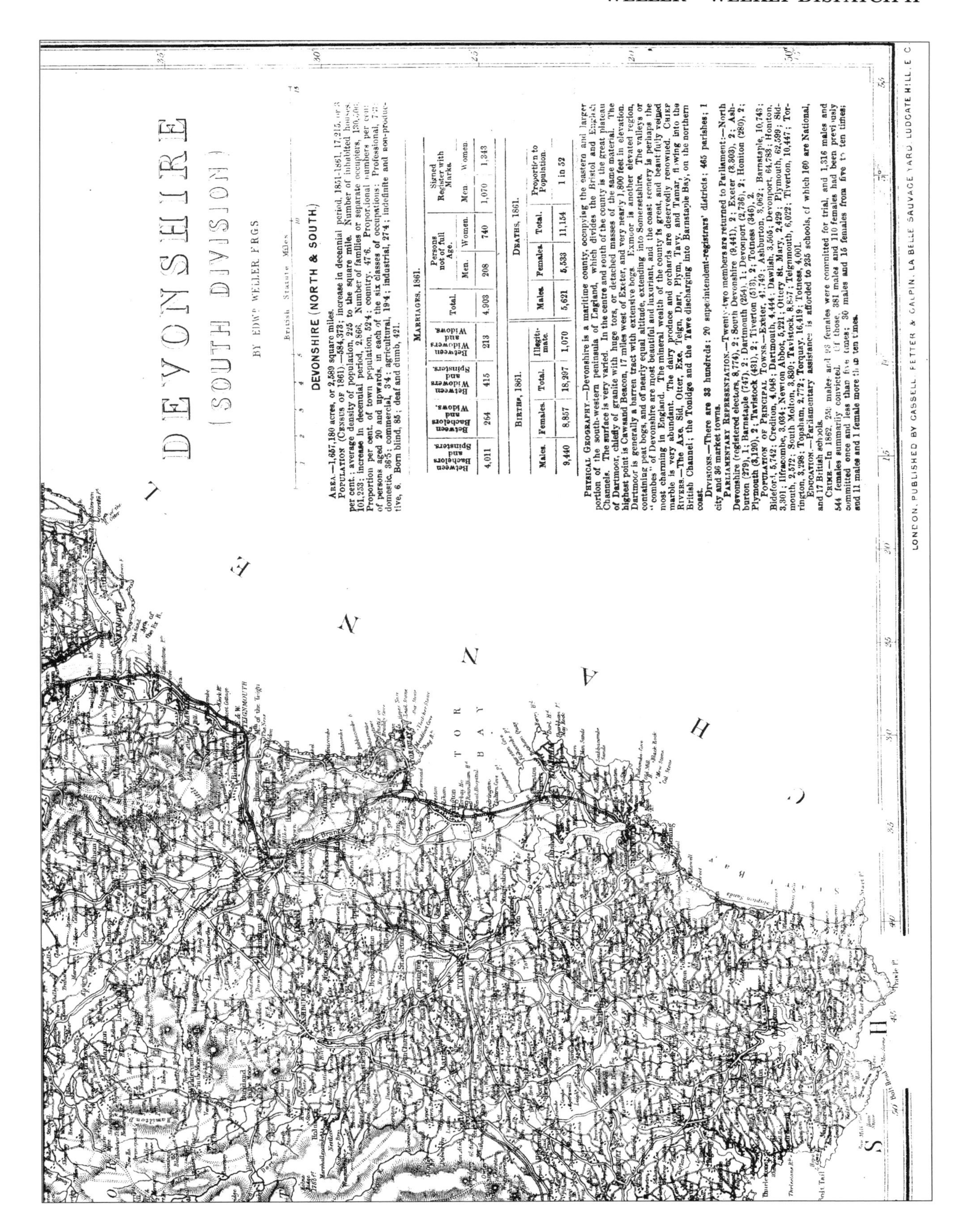

136.4 E Weller *Cassell's British Atlas* – South Sheet with table of statistics

G W BACON & Co. – Weekly Dispatch III

1869–1900

George Washington Bacon (*fl.*1869 - 1932) published a large variety of maps from the late 1860s. He succeeded to the firm of John Wyld III *c.*1893[1], and the firm still survives today with W & A K Johnston. Bacon took over some of the stock of John Cary[2] after it had been used by G F Cruchley (and Gall & Inglis and produced cycling and motoring maps (**see Appendix II**)

Bacon acquired the plates of the *Dispatch Atlas c.*1869 after they had been used in *Cassell's British Atlas.* He exploited these to the full in atlases and also in folding county maps. The atlases were issued into the 20th centruy, and copies of Devon issued after 1901 can be identified by the reference to adjoining maps found in the margins. There is a new imprint; **Bacon's Geographical Establishment** (AeOS) and printer's mark (eg **42B15)** and note **ADJOINING COUNTIES Somerset page 37, Dorset 14, Cornwall 8** (EaOS). The population figures for 1891 and 1901 are included.

Size: 680 x 615 mm. **British Statute Miles** (15=137 mm).

5. 1869 Single sheet with new title: **BACON'S MAP OF DEVONSHIRE BY EDWD WELLER FRGS** (Ea). New imprint: **LONDON, G W BACON & Co. 337, STRAND, OPPOSITE SOMERSET HOUSE.** This is an enlarged lithograph to a larger scale. Lundy is omitted and the Dorset area and Thornecombe are corrected. The previous overlap areas are redraughted. The outline letters of the county name have been filled with an ornate pattern and the map extends into the border on three sides. Signs for railways, stations, canals and roads added. Railways to Seaton, Launceston, Moreton Hampstead, Plymouth-Tavistock-Lidford-Launceston. Planned railways Taunton-Barnstaple, Tiverton-Bampton, Ottery St Mary-Budleigh-Exmouth, Torrington-North Tawton, Kingsbridge, Newton spur to Ashton added, and stations marked by large dots. Parliamentary representation altered, with the map coloured to show the three divisions.[3]

Bacon's County Atlas
London. G W Bacon. 1869. C.

Bacon's Large Print County Atlas Of England And Wales
London. G W Bacon. 1870. TB.

6. 1876 Map on two sheets, each 305 x 430 mm, at a smaller scale (15=115mm): **BACON'S MAP OF DEVONSHIRE BY EDWD WELLER FRGS** (in plainer print on both sheets). Imprint deleted. Note on use of station names added. Railways added, North Sheet: to Torrington, Ilfracombe, Barnstaple-Taunton; South Sheet: Lidford-Okehampton complete, Sidmouth and Brixham deleted. The other division is named but not drawn in. Maps extend over borders at Lundy, Exmouth, Taunton, Otterford, Prawle Point and Churchingford.[4]

Bacon's New Quarto County Atlas
London. G W Bacon & Co. (1876). TB.

1. For an account of the Wyld family see D Smith; The Wyld Family firm; *The Map Collector*, Issue 55; 1991.
2. For an account of the Carys see D Smith; The Cary Family; *The Map Collector*, Issue 43; 1988.
3. Other counties, eg Berks are extant in this state with new address, 127 Strand and dated (1870), or with W C (London) as folding maps with titles such as *Bacon's New Tourist's Map.*
4. Other counties were available in this state as *Bacon's New Pocket Map* mounted and folded.

7. 1876 Ordinary print numbers are added **18** to south and **19** to north (AeOS). The extraneous parts of both Dorset & Devon are removed with the border made good (though erasure is still visible). The railway symbol is changed from 'track' to heavy 'dot-dash'. Railways are Axmouth; through Chard; Moreton Hampstead; Ashburton; Lidford-Launceston. On the North sheet the Devonshire note is rewritten and Key added. A Railway note is added (Ee) erasing the mouth of the Otter.

Bacon's New Quarto County Atlas
London. G W Bacon & Co. (1876). BL.

8. 1883 **a**) South sheet is slightly enlarged (height 330 mm). Maps have new titles: **DEVONSHIRE (NORTH DIVISION)** and **DEVONSHIRE (SOUTH & EAST DIVISIONS).** Both with **REDUCED FROM THE ORDNANCE SURVEY BY EDWD. WELLER. F.R.G.S. Divided into 5 mile squares** (with A-H [&] F-N vertically and 1-13 [or] 1-14 horizontally). The other division is now drawn in in full. Railways:- North Sheet: Holsworthy, Launceston and Hemyock; South Sheet:- Brixham, Chudleigh-Ashton, Launceston and Holsworthy. Border broken at Lyme Regis, Lundy and Exmouth, but is extended below Prawle and ENG**LISH** is completed for the Channel. **IRE** missing in DORSETSHIRE. Page numbers are simply printed **12** and **13** (EaOS).[1]

New Large Scale Ordnance Atlas Of The British Isles
London. George W Bacon. (1883). BL[2], PRO[3].

b) South sheet is slightly smaller (height 310 mm). Border is broken at Prawle Point. The Dartmoor line is drawn as a full railway. ENG**LISH** incomplete.

New Large Scale Ordnance Atlas Of The British Isles
London. George W Bacon. (1883), 1884. W, KB; BCL.

9. 1884 **IRE** added to DORSETSHIRE.

New Large Scale Ordnance Atlas Of The British Isles
London. G W Bacon. 1884. TB, RGS.

10. 1884 The note *Boundary of Boroughs* is changed to *Parliamentary Boroughs*

Bacon's New County Guide And Map of Devon From The Ordnance Survey - folding map
London. G W Bacon & Co. (1884). TB.

New Large Scale Ordnance Atlas Of The British Isles
London. G W Bacon. 1884, (1885). [B]; NLS.

11. 1885 Shows pecked railway Holsworthy-Bude, a further line into Cornwall (Jacobstow) and the Exeter -Bampton line. The note *Railways in Progress* is added.

New Large Scale Ordnance Atlas Of The British Isles
London. G W Bacon. (1885). Leeds, TB, BL.

1. The page numbers were added by stamp and two different type formats were used.
2. The BL copy was delivered in 1883.
3. Lacks title page.

12. 1885 Enlarged to original scale (15=135 mm). Titles are now **DEVONSHIRE (NORTH SHEET)** 425 x 610 mm and **DEVONSHIRE (SOUTH SHEET)** 450 x 650 mm.
IRE is omitted. Line referring to Edward Weller in the title deleted. Explanation redrawn (South Sheet). Railways: Exeter-Bampton (North Sheet); and Kingsbridge, halfway to Salcombe, Brixham Quay, L&SWR toTavistock (but not to Bude) (South Sheet). Cut and folded into soft cover.

Bacon's County Guide Map North Devon from the Ordnance Survey
London. G W Bacon & Co Ltd. (1885). BL.

Bacon's County Guide Map South Devon from the Ordnance Survey
London. G W Bacon & Co Ltd. (1885). BL, W, E[1].

13. 1886 Size and scale as state 11. Ludgate appears in the Railway note - instead of separate boxes for colours the boxes have been drawn together. No colouring for boroughs Tiverton and Barnstaple. Railway from Holsworthy to Jacobstow in Cornwall is shown pecked.

a) In atlas form.

New Large Scale Ordnance Atlas of the British Isles
London. G W Bacon. (1886). TB.

b) Folding map with no page numbers.

Bacon's New County Guide And Map of Devon From The Ordnance Survey
London. G W Bacon & Co. (1887). WM.

c) Page numbers are now EaOS directly above map, not in page corner as before.

New Large Scale Ordnance Atlas of the British Isles
London. G W Bacon. 1888. TB.

14. 1889 New page numbers **DEVON. N. 12** and **DEVON. S. 13**. (both EaOS). The Jacobstow railway in Cornwall is erased.

New Large Scale Atlas of the British Isles from the Ordnance Survey
London. G W Bacon & Co. (1889[2]). BL, NLS, CB.

15. 1890 **a)** South Sheet has the border extension below Prawle Point.

New Large Scale Atlas of the British Isles from the Ordnance Survey
London. G W Bacon & Co., Ltd. 1890. TB.

b) Where Exmouth extends into the border the inner border line has been redrawn to enclose it (previously no border here).

New Large Scale Atlas of the British Isles from the Ordnance Survey
London. G W Bacon & Co., Ltd. 1890, 1891. Hull; PRO.

16. 1891 The reference key is completely redrawn with the railway note moved to the bottom - North sheet (Ee), South sheet (Ae).

New Large Scale Atlas of the British Isles ... Census of 1891
London. G W Bacon & Co., Ltd. 1891, 1892. Liv, BL; TB.

1. Library date of 1895 is too late, the railway network suggests a date of 1890 (Tavistock line).
2. BL copy has receipt stamp, 11 Nov. 1899.

17. 1893 Break in borders at Exmouth and Prawle Point. All pecked lines erased with exception of Totnes-Ashprington. Page Nos added: **DEVON.N.12** and **DEVON.S.13** on map reverse.

New Large Scale Atlas of the British Isles ... Census of 1891
London. G W Bacon & Co. Ltd. 1893. TB.

18. 1895 The Parliamentary Division information replaced by new **REFERENCE** table. REFERENCE TO COLOURS added on both maps, (Aa, North) and (De, South) together with population figures for 1881 and 1891. The grid on the South Sheet is changed, through Borough island (EW) and the b of Bigbury Bay (NS). The Kingsbridge line and Plymouth-Plymstock lines are added. (Note also the pecked line down the centre of the Dart from Totnes to Dartmouth). North and South Sheet have pecked lines. Torrington-Brightly Bridge removed and no pecked lines at Holsworthy. The extended border by Exmouth is altered – now shows 20'.

Commercial And Library Atlas Of The British Isles from the Ordnance Survey[1]
London. G W Bacon & Co., Ltd. 1895. BL, C.

Commercial And Library Atlas Of The British Isles
London. G W Bacon & Co., Ltd. 1896[2], 1897. W, BL, KB; RGS, TB.

19. 1898 South sheet has pecked railway Plymstock-Turnchapel[3].

Commercial And Library Atlas Of The British Isles
London. G W Bacon & Co., Ltd. 1898. W, TB.

20. 1899 The proposed railway Budleigh Salterton is shown and the lines Torrington-Hatherleigh to Brightley Bridge north of Okehampton.

Commercial and Library Atlas of the British Isles Ordnance Survey
London. G W Bacon & Co., Ltd. 1899. TB, BL.

21. 1899 Budleigh Salterton, Plymstock-Yealmpton and Turnchapel lines complete.

Commercial and Library Atlas of the British Isles Ordnance Survey
London. G W Bacon & Co., Ltd. 1899, 1900.
KB, TB; PRO, Liv.

1. Spine and cover title *Bacon's Popular Atlas of the British Isles*
2. Maps of most other counties now have the note on *ADJOINING COUNTIES* and most of these have the new imprint *Bacon's Geographical Establishment* (AeOS).
3. The line from Plymouth to Turnchapel is very unclear as though it has been engraved and then removed.

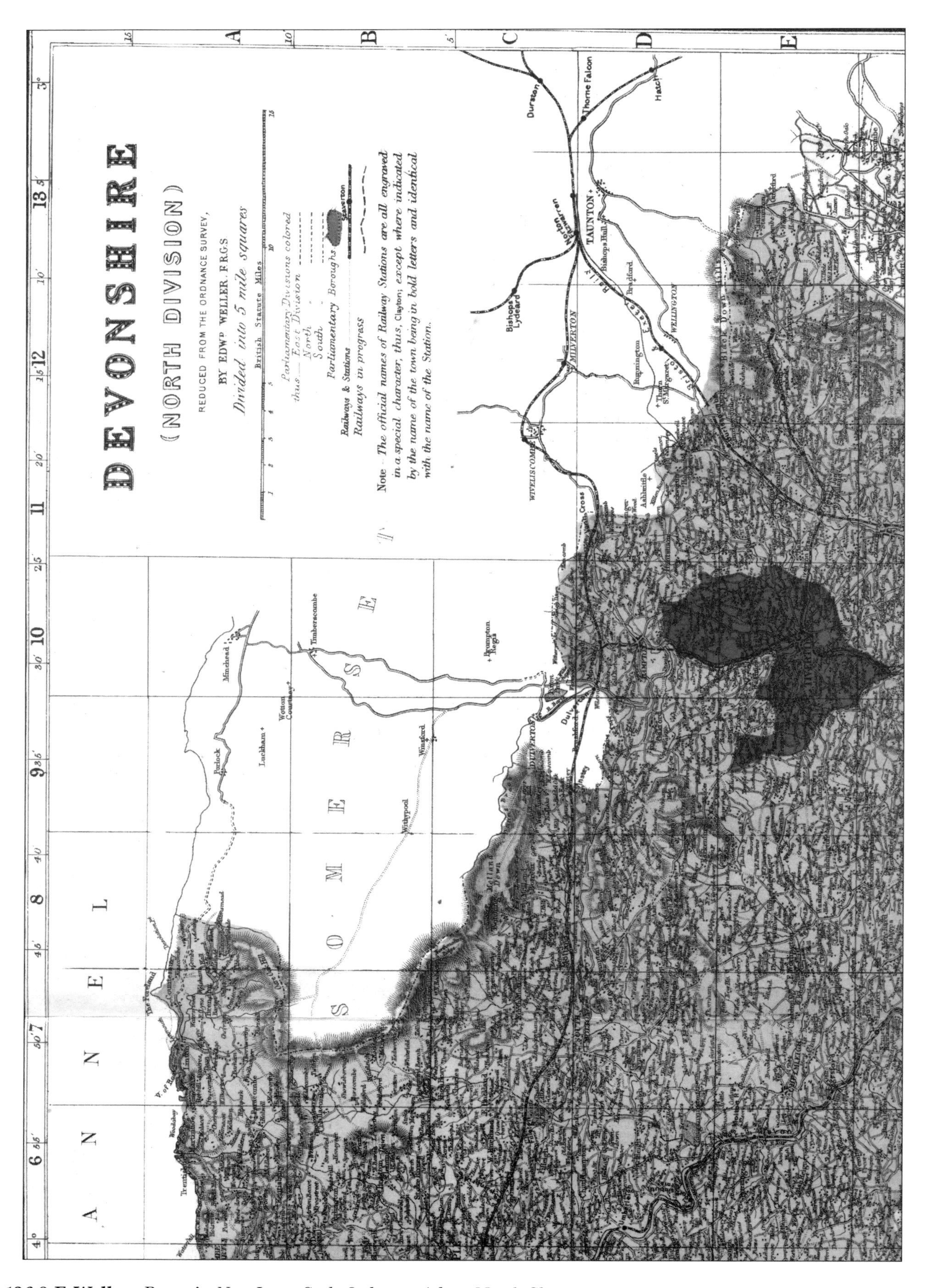

136.8 E Weller Bacon's *New Large Scale Ordnance Atlas* – North Sheet

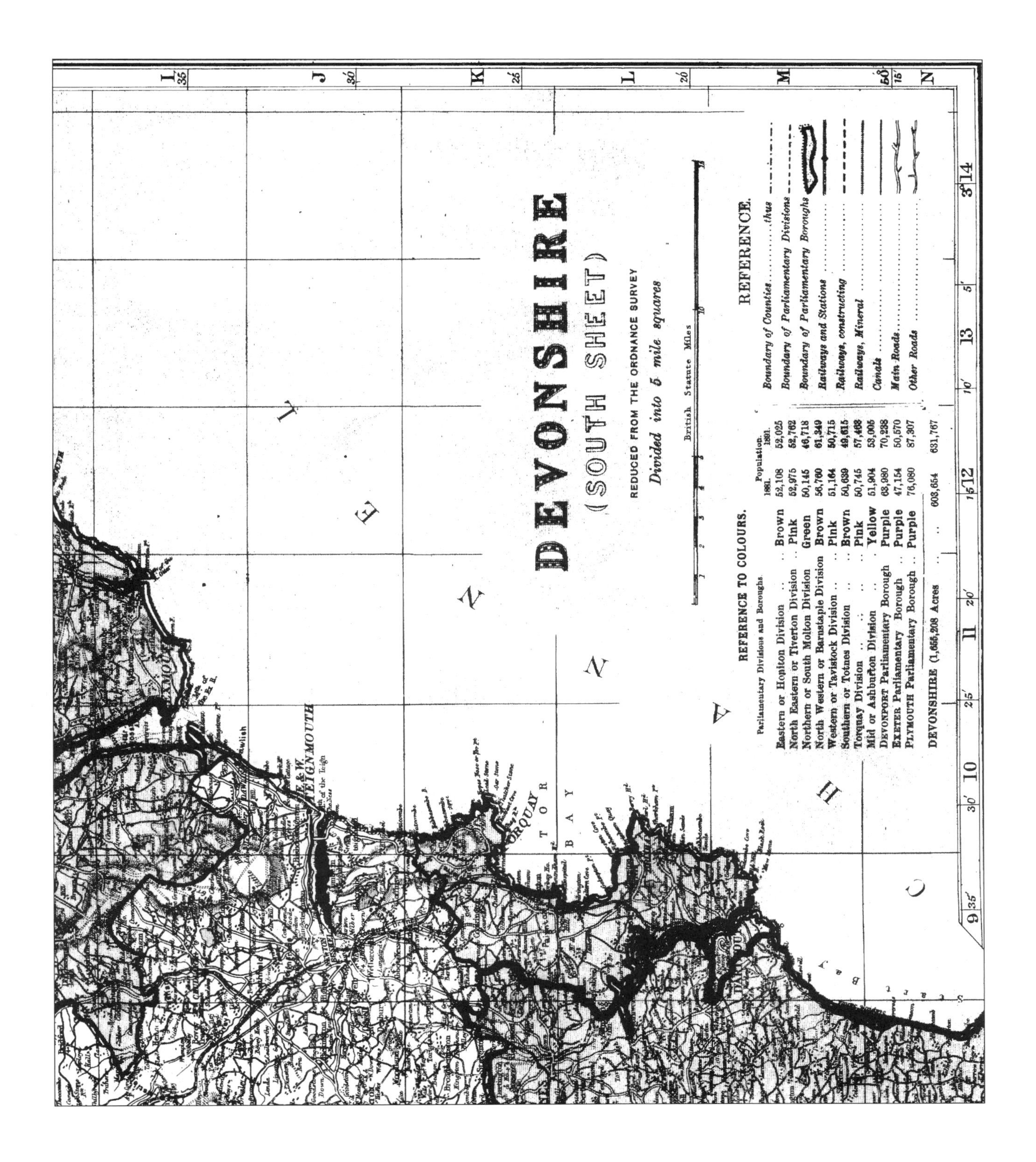

136.18 **E Weller** Bacon's *Commercial and Library Atlas* – South Sheet

137

WALKER/STANFORD

1859

Born in London, Edward Stanford served an apprenticeship as a printer in Malmesbury. He took over the established business of Trelawny William Saunders in 1853 and by the following year was a Fellow of the Royal Geographical Society (proposed by F J Faunthorpe). From 1853 onwards Stanford (1827-1904) had various London addresses, usually in the Charing Cross area.

Although Stanford's was not to become a leading name in map production until the 1880s there are some examples of earlier work including a map for a guide book produced in 1859. The first Stanford map of Devon or of Cornwall is this map of both counties combined. The map appeared in a guide to the coasts by Mackenzie Walcott, published by Stanford and printed by W Clowes of Stamford Street (August 1859). This *Guide to the South Coast of England* was dedicated to Agatha and Constance Walcott. Later when 'Devon and Cornwall' was published separately, this was dedicated to William Brooks King. Part of the first text was reused starting on page 413, following the title page, preface and list of routes. The map was taken from a larger map of the whole country. The original map of England and Wales was engraved by J & C Walker and printed by Standidge and Co. It was issued as Stanford's Railway. And Road Map Of England And Wales in 1862 with added railway to Tavistock. Other coast guides are known.[1] The two counties are coloured and Lundy I(sland) and the Eddystone are shown but not the Scillies.

The Rev. Mackenzie Edward Charles Walcott (*b.*1821) was Precentor of Chichester Cathedral from 1863 until his death in 1880. He wrote other topographical works including the *Cathedrals of England and Wales with their History, Architecture and Traditions* published in 1858 (which according to the *Guardian,* quoted on a page 2 advert, was *singularly free from errors*) and a book on Lincoln's Memorials (1866).

There are known to be at least 21 guides in Stanford's Series of Pocket Guide-Books; the numeration changing during the formation of the series.[2] Originally Vol. 1 was *Paris,* the *South Coast Of England* was Vol. 6 with *Devon And Cornwall* as Vol. 9. Walcott was also responsible for the *East Coast Of England* (in one or three parts -Vols.13-16), the Lakes (Vol.17), the *Cathedrals Of England And Wales* and the *Minsters and Abbey Ruins of the United Kingdom.*

Further Stanford maps of Devon were issued in 1878 (**154**), 1881 (**155**) and 1885 (**161**).

Size: 226 x 298 mm. **English Statute Miles** (10+50=133 mm).

COUNTIES OF CORNWALL AND DEVON (CaOS). Imprint: **Stanfords Geographical Establishment. 6, Charing Cross, London** (EeOS). The railway is shown to Falmouth and to Bideford, Penzance and a false line to Tor Quay. There is a north point (Bc) and scale bar (Ce).

1. 1859 *A Guide To The South Coast Of England, from The Reculvers to the Land's End, and from Cape Cornwall to the Devon Foreland ... By Mackenzie Walcott, M.A.*
London. Edward Stanford. 1859. NLS.

A Guide To The Coasts Of Devon & Cornwall: Descriptive Of Scenery, Historical; Legendary; and Archaeological By Mackenzie Walcott, M. A.
London. Edward Stanford. 1859, 1860[3]. NLS, E, FB, KB; FB.

1. BL has a copy of *A Guide To The Coasts Of Hants & Dorset,* published by Edward Stanford in 1859.
2. The volume numbering was changed in 1860: eg *London,* the *South Coast of England* (Vol.1), (Vol. 3) *Devon and Cornwall* (Vol. 7), *Cathedrals of the United Kingdom* (Vol.18), and the *Minsters and Abbey Ruins of the United Kingdom* (Vol. 19). The series left Britain with *Paris* (now Vol. 21).
3. The front of the cloth binding carried a repeat of the title page but dated 1860 and the back carried a different and fuller list of Stanford's Pocket Guide-Books. The text is an exact repeat but there are no advertisements.

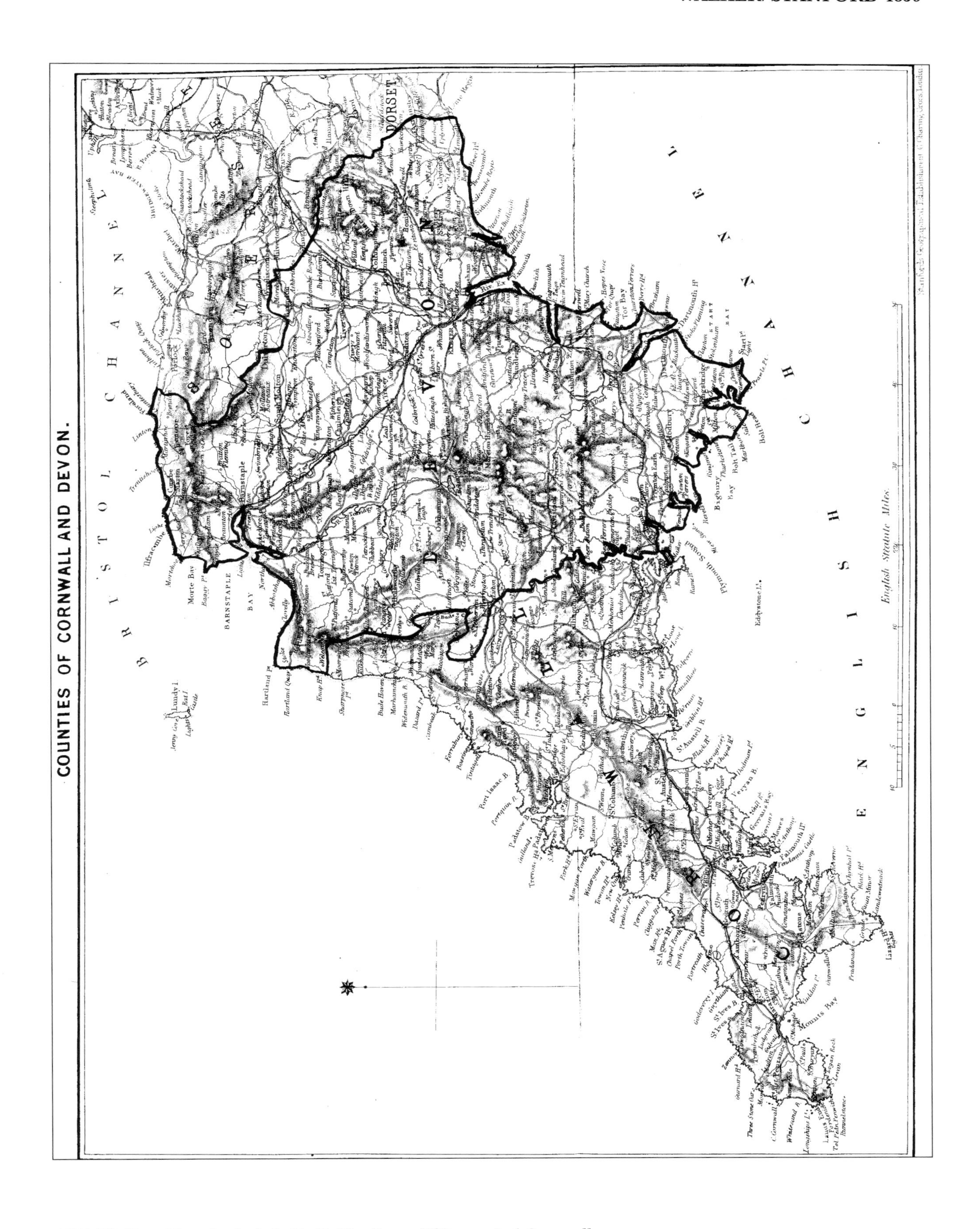

137.1 Walker/Stanford *A Guide To The Coasts Of Devon And Cornwall*

138

GEORGE FREDERICK CRUCHLEY

1859

George Frederick Cruchley (1797-1880) served as an apprentice with the Arrowsmith firm and spent part of his early career working with the Arrowsmith family, but later set up on his own as map-seller, engraver and globe maker at 349 Oxford Street (1823-25), later 38 Ludgate Street (1825-33) and finally 81 Fleet Street (1833-75). He was born in London, the son of John and Ann Cruchley and the story has it that he was christened at St Clement Danes, on St George's Day 1797 - hence his first name.[1]

Cruchley was a jack-of-all-trades: although a trade-card announces his primary business as that of mapseller and publisher, he also offered a complete general engraving and printing service, producing invitation cards, banknotes, cheques, bill-heads and labels. He also engraved seals and brass name-plates. Even coats of arms could be *found* and engraved *on stone, steel, silver and brass.* He was a map wholesaler and advertised a full surveying service.[2]

The first opportunity for Cruchley in his early days came with the production of his maps of London. Two popular maps appeared in the early 1820s: *Improved Environs of London* and *Environs of London Extending Thirty Miles from the Metropolis.* Over the years Cruchley produced many fine plans at large scales. One at the 5-inch to the mile was exhibited at the Great Exhibition of 1851. However, he had also started issuing travel maps and guides much earlier. By 1838 he was offering maps and guides to England, Wales, Scotland, Ireland, and a number of European countries and about two years later his stock included a *Guide to the Levant* (9s), a *Handbook to the East* (15s) and a *Guide to Moscow* (8s 6d).[3]

He reissued some of Arrowsmith's maps and he bought the stock and plates of John Cary *c.*1844[4] and continued to use these throughout his life and it is believed that some plates were still in use in the twentieth century. But by 1860 Cruchley was content in up-dating and reissuing from old plates and resorted to copying or producing lithographic reproductions of others' maps and many of these were mounted on linen and folded into a card slip-case or a book-style cloth binding. The large-scale map, *Cruchley's Reduced Ordnance Map of Devonshire* (cover title), was probably engraved in the late 1850s and is a very close copy of the Palmer/Cary plates produced in 1813 (**80**).

Size: 615 x 950 mm (215 x 140 mm when folded). **Scale of Miles** (12=155 mm).

Map on two sheets with title: **CRUCHLEY'S MAP OF DEVONSHIRE with parts of the adjacent Counties SHEWING ALL THE RAILWAYS AND NAMES OF THE STATIONS, REDUCED FROM THE ORDNANCE SURVEY.** (Ee of South sheet). Imprint: **LONDON, PUBLISHED BY G F CRUCHLEY, MAP-SELLER & GLOBE MANUFACTURER, 81, FLEET STREET.** (Ee of South sheet). Devon is outline coloured red, Somerset green, Dorset and Cornwall are yellow. Thick piano-key border on three sides, broken for Hartland. Railways to Bideford, Tiverton, Tavistock, Exmouth and Plymouth through Liskeard. Line to Kingswear shown red to Paignton station which is named (officially opened August 2nd 1859) and proposed line to Kingswear coloured blue. Proposed line to Watchet.

1. 1859 *Cruchley's Map of Devonshire ...*
London. G F Cruchley. (1859). KB[5].

2. 1860 Now printed on a single sheet, 1230 x 960mm (225 x 150 mm when folded). Folded and hard-backed. The red line (completed railway) extends past Paignton as far as Galmpton.

Cruchley's Map of Devonshire ...
London. G F Cruchley. (1860). KB.

1. See D Smith; George Frederick Cruchley; *The Map Collector*, Issue 49; 1989.
2. Laurence Worms; *Some British Mapmakers*, Ash Rare Books Cat. and Price List; 1992.
3. John Vaughan; 1974; p. 40.
4. D Smith; *op cit.* especially footnote 3 which discusses the possible date of acquisition of Cary's plates by Cruchley.
5. The two maps are designed to slip into a case which is missing.

138.1 Cruchley *Cruchley's Map of Devonshire*

3. 1861 Railways to Brixham Rd Sta. (later renamed Churston), opened 14th March 1861.

a) Single sheet folded and hard-backed

Cruchley's Map of Devonshire ...
London. G F Cruchley. (1861). TQ.

b) Single sheet dissected into 8 parts and laid on linen.

Cruchley's Map of Devonshire ...
London. G F Cruchley. (1861). BL.

4. 1865 Single sheet folded and hard-backed. Railways to Kingswear, and projected lines Yeoford to Tavistock via Oakhampton with branch to Launceston, Moreton Hampstead with branch to Ashton, Sidmouth and Budleigh Salterton. Dissected linen backed and boxed.

Cruchley's Reduced Ordnance Map of Devonshire ...
London. G F Cruchley. (1865). FB.

5. 1868 Two separate sheets (500 x 925 and 510 x 925 mm). Border reduced to three plain lines with tight single line junction and projecting jog for Hartland. Proposed lines: Kingsbridge, Ashton, Oakhampton-Lidford, Barnstaple-Taunton, Ilfracombe, Bideford-N.Tawton, Launceston-Bude.

Cruchley's Map of Devonshire ...
London. G F Cruchley. (1868). B.

6. 1875 Single sheet folded and hard-backed. Brixham Sta. is moved to Brixham and Brixham Rd Sta is now CHURSTON. Barnstaple to Ilfracombe with stops at Pilton and Braunton (completed 1874). The Callington to Calstock line in Cornwall is shown. Lines to Ashton, Holsworthy (to Bude) and Kingsbridge lines are erased.

Cruchley's Map of Devonshire ...
London. G F Cruchley. (1875). E[1].

7. 1875 On two sheets[2] (each 475 x 640 mm) with new title: **CRUCHLEY'S NEW MAP OF NORTH DEVON Shewing all the RAILWAYS & STATIONS, from the ORDNANCE SURVEY.** Same imprint but Cruchley is now **GLOBE MAKER**. Note below the title (Ab) **A MAP OF SOUTH DEVON IS PUBLISHED OF THE SAME SCALE SIZE AND PRICE**. Folded and hard-backed. Railway is shown to Torrington and projected thence to Sampford Courtenay with a branch near Hatherleigh to Holsworthy and on to Stratton in Cornwall and the planned Bampton-Tiverton line shown. The plain border is jogged out 12 mm for Hartland.

Cruchley's New Tourist Map of North Devon from the Ordnance Survey.
London. G F Cruchley. (1875). BL, E[3].

1. Exeter copy is library dated to 1890, but the railway development suggests the earlier date.
2. Although only the north sheet has been seen the note below the title and the evidence of earlier states strongly infers the existence of a south sheet.
3. On the binder of this edition there is an advertisement for a MAP of DEVONSHIRE, with parts of the adjacent Counties: two miles to the inch. Size, 4 feet by 3 feet. (ie approx. 1200 x 900 mm) Rollers varnished, One Guinea.

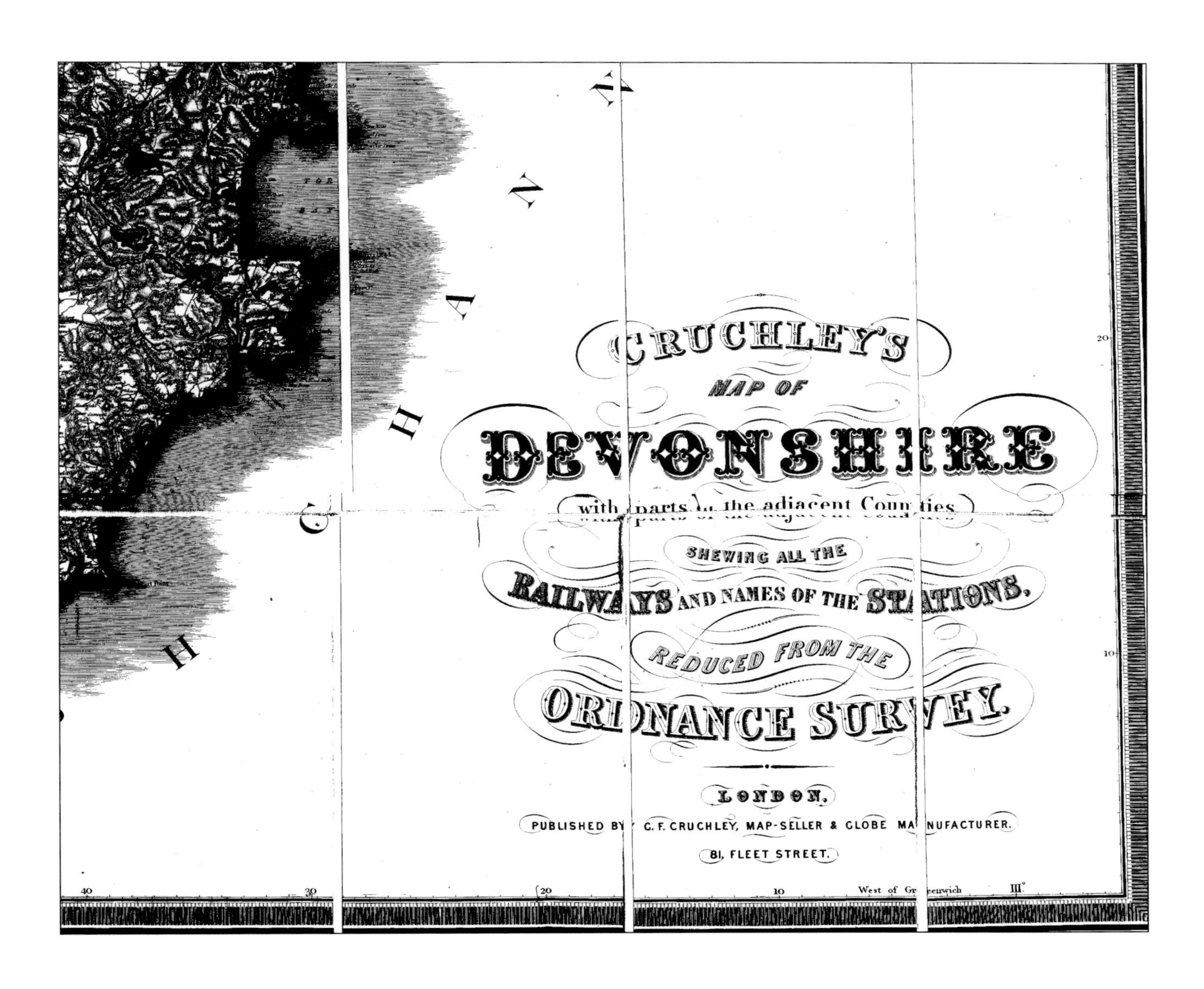

138.4 Cruchley *Cruchley's Map of Devonshire* - detail

139

LEWIS BECKER

1860

An ecclesiastical map that shows the whole of Devon and Cornwall was produced by Lewis Becker for inclusion in the *Transactions* of the Exeter Architectural Society. Becker must surely have been a member of F P Becker's family the inventors of the Omnigraph. The map includes Lundy and Eddystone and is gridded to show Latitude and Longitude. The Archdeaconeries and the Deaneries are coloured in outline. A topographical note is below the title with the Archdeaconeries and Deaneries numbered and coloured. On the left side of the sheet, outside of the map, there is a complete list of the Parishes within the various deaneries and below the lists a note on 'Antiquities'.

The map shows the railways to Penzance, Bideford, Paignton as well as projected railways to Kingswear, Tiverton, and Oakhampton. The L&SWR projected from Exeter to Feniton by then complete to Axminster, Plymouth to Buckland Monachorum and then projected to Tavistock.

Size: 394 x 563 mm. **Scale of Miles** (18 = 107 mm).

AN ECCLESIATICAL MAP OF THE DIOCESE OF EXETER. PREPARED FOR THE DIOCESAN ARCHITECTURAL SOCIETY IN 1860. (Aa). Imprint: **Engraved by Lewis Becker by his Patent Engraving Machines** (Ae). Inset **SCILLY ISLES**: 70 x 80 mm - Scale of Miles (3=18 mm).

1. 1860 *Exeter Architectural Society. Transactions Vol. VI*
Exeter. Wm Pollard. 1861. E.

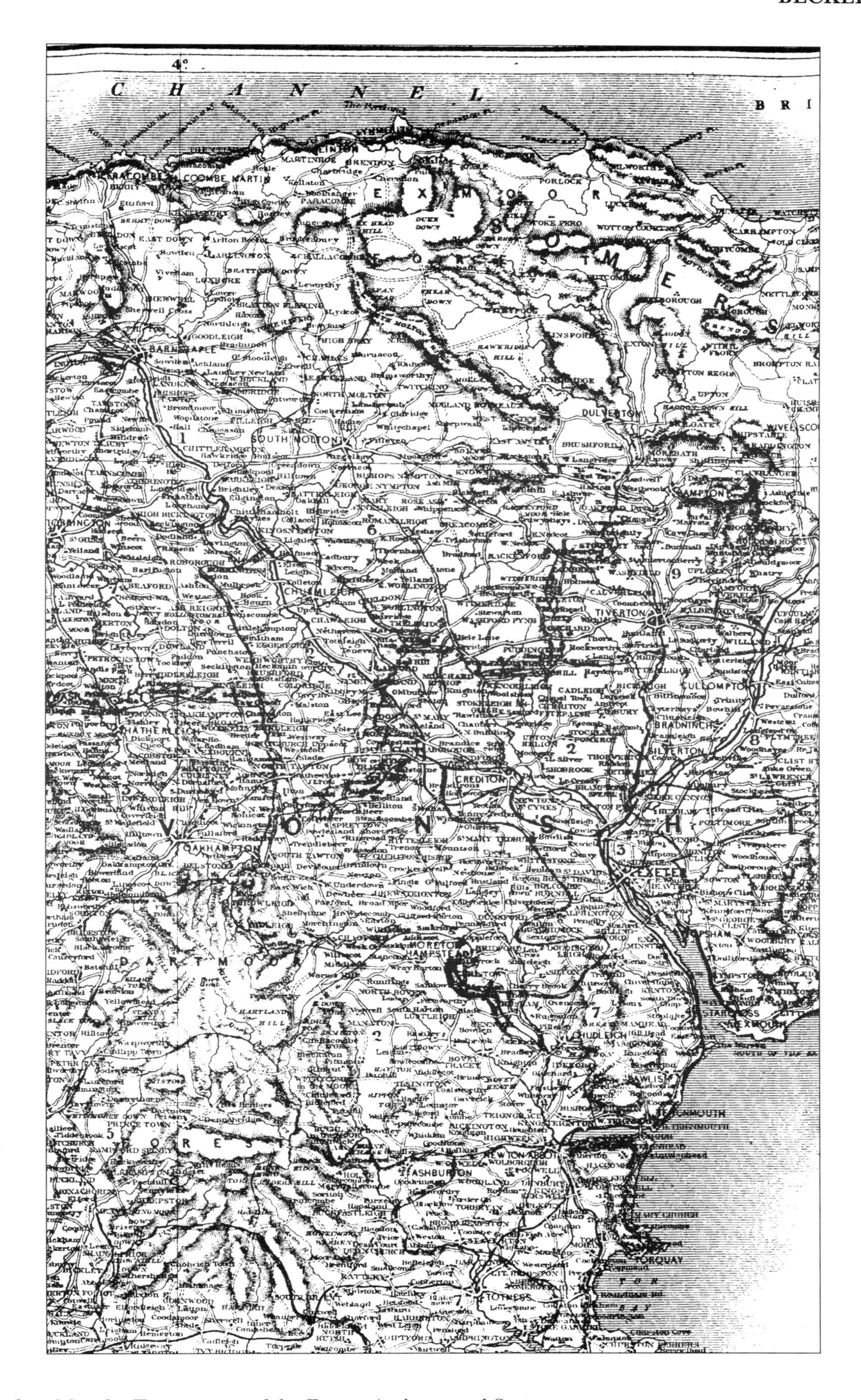

139.1 **Becker** Map for Transactions of the Exeter Architectural Society

140

WALTER McLEOD

1861

Walter McLeod (or M'Leod) was a geographer and headmaster of the Modern School in Chelsea. He prepared several educational atlases between 1858 and 1869 which though interesting do not contain county maps. His one work which did contain maps of the counties was the *Physical Atlas of Great Britain and Ireland* published in 1861 (8vo) by Longman with maps engraved by Edward Weller who had engraved many of the maps in McLeod's other works.

This atlas, published in *Gleig's School Series as the Physical Atlas Of Great Britain & Ireland* (cover title, price 7/6) was aimed especially at schoolchildren *"specifically illustrating the Physical Features of Great Britain and Ireland ... invaluable to those who are directly engaged in the education of youth, whether they are Teachers in Public or Private Schools"*. The first page consists of adverts for McLeod's other school works on geography, reading and spelling, maths and even a grammar for beginners.

The atlas consisted of thirty engravings with several counties on one plate all lithographed at a uniform-scale. The maps include relief geological cross-sections (Devon has a line from just south of Plymouth to the Blackdown Hills and a panorama cross-section view along the east border). This curious little atlas is interesting for its thematic maps.

Size: 131 x 155 mm. No scale bar. [Scale 1M=2 mm]

Title on map **DEVONSHIRE** (Ea): Signature: **E. Weller.** (EeOS). Plate number - **19** - (EaOS). The map has a relief illustration at page top (East) with **Scale of Feet** to show height (1800 ft=21 mm) below (Ee).

1. 1861 *Physical Atlas Of Great Britain And Ireland by Walter M'Leod*
London. Longman, Green, Longman and Roberts. 1861. BL, C, B, NLS.

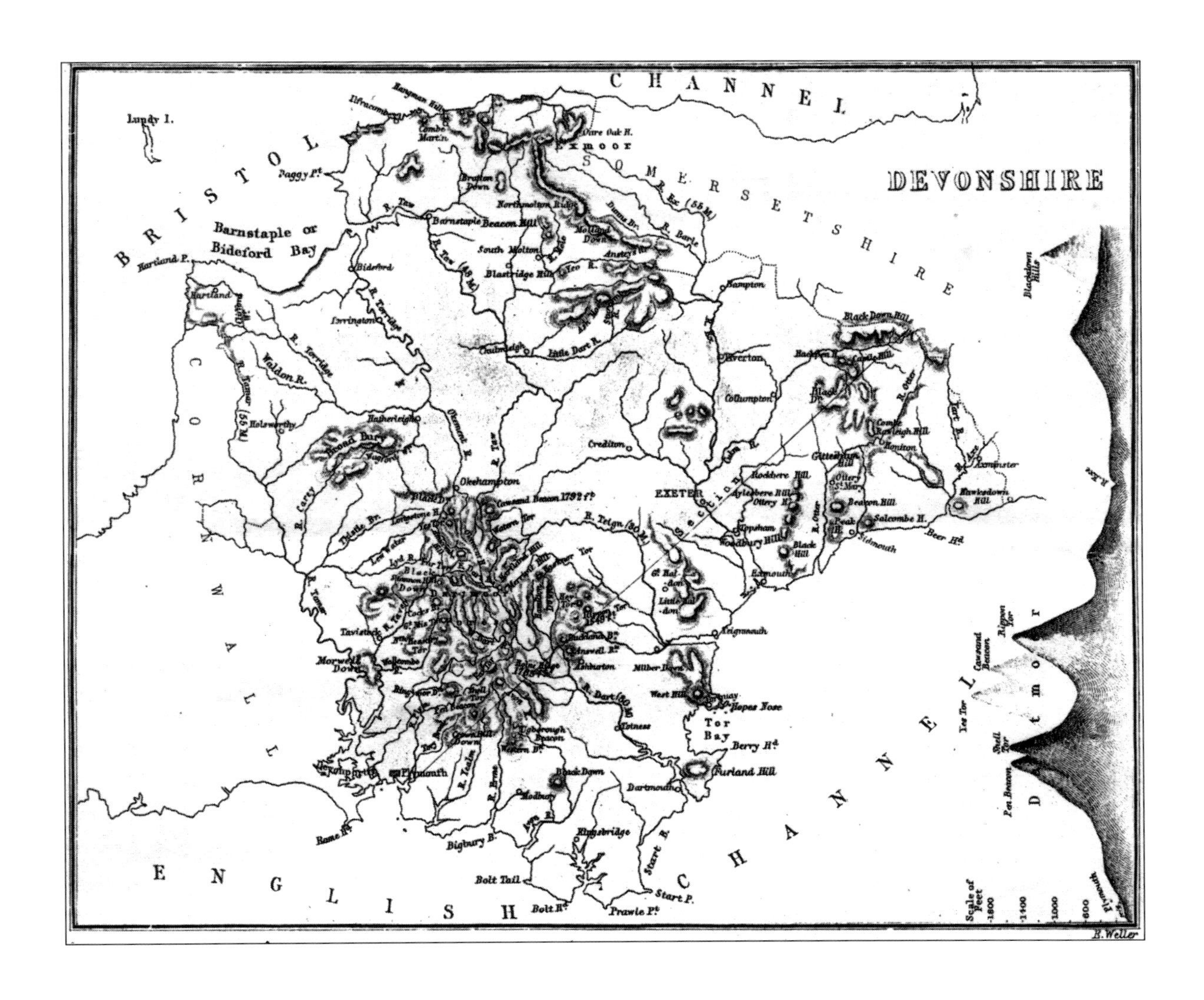

140.1 McLeod *Physical Atlas Of Great Britain And Ireland*

141

GEORGE PHILIP & SON

1862

George Philip (*b.*1799), a Scotsman, established his bookselling business as geographical publisher and globe maker in Liverpool in 1834. His son, George II, joined the business in 1848, a nephew, Thomas, in 1851: and the firm traded as George Philip & Son thereafter. The firm moved to premises at 32 Fleet St, London in 1856. In 1879 there was talk of merging with J Bartholomew, but this did not materialise.[1]

The Philips produced two very popular county atlases, both with many issues up to the end of the century: *Philips' Atlas of the Counties of England* appeared *c.*1862 which was one of the first series of printed coloured county maps; and *Philips' Handy Atlas* (**149**).

Using the latest techniques and up-to-date equipment George Philip and Son constantly updated their maps to include the progress of the railways and often adapted them for other uses, such as cyclists' maps. These included information such as signs for hills, Cyclists' Touring Club agents, repairers and hotels. Seven of these, including Devon, were adapted for C Arthur Pearson (see p.73) and appeared in 1897 in *Pearson's Athletic Record 'every Wednesday morning. Price 1d. Full of interest to cyclists and all other sportsmen. One of these maps will be given each week.*

Exploiting the county map plates, the *Way-About* series of gazetteer guides appeared in the late 1800s. A number of these, including Devon, were based on one county, others were regional. Devon was No. 15 in the series and a folding map was placed opposite the title page. Another guide, *Dodwell's Pocket County Companion*, by Robert Dodwell appeared at approximately the same time. The plan was to produce a series of 41 mainly county guides. Allday's *Guide to Paignton* included a map of most of Devon taken from the same plates.[2]

Philips also produced model county maps printed on a cardboard sheet with a titled cover *Junior Philips' Model County Maps.* A county and relief map it was intended for educational purposes and contained a lot of detail about the county as well as historical notes on the front and back.

Size: 410 x 335 mm. **English Miles** (10=48 mm).

DEVONSHIRE (Ee). Signature: **BY J. BARTHOLOMEW, F.R.G.S.** Imprint: **GEORGE PHILIP & SON. LONDON & LIVERPOOL** (CeOS). Graduated border broken at Hartland and Lyme Regis. Railways to Bideford, Tiverton, Plymouth-Tavistock, Goodrington and Exmouth.

1. 1862 Folded map mounted on card possibly from *Philips' New Series Of County Maps- ..... - From The Ordnance Survey.* See note above. (BL), B.

2. 1865 Railways to Watchet and Churston Ferrers.

Philips' Atlas Of The Counties Of England
London and Liverpool. George Philip & Son. 1865, 1868. GPL, Liv; [TB].

3. 1874 Key added with note: *The colouring represents the Parliamentary Divisions & Parliamentary Boroughs.* Plate number added - **9** - (EeOS vertical). The map is graticuled on a 10' grid with numbers vertical and letters horizontal. Railways to Minehead, Torrington, Dartmoor Loop[3], Launceston, Kingswear, Moreton Hampstead, Ashburton, Kingsbridge, Barnstaple-Taunton and Seaton. Projected railways to Sampford Courteney from Torrington, to Holsworthy and Bude, Ilfracombe, Budleigh Salterton and to Ashton.

Philips' Atlas Of The Counties Of England Reduced From The Ordnance Survey
London and Liverpool. George Philip & Son. 1874, 1875, 1876, 1880.
SGL; BL, Liv, W; BL; W.

1. L Gardiner; 1976; p.24.
2. J L Allday produced *Allday's Paignton and South Devon. Illustrated Guide with maps* (Babbicombe and Birmingham) and used part of the Bartholomew/Philip map.
3. A short stretch from Mary Tavy to Longstone appears incomplete, possibly an engraver's mistake.

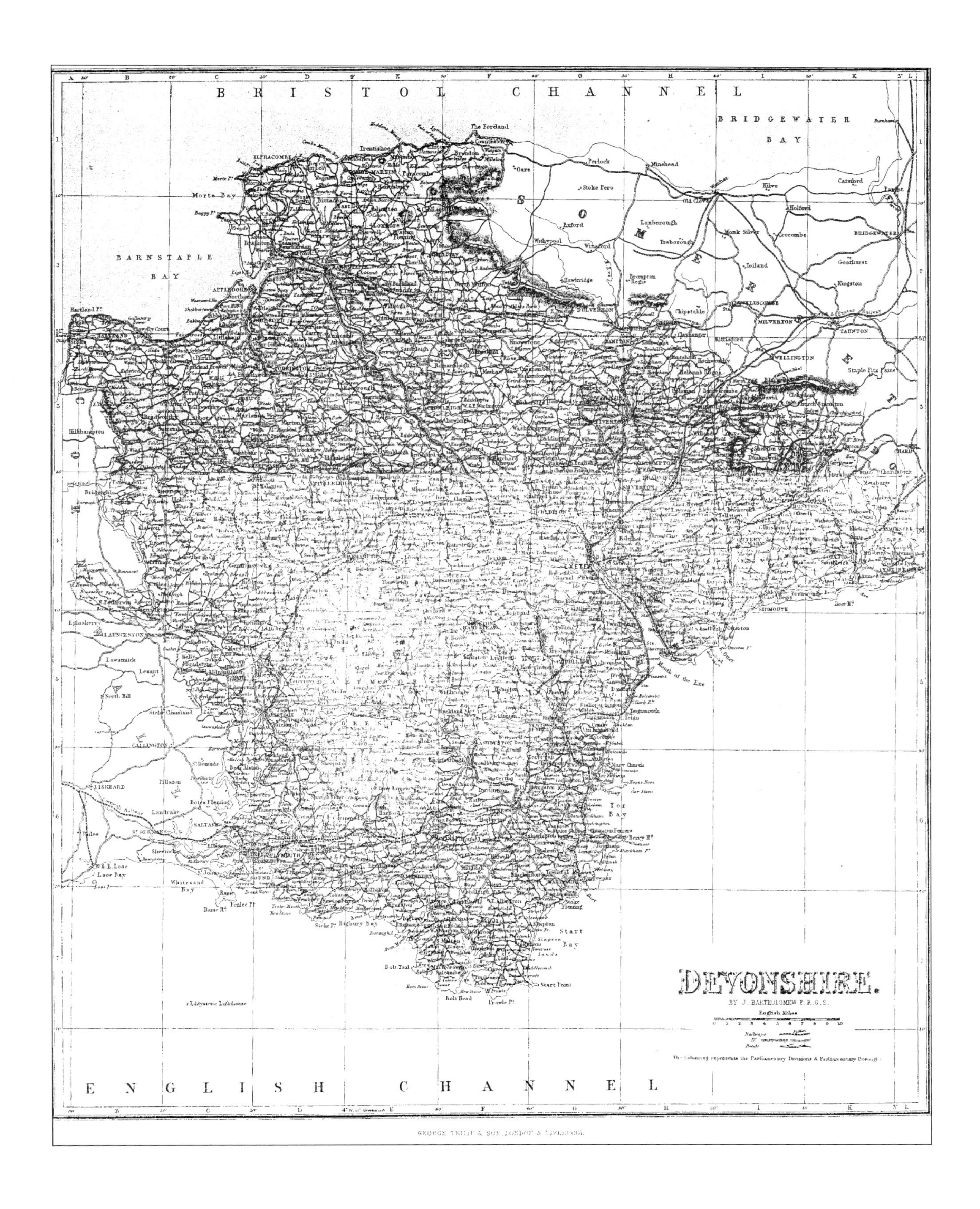

141.6 Philip *Philips' Atlas Of The Counties Of England*

4. 1876 Issued as supplement to paper with statistics of county printed on back. Title: **THE PICTORIAL WORLD MAP OF DEVONSHIRE** (CaOS) and new imprint: **Supplement to the Pictorial World. May 27th 1876** (AaOS). Publisher's imprint and plate No. deleted.

The Pictorial World - An illustrated Weekly Newspaper B[1]

5. 1880 As state 3 without the **Supplement** imprint.

The Pictorial World - An illustrated Weekly Newspape W.

6. 1883 Pictorial World references deleted. Philip imprint reinstated. Projected railway from Tiverton to Bampton. Railway to Ashton (but no station), Kingsbridge, Salcombe Regis near Sidmouth and Ilfracombe. Projected routes to Cornwall deleted.

a) Without DEVON on reverse and imprint is 58 mm long. (FB).

b) With DEVON on reverse and imprint is 65 mm long. (FB).

Philips' Atlas Of The Counties Of England ... New Edition,... Complete Consulting Index,
London and Liverpool. George Philip & Son. 1883. Midd.

7. 1885 Railway from Exeter to Bampton replaces proposed line. Station at Ashton.

Philips' Atlas Of The Counties Of England ... New Edition ...
London and Liverpool. George Philip & Son. 1885. BL, C.

Philips' Atlas Of The Counties Of England
London & Liverpool. George Philip & Son. 1889, (1890). W; [TB, DH].

Philips' New Series Of County Maps Devonshire from the Ordnance Survey Special Edition.
London. G Philip & Son. (1885). EB[2].

Allday's Paignton and South Devon Illustrated Guide with Maps
Babbacombe and Birmingham. J L Allday. (1890)[3]. E.

8. 1885 This issue only: folded in cloth case or mounted on card. Principal roads are marked with red arrows and coded H, HX and HCX according to severity of the hills. Main roads, lakes and reservoirs, Cyclists Touring Club agents, repairers and hotels being marked with overprinted colour. Note on colouring removed.

Philips' Cyclist's Map of the County of Devon,
London & Liverpool. George Philip & Son. (1885). BL, B, E.

9. 1889 As state 7 with proposed railways Launceston-Halwill and L&SWR to Plymouth as pecked lines.

Loose sheet possibly from an atlas. (KB).

1. The British Library has only copies of Berks, Bucks, Essex and Lincs.
2. From *Philips' Popular Series Of County Maps.* Folded 154 x 100 mm.
3. This is a transfer of the south part of the map retaining the portion south of Torrington and re-titled it. *SOUTH DEVON and DARTMOOR* (370 x 410 mm); it retains the old scale bar, 10=58mm, and key but has a North point (Ed).

10. 1893 Railway Launceston-Halwill line complete, railway to Kingsbridge, line continues into Cornwall from Launceston and L&SWR line to Tavistock completed. No plate number.

Philips' New Series Of County Maps Devonshire From The Ordnance Survey Cloth Edition.
London. G Philip & Son. (1893). KB.

11. 1894 New imprint: **ILIFFE & SON. MAP PRINTERS. COVENTRY & LONDON.**[1]. Table of Parliamentary members and colouring note deleted. No DEVON on reverse, no plate number and no note on railways under construction.

The Way About Devonshire - No. 15. ... H S Vaughan ... (Size folded 180 x 95 mm.)
London. Iliffe & Son. (1894[2]). KB, Pl, T.

12. 1896 Imprint: **George Philip & Son, 32 Fleet St. E.C.** (EeOS).

Devonshire Pocket County Companion ... compiled by Robert Dodwell ...
London. Tylston & Edwards and A P Marsden. 1896. T.

13. 1896 Railways as black solid line. Railways into Cornwall from Holsworthy, Budleigh Salterton, Barnstaple-Linton and Plymouth-Yealmpton. Cornish border redrawn to include Borough and Court deleted at Clovelly. Imprint and Key reinstated (as state 10). [KB]

Philips' Atlas Of The Counties Of England
London and Liverpool. George Philip & Son. 1896. [P[3]].

Philips' Atlas Of The Counties Of England
London and Liverpool. George Philip & Son. 1899, 1900[4] [P[5]]; W.

14. 1897 The maps have been photographically reduced in size, have added cycling information, and bear the note *Pearson's 'Athletic Record' County Cycling Maps No...* [6]
assumed

Pearson's Athletic Record
London. C A Pearson. 1897.

1. Although David Smith (1985) reports that only Surrey, Kent, Warwick, Derby, Norfolk, Suffolk, Sussex and Hertford were actually produced, the inside cover of Devon has Norfolk and Suffolk as one guide giving those above as the first seven volumes plus in addition Middlesex, Hereford and Oxford (8 - 10) with guides covering South Wales, the Lake District, Hampshire with the Isle of Wight (11 - 13) and Devon (no 14 listed). An advert at the back also lists Somerset, Essex, Northumberland with Durham, Ireland and East Anglia. Eugene Burden (1991) writes that Berkshire was no. 20
2. Most illustrations are dated 1894 and there is an advert for a Cyclist's Road Book of 1895.
3. A copy once owned by D Kingsley.
4. The atlas was reprinted 1904-5, the maps now have the imprints: *LONDON: THE GEOGRAPHICAL INSTITUTE; 32 FLEET STREET. LIVERPOOL: PHILIP SON & NEPHEW, 45-51 SOUTH CASTLE ST.*
5. Listed in Whitaker, The *Printed Maps of Northampton*; 1948, but no copy has been seen since.
6. David Smith (1985, p. 300) writes that only Surrey, Kent, Hertford, Essex, Sussex, Devon and Derby were issued; Eugene Burden (1991, p. 194) writes that eight were produced.

142

BARTHOLOMEW/BLACK

1862

Black's guides appeared from 1855 (**130**) when the *Tourists' Guide to Devonshire and Cornwall* appeared. In 1862 the work was revised with new maps. John Bartholomew drew and egraved a map of the country and a transfer was taken for *Blacks Guide to the South Western Counties of England* (1862).[1] This was a map of *Dorset, Devon & Cornwall.* From *c.*1869[2] the map of Dorset was separated (there were now two maps, Dorset and Devon with Cornwall) in the three county volume. Concurrently the guide to Devon was issued with the larger map now cut down and rebordered, showing Devon separately, and until *c.*1878 *Black's Guide To Devonshire* always included this separate map of Devon. For the Tenth Edition (1879), a new large Bartholomew map was used (see **150**).

Although McFarlane's name appears on the first map, the guides were printed by R & R Clark of Edinburgh, and were reprinted almost every year with little or, more often, no change to the text as shown by the fact that the pagination for *Devonshire* remained that of the combined volume. The early *Black's Guides* had an Index Map on the inside front cover which covered Dorset, Devon & Cornwall (also regularly updated) and was used in all three county guides. In 1882 this was replaced by a key map of Devon only (**156**).

Size: 280 x 260 mm. **Scale of English Miles** (20=63mm).

1. 1862 a) **DORSET, DEVON & CORNWALL** issued as one map (270 x 540 mm). Imprints: **Published by A & C Black, Edinburgh** (CeOS, 40 mm) Signatures **Drawn & Engraved by John Bartholomew Edin^r^. F.R.G.S.** (EeOS) and **Printed by W H M^c^Farlane, Edin^r^** (AeOS). Border broken at The Foreland, Lands End, Lizard Head and at Ringwood and Salisbury (east). Inset of the **SCILLY ISLES** (Ac).

Black's Guide To The South-Western Counties ... Dorsetshire, Devon, And Cornwall
Edinburgh. Adam and Charles Black. 1862. BL.

b) **DEVON** issued as one map (280 x 260 mm). Imprint and signatures as above. Border broken at The Foreland, Widemouth Bay and Whitechurch (east). Shows GWR to Plymouth and L&SWR to Exeter, with lines to Goodrington, Bideford, Tavistock, Tiverton and through Liskeard.Ttile is set below a horizontal line through Prawle Point. English Channel written along the coast (just entering east border). Scale bar 20 mm from lower border.

Black's Guide To The South-Western Counties Of England Devonshire
Edinburgh. Adam and Charles Black. 1862. KB.

2. 1862 a) **DORSET, DEVON & CORNWALL.** Printer's signature removed. Railways to Chard, Watchet and Exmouth.

Black's Guide To The South-Western Counties ... Dorsetshire, Devon, And Cornwall
Edinburgh. Adam and Charles Black. 1864, 1866. E; KB.

b) **DEVONSHIRE** (280 x 260 mm). Scale bar 15 mm from lower border. Imprint and signature as above.

Black's Guide To Devonshire
Edinburgh. Adam and Charles Black. 1863, 1863 (1864). C; Pl.

1. The three counties volume was probably written in 1861 and published the following year. A note on page 152 begins *Even while we write (January 1861)* and refers to a Spanish ship cast onto Morte Stone.
2. The earliest known work so far discovered by the authors; there may be earlier versions.

142.4 **Bartholomew/Black** *Black's Guide To Devonshire*

3. 1864 **DEVONSHIRE**. Cornwall named vertically (north-south) just inside map border.

Black's Guide To Devonshire
Edinburgh. Adam and Charles Black. 1864. E.

4. 1865 **DEVONSHIRE**. Railways shown to Launceston and Kingswear. Cornwall now reads south-north.

Black's Guide To Devonshire
Edinburgh. Adam and Charles Black. 1865 (1866). E, FB.

5. 1867 **DEVONSHIRE**. Railways continued to Ashburton, Kingsbridge and Moreton Hampstead and planned Launceston to Bodmin. Signature change: **FRGS** deleted.

Black's New Guide To Devonshire
Edinburgh. Adam and Charles Black. 1867. E, KB.

6. 1868 **DEVONSHIRE**. Railway Chard to Taunton and planned line Colbrook via Oakhampton to Lidford.

Black's New Guide To Devonshire
Edinburgh. Adam and Charles Black. 1868. KB.

7. 1869 **DEVON & CORNWALL.** above the scale bar (Ee) (277 x 400 mm). The border is broken at the Lizard, the Foreland and Land's End. Eastern border stops at Lyme Regis. Signature: **John Bartholomew Edin[r].** Inset for the **SCILLY ISLES** (Ac). Railways in Cornwall (at New Quay and Bodmin).

Black's Guide To The Counties of Dorset, Devon & Cornwall
Edinburgh. Adam and Charles Black. 1869. Pl.

8. 1869 **DEVON & CORNWALL.** Signature: **Drawn & Engraved by John Bartholomew Edin[r].** Railway extension at Watchet (into Brendon Hills). The 'English Channel' repositioned with the 'H' just left of the title.

Black's Guide To The Counties of Dorset, Devon & Cornwall
Edinburgh. Adam and Charles Black. 1869. KB.

9. 1869 **DEVONSHIRE**. Railway to Brixham.

Black's Guide To Devonshire
Edinburgh. Adam and Charles Black. 1869. KB.

10. 1871 **DEVONSHIRE**. Imprint and signature erased (apparently only for this issue).

Black's Guide To Devonshire
Edinburgh. Adam and Charles Black. 1870 (1871). Pl.

11. 1871 Imprint and signature reinstated. Railways added Barnstaple-Taunton and Seaton.

a) **DEVON & CORNWALL**. ENGLISH CHANNEL is rewritten straight along the lower border (240 mm).

Black's Guide To The Counties of Dorset, Devon & Cornwall
Edinburgh. Adam and Charles Black. 1871. E.

b) DEVONSHIRE. The title and scale are raised, the title is in line with Start Pt. and scale bar 30 mm from lower border.

Black's Guide To Devonshire
Edinburgh. Adam and Charles Black. 1872, 1873. KB; Pl.

12. 1873 **DEVONSHIRE**. Height reduced to 255 mm. Title is raised and is now in a line with Stoke Fleming and Street. Signature shortened to: **John Bartholomew, Edinr**. (EeOS).

Black's Guide To Devonshire
Edinburgh. Adam and Charles Black. 1872 (1873). KB.

13. 1874 **a) DEVONSHIRE.** The word **CHANNEL** added in Bristol Channel.

Black's Guide To Devonshire
Edinburgh. Adam and Charles Black. 1874. KB.

b) DEVON & CORNWALL. [1] Signature: **John Bartholomew, Edinr**.

Black's Guide To The Counties of Dorset, Devon & Cornwall. Seventh Edition
Edinburgh. Adam and Charles Black. 1874. [P].

Black's Guide To The Counties of Dorset, Devon & Cornwall. Eighth Edition
Edinburgh. Adam and Charles Black. 1875. FB.

14. 1875 **DEVONSHIRE**. Shorter signature: **J. Bartholomew. Edinr**. Railways to Torrington, Sidmouth, Ilfracombe (wrong route), Minehead, Oakhampton-Tavistock line completed. The word **CHANNEL** deleted in Bristol Channel. ENGLISH CHANNEL is along the lower border (140 mm).

Black's Guide To Devonshire
Edinburgh. Adam and Charles Black. 1875. E, TQ, FB, KB.

15. 1876 **a) DEVONSHIRE.** Shorter imprint: **A & C Black, Edinburgh**.

Black's Guide To Devonshire
Edinburgh. Adam and Charles Black. 1876, 1876 (1877), 1877, 1878[2].
E, KB; E; E; TQ, KB.

b) DEVON & CORNWALL The title and 'English Channel' revert to that found in state 11, but with the scale bar closer to the title. Shorter imprint: **A & C Black, Edinburgh.**

Black's Guide To The Counties of Dorset, Devon & Cornwall. Tenth Edition
Edinburgh. Adam and Charles Black. 1878. E.

Black's Guide To The Counties of Dorset, Devon & Cornwall. Eleventh Edition
Edinburgh. Adam and Charles Black. 1879. C.

16. 1881 **DEVON & CORNWALL**. The two counties are now coloured. Railways extended to Holsworthy and Hemyock.

Black's Guide To The Counties of Dorset, Devon & Cornwall. Eleventh Edition
Edinburgh. Adam and Charles Black. 1881. FB.

1. Devon & Cornwall is a transfer of the regional map and thus CHANNEL is always present.
2. From 1879 copies of Black's guide to *Devonshire* had a new Bartholomew map (**149**).

143

W J SACKETT
1864

In 1864 W J Sackett lithographed *A New Set of Diocesan Maps* which was issued in parts. This work was by James Thomas Law (1790-1876) prebendary of Lichfield 1818, and Chancellor 1821, and William Francis, who (according to the introduction) took over much of the pre-press work when Law's eyesight started to fail.

The various dioceses were each bound into a blue file - Devon being *No.9. Diocese of Exeter.* On the reverse of the title page there is an introductory letter to the Students of Lichfield Technical College and a preface by J T Law dated Nov 1st 1864. This is followed by a *Synopsis of the Diocese with half of the Arch Deaconery* and an index on the reverse, the remainder being stuck to the west side of the map with the east side stuck to the blue cover.

The map extends into the border at four places (including Ilfracombe and Land's End) and at Lizard Point the map has been curiously trimmed to extend part of the sheet by approximately 2 cm in order to include part of the labelling.

The map is preceded by two pages listing the various parishes under curacies, vicarages, parsonages and rectorates. On the page backing the map (actually 2 pages pasted back to back) there is a further index map. This is also titled *Exeter* is approximately 80 x 60 mm and has no border (Fig 5. p.xxxi). The same imprint of Sackett is found under the listing.

James Thomas Law also published a set of lectures: *Lectures on the First and Subsequent Divisions of the Kingdom into Provinces and Dioceses: and the rearrangement of Dioceses by the Ecclesiastical Commissioners, delivered to the Students of the Theological College, Lichfield.* This was published by Stevens and Sons in London as well as the local Lichfield firm of Thomas George Lomax (1868). This work included four maps of England and Wales: Map of the Saxon Heptarchy; Map of 1835 Dioceses; Proposed Dioceses; and Present Boundaries of the Dioceses. Three of these are signed *Drawn by Revd W F Francis.* The work was dedicated to Dean Howard of Lichfield.

Size: 365 x 500 mm.

No scale bar.
]Scale 1M= 2 mm]

Exeter: (Aa). Imprint: **W J Sackett, Printer, & Lithographer, II, Bull St. Birmm.** (CeOS). The map extends beyond the border in four places (see text). Inset map of the **SCILLY ISLANDS** (Ac). Arms of the Bishop (Ba).

1. 1864 *A New Set Of Diocesan Maps. By James Thos. Law, and William F. Francis* (Birmingham). (Sackett). 1864. BL, C[1], W.

1. The collection of the 27 dioceses has no title.

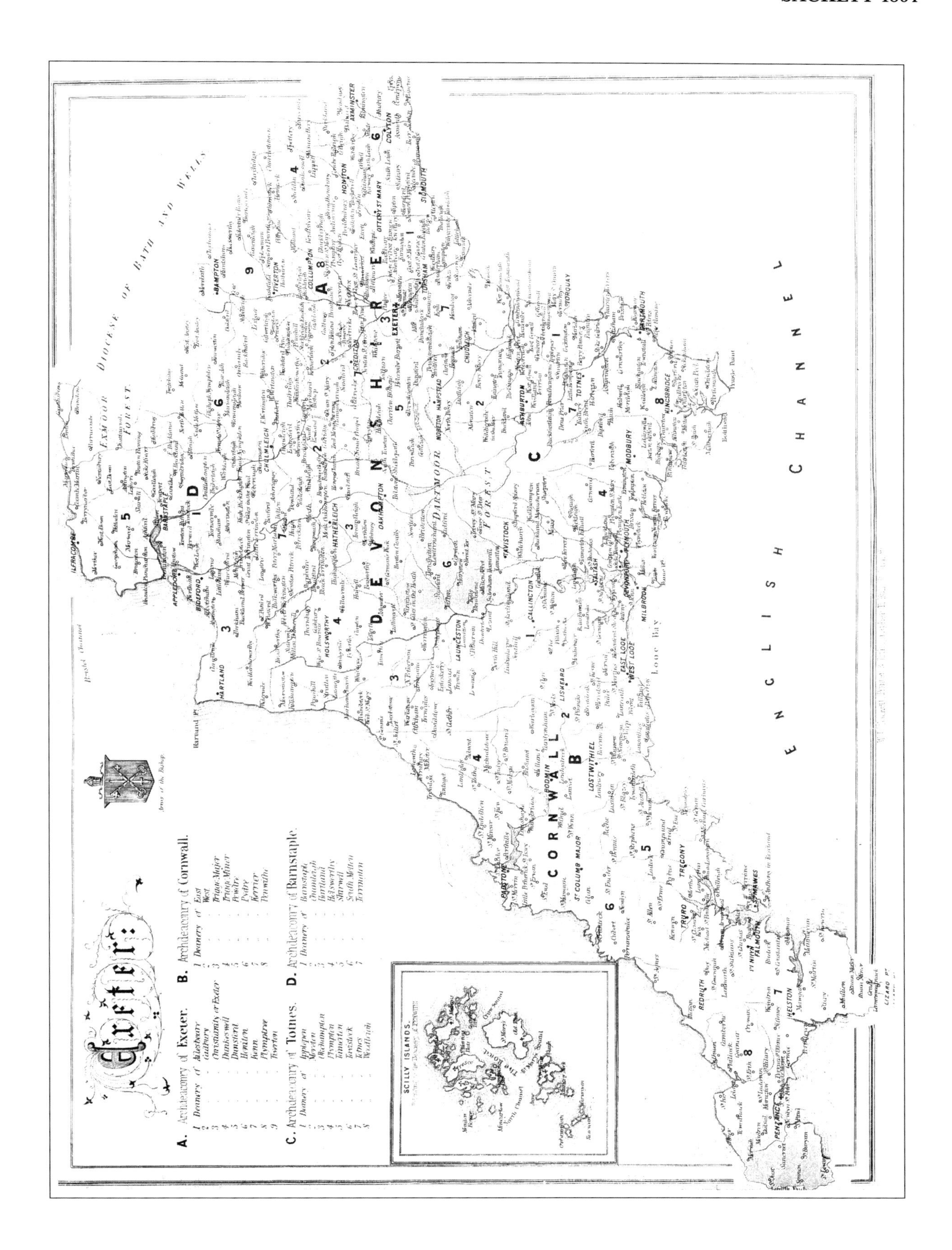

143.1 **Sackett** *A New Set Of Diocesan Maps*

144

WILLIAM HUGHES

1864

William Nightingale Hughes (1817-76) was a geographer and engraver with premises at Aldine Chambers, Paternoster Row, London. He engraved the first maps that were later printed in colour; the *Journey-Book* series published by Charles Knight (four counties were engraved but only three maps were printed in colour). Hughes engraved a large number of maps of all areas of the world including maps for Knight's *Illustrated Atlas of Scriptural Geography* (1840), *Atlas of Constructive Geography* 1841 for student use, maps for A & C Black 1840-53, including *Black's Tourist's and Sportsman's Companion* of 1852 and *Black's General Atlas* (with maps engraved on steel by Sidney Hall), Stanford and Mackenzie's *Modern Geography* 1866 (also issued as *Hughes' New Comprehensive Atlas*) and the London Environs map for *Barclay's Dictionary.* He produced an industrial map of the *British Isles* and a Map of England and Wales (700 x 1000 mm) for Philips.

Hughes prepared *A Topographical Dictionary of the British Islands* which was published in *The National Gazetteer* issued in parts from 1863 to 1868.[1] It contained sixty-eight maps including 39 counties plus other maps of the British Isles, all with his imprint. The maps were gathered together *c.*1868.[2] The maps were then used again for *Hughes' New Parliamentary and County Atlas* (1886).

William Hughes was Professor of Geography at King's College 1863-75. He had also been cataloguer of books at the British Museum between 1841-43 and a number of geographical works are assigned to a W Hughes between 1840 and 1870.

A later, loose copy of a map of Surrey has been seen. This was titled *Map Of Surrey showing the Roman Road and Places of General Interest.* There was an added reference added above the map: *The Coloured portions represent the Old Parliamentary Divisions. The Blue Lines show the present divisions with their Titles.* Presumably printed after the boundary revisions (1885) Surrey may have been the only county so issued.

Size: 240 x 300 mm. **English Miles** (15=50 mm).

DEVONSHIRE (Ee) with key to **Boundary between North and South Devon** and railways. Signature: **W. Hughes** (EeOS). Publisher's imprint: **LONDON: JAMES S. VIRTUE** (CeOS). Railways to Tavistock, Bideford, Tiverton, Watchet, Exmouth and Goodrington (near Paignton).

1. 1864 *The National Gazetteer: A Topographical Dictionary of the British Islands*
London. James S Virtue. (1864). BL[3].

The National Gazetteer of Great Britain and Ireland
London. Virtue & Co. 1868. W, NLS, CB, GUL.

2. 1868 New imprint: **LONDON: VIRTUE & Co**. Railways to Kingswear, Tavistock-Launceston and Newland Bridge. (E).

The National Gazetteer of Great Britain and Ireland
London. Virtue & Co. 1868. Leics, C.

The National Gazetteer of Great Britain and Ireland
London. James S Virtue. 1868. [TB].

The National Gazetteer Of Great Britain And Ireland
London. Virtue & Co. 1870. BL[4], E.

1. Devon may have been issued shortly after Berks (*c.*1863) which agrees with the Paignton railway construction.
2. Although the map of Devonshire is usually found in the volume covering ABB to EYW, it has also been noted in the volume for KNE to MAI.
3. Atlas with two title pages. One title page is James Virtue, the other Virtue & Co. One is probably from the parts issue, the second given away later for binding purposes.
4. Volume with no title page.

3. 1873 Altitudes on hills added eg Cut Hill 1971 (Bc). Key (Ee) refers now to **Boundary between Parliamentary Divisions, North, East & South Devon**. Railways to Moreton Hampstead, Seaton, Taunton-Chard, and incorrectly to Sth Tawton (near Oakhampton). **Heights in feet**.

A New County Atlas Of Great Britain And Ireland Containing Sixty-eight Coloured Maps
London. Virtue & Co. (1873). BL.

The National Gazetteer
London. Virtue & Co. (1875). W.

4. 1875 Station symbol added to the railway key and also on the map.

A New County Atlas Of Great Britain And Ireland Containing Sixty-eight Coloured Maps
London. Virtue & Co. (1875). W.

5. 1886 Imprint: **LONDON. J. S. VIRTUE & CO. LIMITED**. Railways to Oakhampton and on to Tavistock, Holsworthy, Torrington, Ilfracombe (incorrectly), Bampton-Tiverton, and Taunton-Barnstaple. The county is coloured yellow with the new divisions in red (Barnstaple, South Molton, Tiverton, Tavistock, Honiton, Teignbridge, Torquay, Totnes).

A New Parliamentary And County Atlas Of Great Britain and Ireland containing 72 coloured maps by W Hughes, Esq, FRGS and others edited by Professor A H Keane BA
London. J S Virtue & Co. Ltd. (1886). Leeds, Hull, BL, BRL.

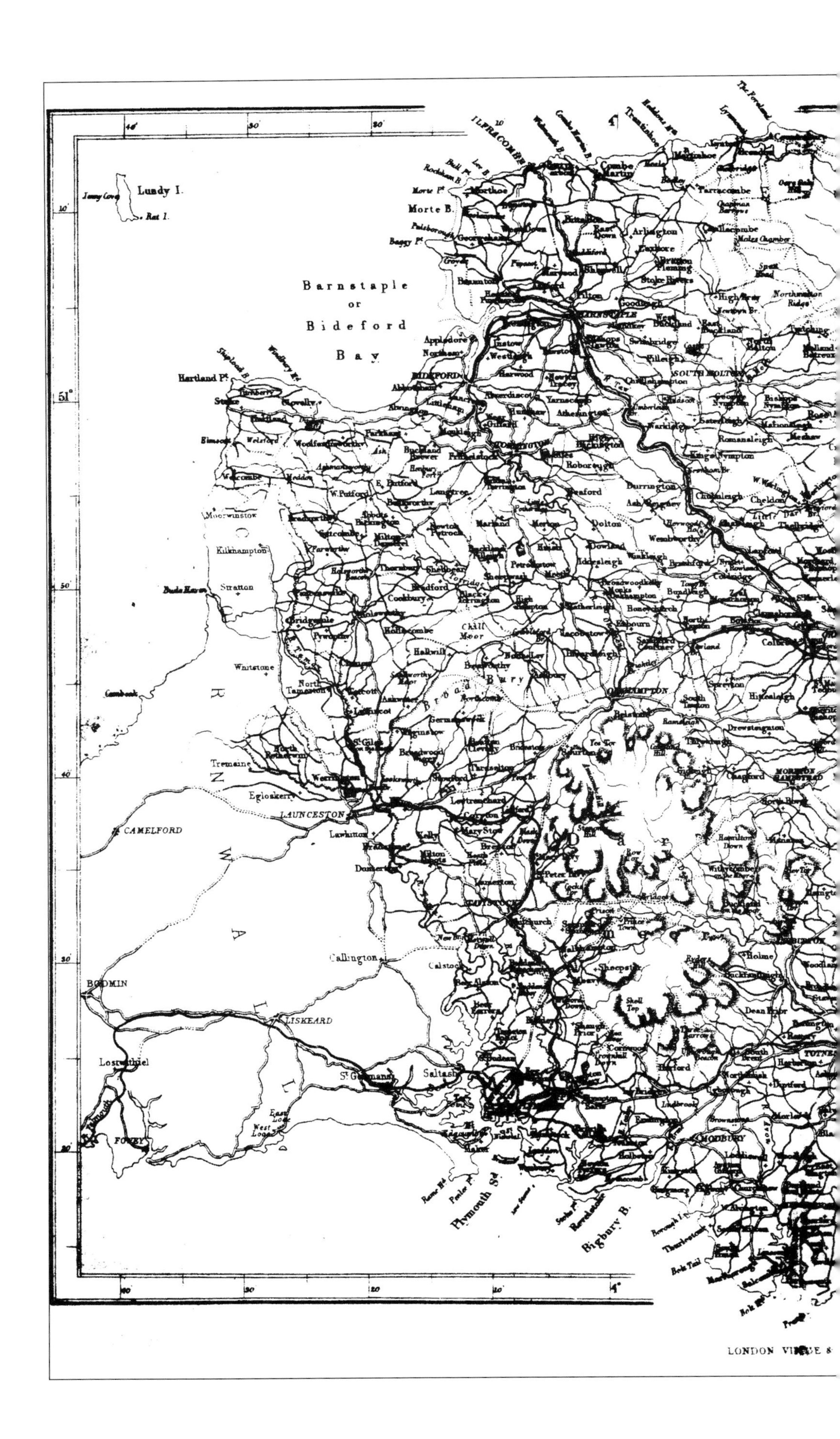

Lundy I.
Barnstaple
or
Bideford
Bay
Hartland Pt.
LAUNCESTON
CAMELFORD
BODMIN
LISKEARD
Lostwithiel
FOWEY
Callington
Saltash
Plymouth Sd.
Bigbury B.
OKEHAMPTON
Kilkhampton
Stratton
Whitstone
Tremaine
Egloskerry
Morte B.
LONDON VIRTUE &

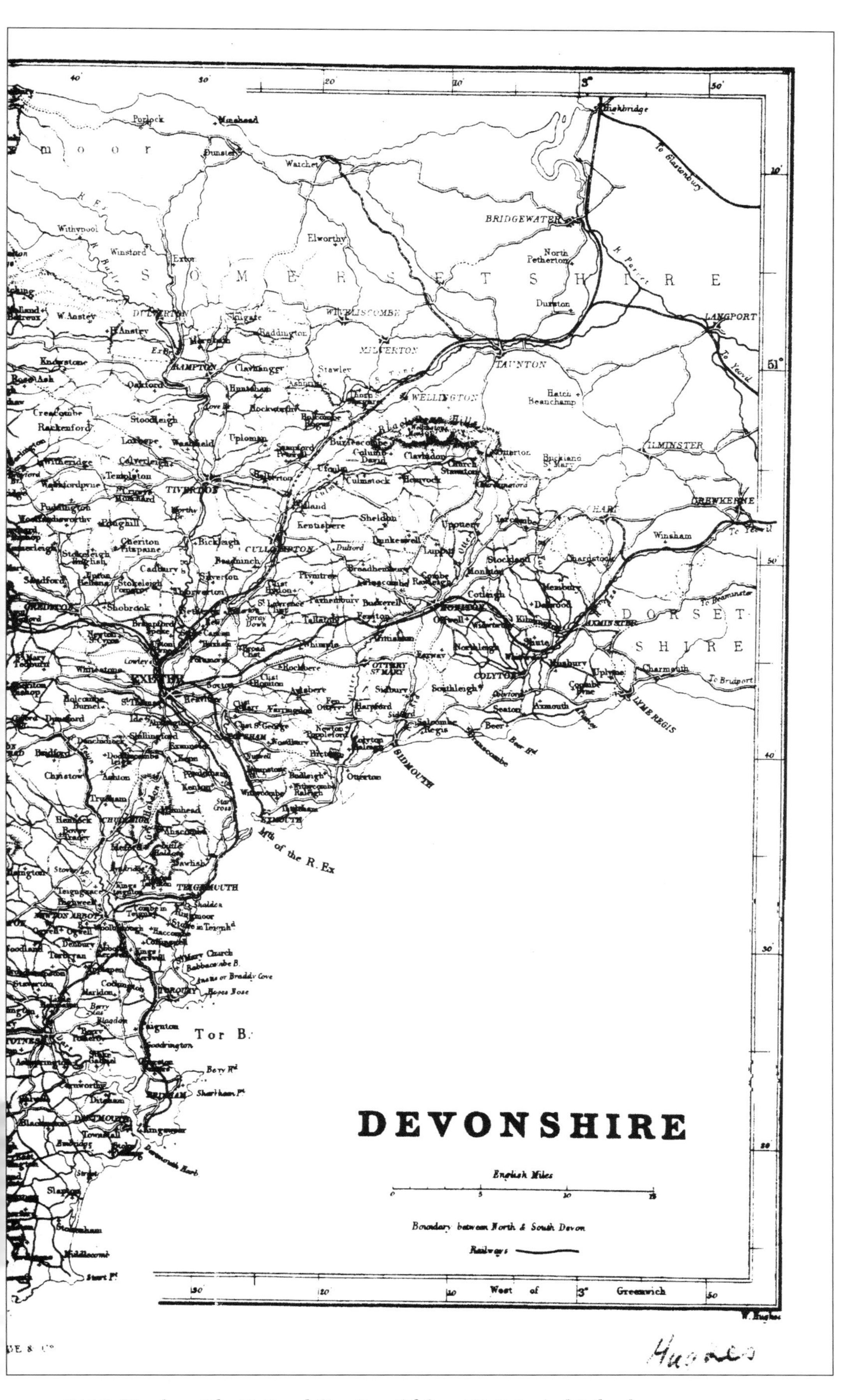

144.2 Hughes *The National Gazetteer Of Great Britain And Ireland*

145

THOMAS SPARGO

1865

Thomas Spargo is noted for his interest in the mining activities of Devon and Cornwall, and he called himself both a Mining Engineer[1] and Share Broker. In his book *The Mines of Cornwall and Devon* he included a map of each county with the divisions in the county outlined and coloured. An earlier edition of 1864 did not contain the map of Devon.[2] The equivalent map of Cornwall is by Valentine Smith, 5 London Wall whose imprint appears on many of the plans; other companies used were Newbery and Alexander of Holborn and Varty of Camomile Street in EC London. Cornwall was drawn to a scale of 12½ miles to the inch and showed only the divisions in the county. In 1870 he published *The Mines of Wales, their present position and prospects.*

The *Mines of Cornwall* was published by Emily Faithfull, *Printer in Ordinary to Her Majesty,* at the Victoria Press, 83a Farringdon Street, EC, London. Emily was born in Headley, Surrey on May 27th 1835, the daughter of the rector. A member of the affluent middle class she was soon engaged in movements such as the National Association for the Promotion of the Social Sciences. She did not totally support the suffragette movement as she felt employment was more important than the right to vote and became involved in the Society for Promoting the Employment of Women in the late 1850s. In 1860 she opened the Victoria Press employing female compositors in the face of criticism and malicious damage by male workers. She founded *The Victoria Magazine* in 1863 and *Work and Women,* a penny weekly, in 1865. Both publications focussed on the improvement of women's knowledge. From 1872 she visited America three times giving talks on women's employment. In 1886 she received £100 from the Royal Bounty and in 1889 a Civil List pension of £50. She died on May 31st 1895 in Manchester.[3]

Size: 232 x 353 mm. **Scale** (12=155 mm).

PLAN OF THE COUNTY OF DEVON Shewing its Divisions &c. (Ac). There is a north point and key to railways. A table of Devonshire Districts shows area and population statistics for 1861. Railways are shown to Bideford, Kingswear, Exmouth and Kingsbridge and the Dartmoor Loop is closed.

1. 1865 *The Mines Of Cornwall And Devon: Statistics And Observations by Thomas Spargo*
London. Emily Faithfull. 1865. BL, B.

1. Spargo titled one of the Cornish maps: *Map of Fowey Consols and Par Consols Mining District - Sept 1863 - By Thomas Spargo Mining Engineer - at a scale of 600 fathoms to 1 inch.*
2. A copy is in the Devon and Exeter Institute in Exeter. Facsimiles were issued (without maps) between 1959 & 1961.
3. Eric Ratcliffe; *The Caxton Of Her Age*, Images Publishing (Malvern) Ltd.; 1993.

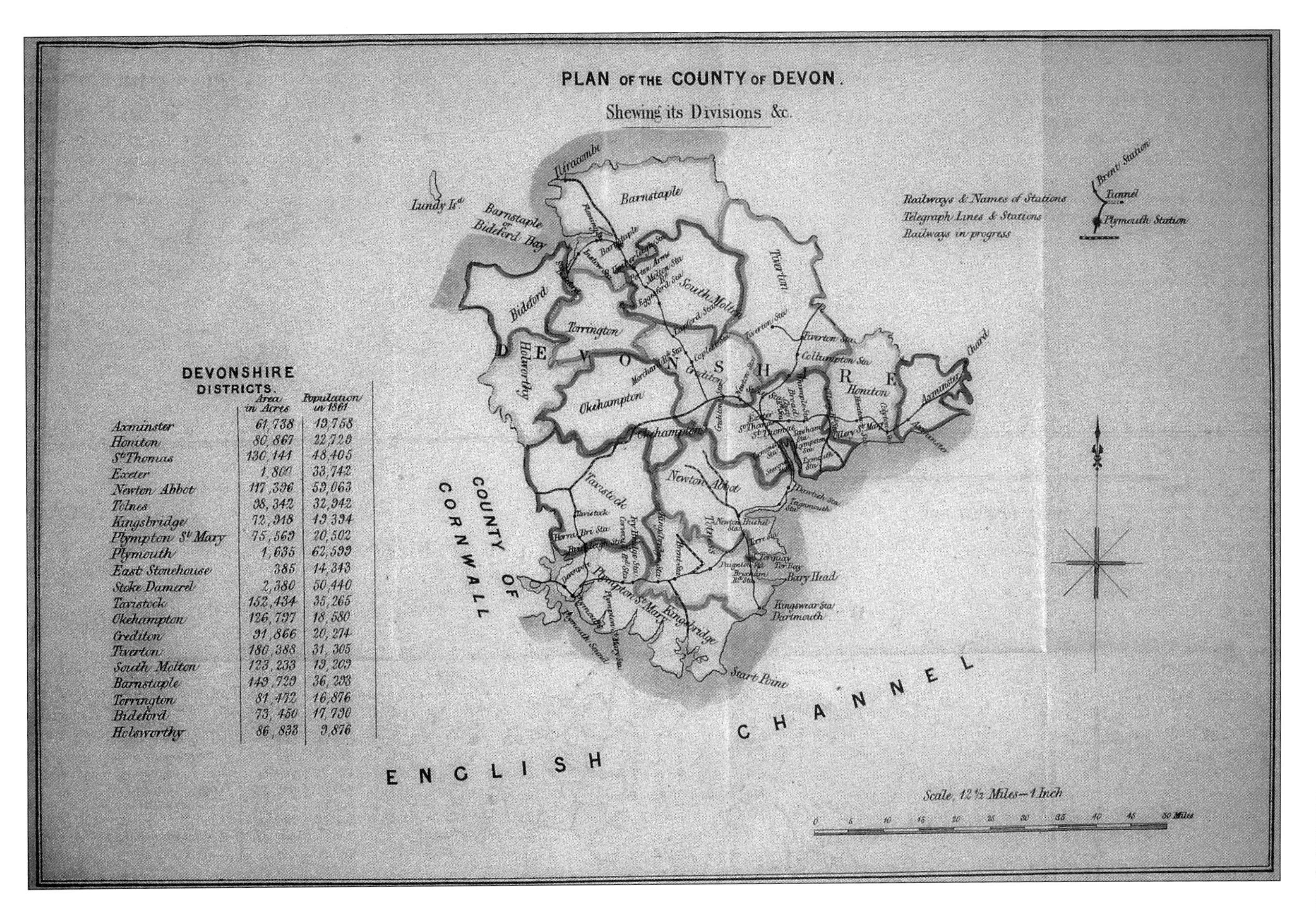

	Area in Acres	Population in 1861
Axminster	61,738	19,758
Honiton	80,867	22,729
St Thomas	130,141	48,405
Exeter	1,800	33,742
Newton Abbot	117,396	59,063
Tolnes	98,342	32,942
Kingsbridge	72,918	19,394
Plympton St Mary	75,569	20,502
Plymouth	1,635	62,599
East Stonehouse	385	14,343
Stoke Damerel	2,380	50,440
Tavistock	152,434	35,265
Okehampton	126,797	18,580
Crediton	91,866	20,274
Tiverton	180,388	31,305
South Molton	123,233	19,209
Barnstaple	149,729	36,293
Torrington	81,472	16,876
Bideford	73,450	17,790
Holsworthy	86,833	9,876

145.1 Spargo *The Mines Of Cornwall And Devon*

146

HENRY JAMES – Boundary Commission

1868

In 1867 the Boundary Commission produced 196 plans of English Boroughs (Newport, Isle of Wight, is represented by two plans) and 51 of Welsh Boroughs. Although the report purported to include each Borough and County, very few county or part-county maps were included.[1] Those that were, were reproduced from the Ordnance Survey 4 inches to the mile with title NEW DIVISIONS OF COUNTY. But in the case of Devon it was probably reduced from the 1809 survey carried out by Colonel William Mudge and Robert Dawson.

Henry James (1803-1877), whose name appears on the map of Devon, had overseen the initial stages of the later surveying of towns and boroughs in the West Country. He joined the Ordnance Survey in 1827 and became the director-general in 1854, a post he held until 1875. He had taken over from Lt-Colonel Lewis Hall as head of the Ordnance Survey and it was largely his efforts that led to the standard 1:2500 scale being adopted. James was a supporter of lithography and his enthusiasm for the new technique of photo-zincography (in 1860 he wrote a treatise *Photo-zincography*) led to it becoming a standard process in the production of Ordnance Survey maps. He became a Fellow of the Royal Society, was knighted in 1860 and promoted to Lieutenant-General in 1874.

The boundary maps were prepared under the direction of Captain R M Parsons by lithographic transfer with overprinted colour representing the parliamentary boundary of 1832, the proposed Parliamentary boundary of 1868, the municipal boundary, and the parish and township boundaries. A note explains that: *A map of each Borough and County taken from the Ordnance Survey plans is appended to the Reports for the purpose of illustrating the existing and proposed Boundaries. These maps, however, many of which are of old date, are far from conveying an adequate idea of the extension of building which has taken place in recent years, and must not be considered as indicating the character of the Districts within the new proposed Boundaries.*[2] the railways, for example, are hopelessly out of date, eg the railway to Paignton is included (1859) but not that to Kingswear (1864). The report was printed by George Edward Eyre and William Spottiswoode for HMSO.

Size: 480 x 335 mm incl. panel. **Scale Five Miles to an Inch** (5+20=130 mm).

DEVONSHIRE (NEW DIVISIONS OF COUNTY) in frame (Ea). Map is orientated slightly north-east with compass (Ba). Below map is a panel with **REFERENCE** to the boundaries and places of elections and 'handwritten' signature **Henry James. Colonel Royal Engineers.** Imprint: **Zincographed at the Ordnance Survey Office Southampton under the superintendance of Captn R M Parsons R.E. F.R.A.S. Col. Sir H. James R.E. F.R.S. &c. Director.** Date **1868** is below the imprint (both CeOS).

1. 1868 *The Representation Of The People Act, 1867 ... Report Of The Boundary Commissioners For England And Wales*
London. HMSO. 1868. BL, (KB).

1. Besides Devonshire only 8 county maps were included: Cheshire, Derby, Essex, (North and South) Lancs, Lincs, Norfolk, Somerset and Staffs. There were further maps of West Kent, East Surrey and the West Riding of Yorkshire.
2. *The Representation of the People Act, 1867*; pp. xiii-xiv.

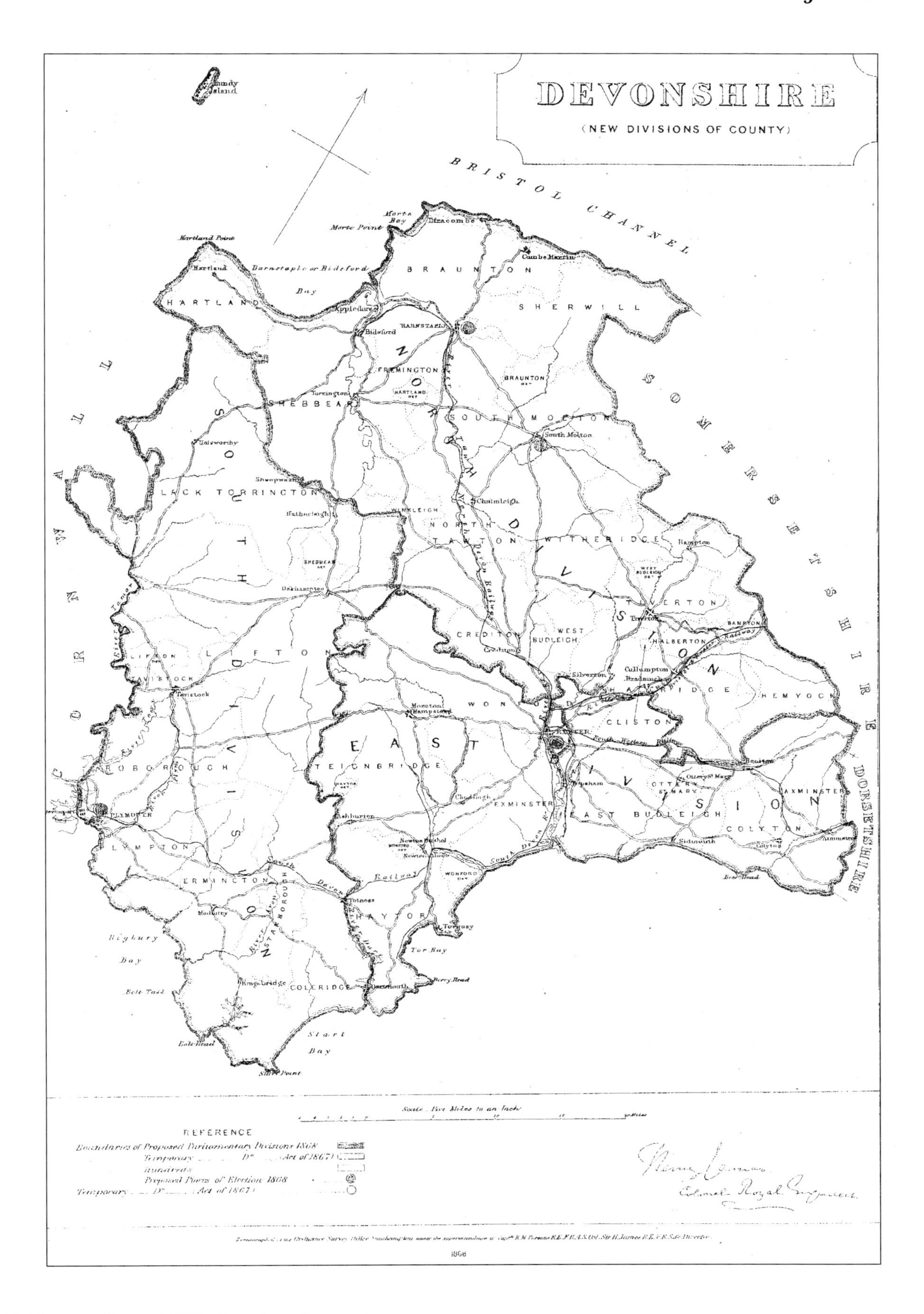

146.1 James *Report Of The Boundary Commissioners*

147

JACKSON & PARTRIDGE
1868

A map series produced for children's amusement as well as their education appeared in *c.*1868. The simple outline woodcut maps have little detail and show only unnamed rivers and a few towns. Devon appears on an illustrated page in *The Children's Friend*, published by Seeley & Co. of Fleet Street, Jackson & Co. and Samuel W Partridge & Co. of Paternoster Row. The map is part of a simple puzzle in which the child has to decipher the county's descriptions by using the adjoining delightful picture.

The first map puzzle was of Anglesea and Holyhead (unnumbered) on March 2nd 1868. Other British and Irish counties appeared regularly, dated the 1st of each month, with these subsequent issues being numbered consecutively. The counties published in 1868 were Cambs, Beds, Devonshire, Berks, Brecknock, Bucks, Caermarthen, Banff and Antrim. The reverse of Devon is dated *June 1, 1868* and titled *THE CHILDREN'S FRIEND.* (Ca) and *83* (Ea) with the finish of a moral story, two Christian morals and the *SOLUTION TO NO. III - BEDFORDSHIRE.*

Many of the same maps and pictures, but with different verso, appeared some sixteen years later in another children's publication, *Early Days* in 1883. Other counties including Devon had new maps (see **158**).

Size: 36 x 40 mm. Scale is irrelevant.

The map is (Bb) on a page 175 x 127 mm entitled **HIEROGLYPHICAL** (Ca) **COUNTY** (Ba) **READINGS No. IV.** (Da) with the county arms (Cb). The page number **84** (left) and running title **[THE CHILDREN'S FRIEND, June 1, 1868.** (right) are at top of page. The map title is **DEVONSHIRE** (Ce). The map and illustrations form part of the text, or puzzle; *The scenery in* (map) *is very varied and beautiful; including* (vales) *and* (waterfalls) *and remarkable (*coastal) *views ... .*

1. 1868 *The Children's Friend. Price One Penny*
London. Seeley, Jackson And Co. and S W Partridge And Co. 1868. BL, (FB).

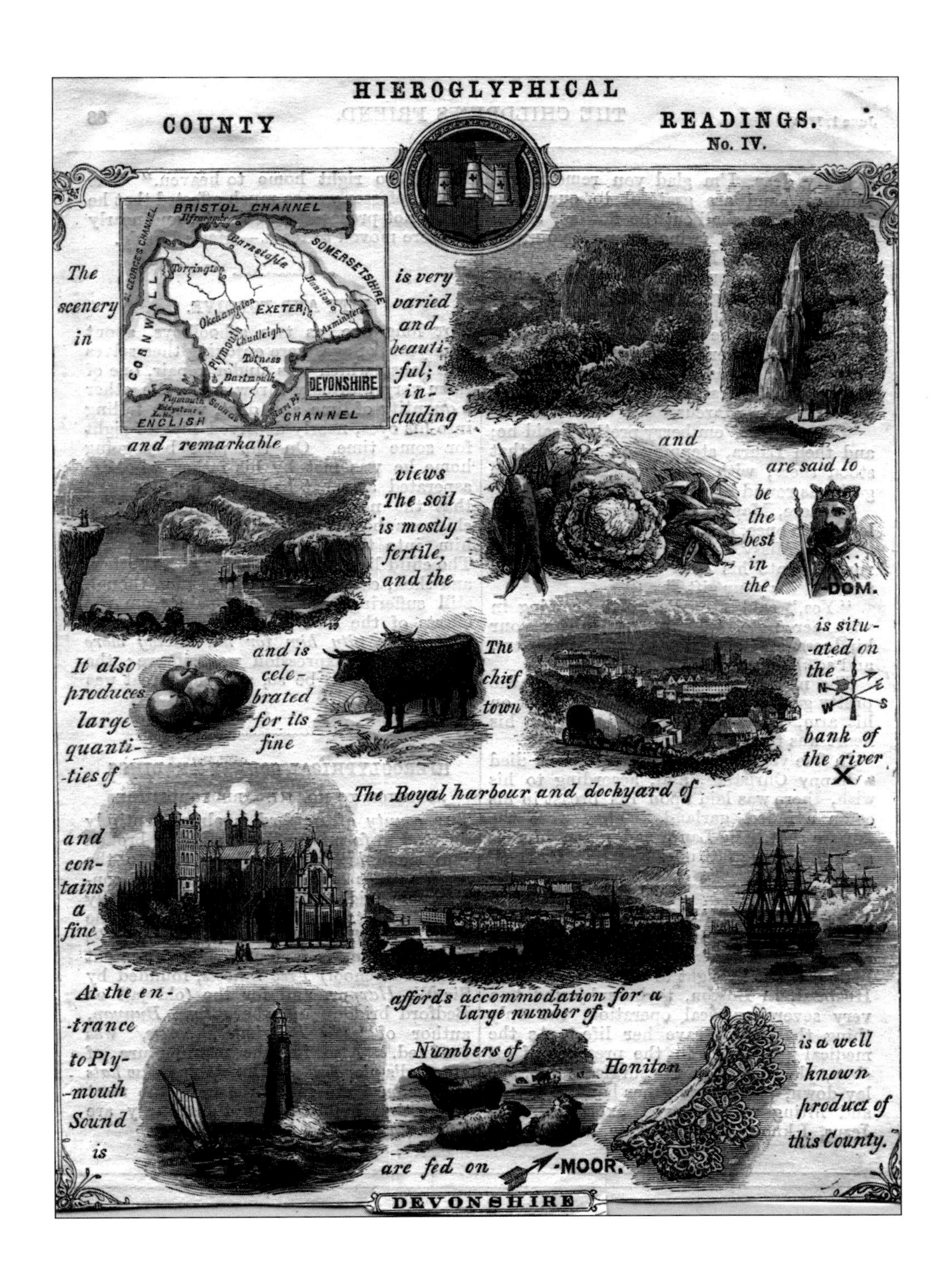

147.1 **Jackson and Partridge** *The Children's Friend*

148

BARNES & HOWE

1870

A number of maps of Devon are to be found in works on the dioceses: one was issued by Sackett (see **143**); one a copy of Arrowsmith by Oliver (see **129** and **80a**); and another was issued by Reginald Henry Barnes and Newton E Howe. This map shows principal towns and the rivers, though no roads are shown the railways are included: lines to Ashburton and Brixham are drawn but not the line to Kingswear and some proposed lines are shown with a pecked line, eg Okehampton to Tavistock, Bideford to Sampford Courtney, and Taunton to Barnstaple and Ilfracombe which help date the map to 1870-72. In 1896 Bishop Frederick became Archbishop of Canterbury.

The map was dedicated to Frederick Temple (1821-1902), formerly Headmaster at Rugby, who was consecrated as Bishop of Exeter on December 21st 1869 and installed eight days later as Exeter's 61st bishop. The map may have been produced to commemmorate this event.

The episcopal see for Devonshire was first established at Crediton in 909 AD. The ancient Cornish see is believed to have been founded at Petroc-Stow in 905, and when that place was sacked by the Danes in 981 it moved to St Germans. In 1050 the bishoprics of Tawton, St. Germans and Crediton were united by Edward the Confessor into one diocese in the church of St Peter at Exeter. Seven years after this map was completed, the Cornish see was separated and returned to Cornwall and to Truro. The diocese of Exeter was then limited to the county boundaries.

Reginald Henry Barnes M A was prebendary of Exeter and vicar of Heavitree where he lived. He produced a number of ecclesiastical works; *Asa's Victory* in 1869; *sermons given by Henry, Lord Bishop of Exeter in 1863* together with Christopher Churchill Bartholomew; and wrote a preface for Henry George Tomkins' *Church of the First Born*, 1863. He might have been related to Samuel Barnes who together with Arthur Burch had a solicitor's practice in Palace Gate, Exeter. Arthur Burch was proctor, notary public, secretary to the Bishop of Exeter and the Bishop of Cornwall. In addition his duties included registrar of the diocese and in his legal role he was solicitor to Exeter School.

According to J T White's *History of Torquay* (Directory Office, Torquay; 1878), Temple's selection was not popular. *When it became known that Dr Temple would be nominated, there was a general outcry, and on the 25th of October, at a meeting of the clergy held in Torquay, it was resolved to present a memorial to Her Majesty against the contemplated appointment of Dr Temple as Bishop of this diocese, on the ground that he was one of the authors of the famous "Essays and Reviews"*. However, White goes on to recount that at the time of writing Dr Temple had been accepted *by his lofty example and the devotion with which he labours in the interest and welfare of the diocese.*

Size: 587 x 461 mm. **Scale of Miles** (11 =68 mm).

A PAROCHIAL BOUNDARY MAP OF THE COUNTY OF DEVON. Reduced & corrected from the Tithe Maps in the Registry of the Diocese, by the Permission of ARTHUR BURCH, ESQ[R]. & dedicated to the RIGHT REVEREND FREDERICK LORD BISHOP OF EXETER by his faithful servants REGINALD HENRY BARNES AND NEWTON E HOWE. With below a list of parishes not named on the map. No imprints.

1. 1870 Loose map. E.

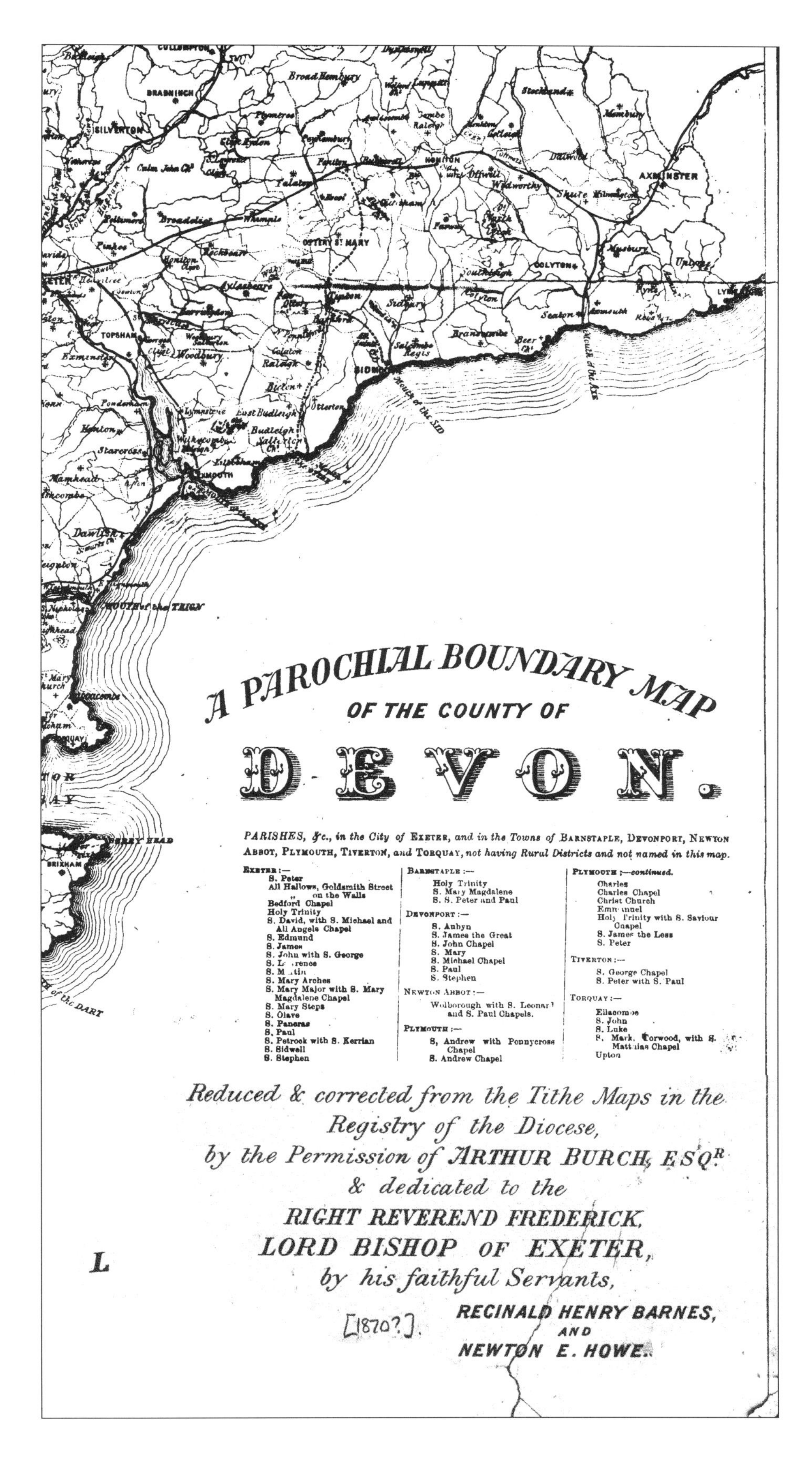

148.1 Barnes & Howe Loose map

149

GEORGE PHILIP & SON

1872

The family firm of George Philip & Son moved to London from Liverpool in 1856 but they continued to issue their important works from both cities. In addition to the atlas published in 1862 (see **141**) they produced a set of smaller county maps that appeared in *Philips' Handy Atlas.* Never used for direct intaglio printing the plates continued to be used until *c.*1938 (in *Philips' Handy Administrative Atlas of England & Wales*).

Philips' Handy Atlas contained 35 English county maps (some maps with two counties, for example Cambridge and Huntingdon), maps of the English railways, North and South Wales. Many of the maps were issued, before the atlas appeared, as individual county maps with a special text and cover for school use; *The Geography of Devonshire for use in Schools* was written by Rev. J P Faunthorpe MA, FRGS and was issued in 1872. John Pincher Faunthorpe graduated from London University in 1865, was ordained priest three years later and was Vice-Principal of St. John's College in Battersea and later the principal of Whitelands College in Chelsea. He wrote other geographies for schools, including *Elementary Physical Atlas* and *Geography of the British Colonies.*

Many of Philips' maps appeared in directories: in 1873 some county maps were published as *Butcher & Co`s Series of Directory Maps* and in 1878 *Eyre's Shilling County Guides* (see below). Devon may also have appeared as one of - *Steven's Series of Directory Maps* (London. G Stevens, 1881-1891) proceeded by Stevens' Postal Directories Publishing Co.using modified Philips' maps.

The *Handy Atlas* was re-issued post-1901 with a new map imprint: *London Geographical Institute. George Philip & Son, Ltd.* (London and Liverpool. George Philip, Son & Nephew.).

H Rider Haggard of *King Solomon's Mine* fame wrote *Rural England* in 1902 which included a late edition of the Devon map with new overprinting of geological information (London, New York, Bombay. Longmans, Green, And Co.). The Devon map appeared in Volume One opposite page 175 and had an interesting note in the top left corner: *Average weekly wage of ordinary labourers is frequently augmented by a cottage and garden, cider, and potato ground, all free.*

In addition the map was adapted to include physical and geological features and given new titles, *Physical Map Of Devon* and *Geological Map Of Devon* and was printed on the inside covers of the *Cambridge County Geographies - Devonshire* by Francis Knight and Louie Dutton published from 1910.

Size: 194 x 145 mm. **English Miles** (12=25 mm).

THE COUNTY OF DEVON with scale, note on 3 Parliamentary areas and key (Ee). Imprint: **GEORGE PHILIP & SON LONDON & LIVERPOOL** (CeOS). Imprint [for this issue only]: **PHILIPS' EDUCATIONAL SERIES OF COUNTY MAPS** (CaOS). No page number. Railways: Plymouth and Cornwall, Bideford, Plymouth-Launceston, Crediton-Oakhampton-Lidford, Brixham, Tiverton, Barnstaple-Taunton, Seaton, Exmouth, Moreton Hampstead, Kingswear, Kingsbridge (along the line of the Avon river), Ashburton and L&SWR to Exeter.

1. 1872 *The Geography of Devonshire for use in Schools - by Rev. J. P. Faunthorpe MA, FRGS*
London and Liverpool. George Philip & Son. 1872[1], 1877[2]. C; T.

2. 1873 Top imprint removed. Plate number **9** (EaOS) and on reverse.

Philips' Handy Atlas Of The Counties Of England, By John Bartholomew
London and Liverpool. George Philip & Son. 1873. C, TB, EB, BL, W, B.

3. 1874 Plate number now EeOS, vertically. The railway Bideford-Torrington shown unclearly.

Philips' Handy Atlas Of The Counties Of England, By John Bartholomew
London and Liverpool. George Philip & Son. 1874. NLS, TB, EB, W.

1. There were three very small key maps included showing a) Geology, b) Rivers and c) Mountains and Tablelands. Each map is approx. 65 x 70 mm. (See Fig.3. p.xxx)
2. *Second Edition, Revised, Enlarged and Corrected.*

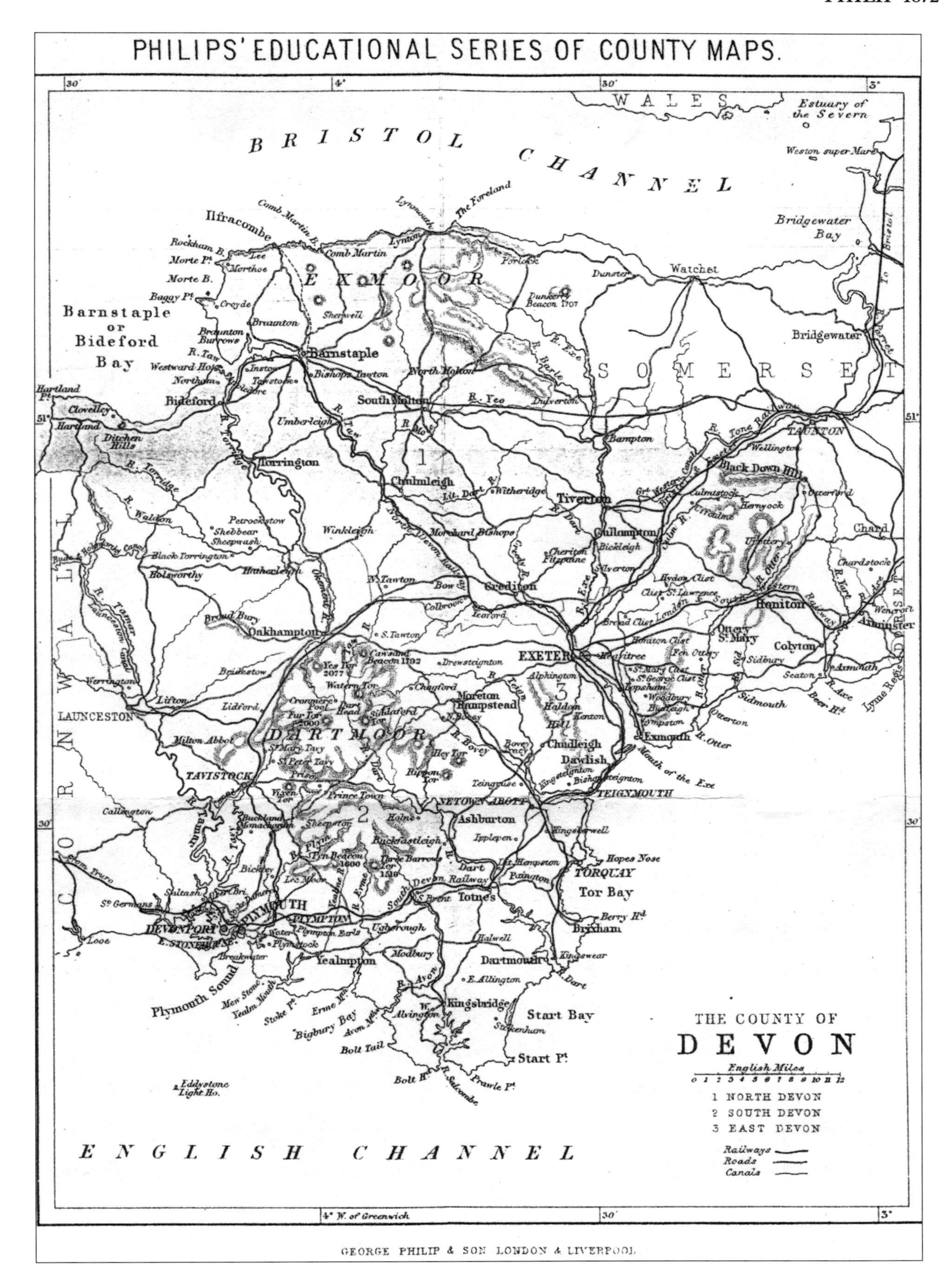

149.1 Philip *Philips' Handy Atlas Of The Counties Of England*

4. 1876 Sign for railway stations added in key, frame has graticule based on 10 minutes of a degree. Letters A-G horizontally and numbers 1-6 vertically in the borders. Railways added: Ilfracombe (curved westwards), Sidmouth. Some place names added, eg Portsmouth Arms (Torrington), Castle Hill (Sth Molton) and Eggesford near Chulmleigh.

Philips' Handy Atlas ... New And Revised Edition
London and Liverpool. George Philip & Son. 1876, 1877. BL, W, C, EB, TB; TB, EB.

5. 1878 Railways from Watchet to Minehead and to Hemyock added.

Philips' Handy Atlas ... New And Revised Edition
London and Liverpool. George Philip & Son. 1878, (1878), 1879. TB; EB; EB.

6. 1878 Imprint erased and new added [for this state only]: **EYRE BROTHERS' SERIES OF GUIDE MAPS**. (CaOS). **10, PATERNOSTER SQUARE, LONDON**. (CeOS). No plate number, divisions tinted. Reverse has adverts for Torquay College etc.

Eyre's Guide to the Seaside and Visiting Resorts of Devon & Cornwall
London. Eyre Bros. 1878. E[1].

Eyre's Hotels Of The United Kingdom
London. Eyre Bros. (1878). [P[2]].

7. 1879 PHILIP imprint. Extensive revisions with added towns, eg Halwell, Ashwater and Virginstow near Oakhampton. Railway to Holsworthy, Ilfracombe line corrected and road Torrington-Crediton added.

Philips' Handy Atlas ... New And Enlarged Edition[3]
London and Liverpool. George Philip & Son. 1880, 1882. EB; TB.

8. 1880 Railway in Cornwall added with loop through Callington.

Philips' Handy Atlas ... New And Enlarged Edition
London and Liverpool. George Philip & Son. 1880, (1881), 1882. TB; EB; EB.

9. 1882 Railway to Ashburton upgraded (thicker), line through Chudleigh added. Kingsbridge line deleted.The Sidmouth line has been slightly altered west of the River Otter before turning eastwards.

Philips' Handy Atlas ... New And Enlarged Edition
London and Liverpool. George Philip & Son. 1882, 1884, 1885. W, TB, EB; TB, EB; TB.

10. 1885 Railway to Princetown changed to solid line. Cornwall loop has short stretch deleted below Callington. Chudleigh line realigned.

Philips' Handy Atlas ... New And Enlarged Edition
London and Liverpool. George Philip & Son. 1885. EB.

Philips' Handy Atlas ... New And Revised Edition
London and Liverpool. George Philip & Son. 1886. NLS.

1. Spine title is *Eyres' Guide to the Watering and Visiting Places of Devon & Cornwall.*
2. The Eyre brothers also published, *Eyre's Hotels Of The United Kingdom* c.1879-1880. This guide (240pp) was: a directory of hotels, clubs and hydropathic establishments, and a gazetteer of every important place in Britain; showing the best means of travelling thither; distance from London; fares; postal intelligence &c. It included 42 coloured county maps, engraved views of the hotels and many illustrated adverts. The map of Devon was presumably that described here. Advertised in *Messenger's Auction Catalogue* 9.1.97.
3. *New And Enlarged Edition* includes Isle of Wight, Isle of Man & Channel Isles.

11. 1885 **a**) Coloured and titled to show the new political divisions according to the Redistribution Bill, 1885 and Note added: **The colouring represents the Parliamentary Divisions each returning 1 member** (Ee). Names and boundaries of the new (1885) Parliamentary divisions added. Railway Exeter-Bampton added.

Philips' Handy Atlas of the Counties of England including maps of North & South Wales, the Channel Islands, and the Isle of Man. Reduced from the Ordnance Survey
London and Liverpool. George Philip & Son. 1885, (1885), 1886.
TB, BL, RGS, KB; TB[1]; TB.

b) Has **DEVON** printed on the reverse.

Philips' Handy Atlas
London and Liverpool. George Philip & Son. 1886. BL, EB.

12. 1887 Railway to Launceston from Halwell added.

Philips' Handy Atlas
London and Liverpool. George Philip & Son. 1887, (1887[1]), 1888, 1889.
BL, C, TB; EB; TB, EB; EB.

13. 1892 L&SWR to Tavistock railway. Dotted boundary lines to Plymouth and Devonport. Some maps in the Atlas show new coloring: dots replaced by cress-hatching, vertical bars and wash; others still have dotted technique.

Philips' Handy Atlas
London and Liverpool. George Philip & Son. 1891, 1892, 1893, 1895.
CB; TB; EB, W; NLS.

14. 1895 Note added (Ee): **NOTE. Railway stations marked 'Sta' bear the same names as their nearest town or village.** Names of stations added on map. Kingsbridge line redrawn (no longer follows the line of the River Avon) and Launceston line into Cornwall added.

Philips' Handy Atlas ... with maps of the County of London, North & South Wales, the Channel Islands, the Isle of Man, ... The maps are coloured to show the political divisions ... shewing every railway station in England and Wales. ... New and Enlarged Edition
London and Liverpool. George Philip & Son. (1895), 1895, (1896), 1898.
TB; EB, BL, C, W; EB; B.

15. 1898 Railway to Budleigh Salterton.

Philips' Handy Atlas
London: G Philip & Son. Liverpool: Philip, Son & Nephew. (1898), 1898, 1900.
BCL; TB, EB; EB.

1. Published London, Brighton, New York. Society for Promoting Christian Knowledge.

150

BARTHOLOMEW/HEYDON

1872

John Bartholomew I (1805-1861) founded the company and John Bartholomew II (1831-1893) took over in 1859. One of their most important works was the engraving of the *Imperial Map of England & Wales,* published in 1868 in atlas form on 16 sheets. Although originally only sectional maps of England and Wales were prepared from the plates, later they were used to produce county maps. The publisher of the *Imperial Map* was Archibald Fullarton, who commissioned it to replace the earlier maps from his *Parliamentary Gazetteer* (**107**). Besides the large sheet map of England and Wales, with its instructions on how to form a wall map the Gazetteer also included many plates of scenery and architecture. Throughout the life of these plates updated transfers in different sizes, in similar style but newly titled, were taken to produce county maps, excursion maps and local guides by publishers such as Houlston & Son, Houlston & Wright, G W Bacon & Co.,[1] Charles Pearson[2], Varnan Chown & Co[3]. and *Darlington's Devon and Cornwall* guide (*c.*1908) used a transfer of North Devon.

The first county map of Devon taken from the *Imperial Map* was a folding lithograph map by John Heydon; printer, stationer and book-seller in Devonport, according to White's *Gazetteer* of Devonshire, 1850. He published a print of Brunel's bridge at Saltash in 1859.[4] Whether Heydon had the map specially printed or why he added the Plymouth Forts is not known. He possibly sold it as a tourist's map of Plymouth.

The map of Devon was next issued by W H Smith, a company started by two brothers *c.*1820, Henry Edward and William Henry, who actually ran the company. But it was not until his son, also William Henry (1825-1891), became a partner in 1846 that the firm became the well-known booksellers and stationers. In 1848 he started a newsagent shop at Euston with a concession from the L&NW Railway. This was soon to be converted into a virtual railway book-stall monopoly. The firm soon became the largest newsagent in the country and expanded further into circulating libraries. William Henry later entered Parliament and became the First Lord of the Admiralty in 1877 (Gilbert & Sullivan's HMS Pinafore was supposedly based on him).[5] W H Smith used to have an estate in Rewe in east Devon where he had a stable built opposite the church of St Mary the Virgin church for the use of the parishioners so that they could tie up their horses.[6]

Black's Guide To Devonshire included Bartholomew's map in all editions after 1882. Originally the guide, printed by R & R Clark of Edinburgh, contained two maps: a sketch, or index map on the inside front cover[7]; and the county map, taken from the *Imperial Map.* This was originally on two sheets inserted between text pages, in the fourteenth edition it was on four sheets and later it was a single sheet, folded and located in a pocket formed carefully on the back cover. In 1892 Black's guide was extensively revised by "C W", probably Charles Worthy who had also edited some of Murray's guides. The fifteenth edition of *Black's Guide* was revised by A R Hope Moncrieff who was the editor of a number of topographical guide books for Blacks into the 1920s. A seventeenth edition was issued in 1902.

The *Imperial Map* was continually updated and W H Smith used the 1884 state for their general *Series of Tourists' Maps & Plans of England.* The series included both *North* and *South Devon, Environs of Plymouth* and *Environs of Exeter*[8]. Transfers were taken for sectional maps for guides, eg Dulau and Co's *Thorough Guides* series (**157**); regional maps; maps published in *The Royal Atlas of England and Wales* (see **175**); and John Murray used this map in the 11th *Edition* of his 1895 guide 1895 (printed by Spottiswoode & Co.). Bartholomew's own *Road Map of England and Wales* (*c.*1903 with Devon and Cornwall on sheet 10) was from the Imperial Map plate.

1. For example *Bacon's County Map of Devonshire with parts of adjoining counties* (*c.*1910).
2. The South section only was used for *Pearson's Gossipy Guide to South Devon.* London. C Arthur Pearson Ltd. (1901). Title: ***PEARSON'S MAP OF SOUTH DEVON*** (CaOS) on one sheet (257 x 385 mm). Imprints: *C Arthur Pearson, Ltd., London, W.C.* (CeOS). *Copyright* added to signature (EeOS). Area covered is from Whitesand Bay to Lyme Regis and inland to Crediton.
3. Varnan, Chown & Co's "Half-Inch To Mile" Map Of North Devon District. Signatures: The Edinburgh Geographical Institute (AeOS) and Copyright – John Bartholomew & Co. (EeOS). 290 mm x 560 mm..Scale 2 Miles to an Inch (5=65 mm) (CeOS). Inset PLAN OF ILFRACOMBE. Railways to Lynton and Northam and planned to Appledore. Map reverse covered in adverts. Folded into yellow cardboard covers *c.*1903.
4. Lithograph by J Needham after C A Scott; see Somers Cocks; 1977.
5. Pocklington, G R; *The Story of W H Smith.* London; 1921.
6. Anthony Taylor; *Culm Valley Album*; private printing.
7. In the volumes before 1882 this was a map of Devon, Cornwall and Dorset.
8. Launceston to Bridport - Plymouth to Barnstaple (345 x 450 mm) with Smith imprint and J Bartholomew signature, folded into a small booklet with index and index map to the series.

150.3 Bartholomew/Heydon *Black's Guide To Devonshire*

Size: 473 x 568 mm. **SCALE 4 MILES TO AN INCH** (12=75 mm).

JOHN HEYDON'S MAP OF THE ENVIRONS OF PLYMOUTH, DEVONPORT AND STONEHOUSE, OR PEDESTRIANS COMPANION. Signature: **J Bartholomew Edin^{r}** (EeOS). Shows all but the extreme south east corner of the county with Cornwall to Gerrans Bay. Key includes railways, railways in progress and symbols for forts and batteries around Plymouth. Devon railways: Kingsbridge, Seaton, Watchet, Exmouth, Tiverton, Bideford, Kingswear with Brixham, Ashburton, Moreton Hampstead, Plymouth to Launceston and Exeter to Greenslade. Proposed lines: Sidmouth, complete Dartmoor Loop with branches Launceston to Bodmin, Greenslade-Hatherleigh-Holsworthy-Bude and Bideford to Torrington and Hatherleigh, Ilfracombe via Bittadon. Village of Elmdon named near Sourton.

1. 1872 *John Heydon's Map Of The Environs Of Plymouth ...*
Plymouth. J Heydon. (1872). Pl.

2. 1873 Map on two sheets: **NORTH DEVON**[1] (Aa) and **SOUTH DEVON** folding into red covers. New imprint: **London. W.H.Smith & Son. 186 Strand**. (CeOS). Size: 340 x 470 mm. **Scale of Miles** (8=50 mm). Covers an area from Bude Bay (Lundy breaking border) to Bridgewater and Ashburton to The Foreland. Railways to Ilfracombe via Braunton (replaces proposed route leaving traces), Torrington, Sidmouth, Minehead and Brendon Hills from Watchet, Taunton-Barnstaple and Greenslade to Launceston. Westward Ho! and Sandridge in sea by the Foreland are added. Division letters. Forts etc. deleted.

W H Smith and Sons Reduced Ordnance Map of North Devon
London. W H Smith & Son. (1873 [2]). KB.

W H Smith and Sons Reduced Ordnance Map of Devon, South, and Dartmoor
London. W H Smith & Son. (1874). EB.

3. 1879 Map on one sheet with title: **DEVONSHIRE**. New imprint: **Published by A & C Black. Edinburgh.** (CeOS). Size: *c.*460 x 455 mm. Railway shown incorrectly Okehampton to Holsworthy (from Elmdon).

Black's Guide To Devonshire ... Tenth Edition
Edinburgh. Adam and Charles Black. 1879. FB.

4. 1881 Two sheets, *c.*235 x 450 mm: **DEVONSHIRE (NORTH SECTION)** (Aa) with some loss, eg everything below Exeter, Sandridge cut at The Foreland and (Pen)kuke (Ae) and Whitland (Ee) incomplete. Removals, eg Compass Pt and Cleave Cross Pt (Ab). **DEVONSHIRE (SOUTH SECTION)** (Ee) south from Exeter and Lansallos to Lyme Regis (breaking border). Imprint (40 mm).

Black's Guide To Devonshire ... Tenth Edition
Edinburgh. Adam and Charles Black. 1881[3]. KB.

5. 1882 **a)** Changes to north sheet only. Imprint shorter (31 mm). Penkuke (Ae) and Whitlands (Ee) now removed. Old Ilfracombe route now entirely erased.

Black's Guide To Devonshire ... Eleventh Edition
Edinburgh. Adam and Charles Black. 1882[4]. KB.

b) South sheet imprint shorter (31 mm).

1. Houlston & Wright published smaller transfers of these maps, folded into green covers and titled: *Tourists Handy Maps from the Ordnance Survey. North Devon* (South Devon) *&c. Price Fourpence.* Size 250x350mm. Scale of miles (8=50mm) Imprint: *J. Bartholemew Edinburgh* (EeOS). The series included *Torquay and Neighbourhood* and *Plymouth and Dartmoor.*
2. The North sheet was advertised in *Black's Guide To Devonshire* 1872 (1873). South Sheet lists both inside cover.
3. Addendum added with population figures for 1871 and 1881.
4. Addendum is expanded to become an introduction: General Description.

Black's Guide To Devonshire ... Eleventh Edition
Edinburgh. Adam and Charles Black. 1883, 1883 (1884). E, Pl[1]; KB.

6. 1884 Map similar to state 2 in size and imprints. Title and scale now (Ee). Railway to Hemyock. Depth lines added at Lundy. Budleigh Salterton now in capitals. Compass Pt and Cleave Cross Pt (Ab) both reappear.

W H Smith and Sons Reduced Ordnance Map of North Devon
London. W H Smith & Son. (1884). BL.

7. 1884 Title and imprints as state 5 but with added changes as state 6. The borders are widened, eg only Lyme Regis breaking the outer line on South sheet. South Sheet has names added (Axe Landslip and Ramillies Cove and Ralph's Hole at Bolt Tail and Sands Hotel near Slapton). Line to Ashton. Elmdon is now Meldon.

Black's Guide To Devonshire ... Twelfth Edition
Edinburgh. Adam and Charles Black. 1884, 1885, 1885 (1886). MW; GUL, T[2]; Pl.

8. 1888 Division letters removed. North sheet has names added on west coast (eg Duck Pool, Lighthouse, Blackmouth Mill, Hennacliff); Launceston to Halwell (straight line ending in T-junction), Exeter to Bampton completed.

Black's Guide To Devonshire ... Twelfth Edition
Edinburgh. Adam and Charles Black. 1886 (1888). C, NDL, Pl , KB.

Black's Guide To Devonshire ... Thirteenth Edition
Edinburgh. Adam and Charles Black. 1889, 1889 (1890). NDL, FB; KB.

9. 1891 New signature: **John Bartholomew & Co., Edinr**. The maps show changes, eg North sheet has railway lined curved (west) round Halwell; South Sheet has L&SWR Plymouth-Tavistock line. Fathom lines (Ae) deleted.

Black's Guide To Devonshire ... Thirteenth Edition
London and Edinburgh. Adam and Charles Black. 1889 (1891). KB.

Black's Guide To Devonshire ... Fourteenth Edition ... Revised and Corrected[3]
London and Edinburgh. Adam and Charles Black. 1892. TQ, Pl, KB.

10. 1892 Map similar to state 6 in size and layout, but no imprint. Signature and railways as state 9.

W H Smith and Son's Reduced Ordnance Maps For Tourists North Devon
London. W H Smith & Son. (1892). KB.

11. 1892 Four sections: (*c*.240 x 250 mm): **DEVONSHIRE NORTH EAST SECTION** or **NORTH WEST SECTION** or **SOUTH WEST SECTION** or **SOUTH EAST SECTION**. Imprint: **Published by A & C Black, London** (CeOS) and signature: **J Bartholomew, Edinr**. (EeOS). Overprinted **DEVON** erased. **SCALE 4 MILES TO AN INCH**.

Black's Guide To Devonshire ... Fourteenth Edition Revised and Corrected
London and Edinburgh. Adam and Charles Black. 1892 (1893). E; Pl, T[4].

1. North sheet only.
2. North sheet only.
3. With new Preface signed C.W.
4. Torquay Library copy with only SW and NW sections, both torn. NW section has signature J Bartholomew & Co.

12. 1894 **a**) Four sections: the earlier projected railways to Hatherleigh are erased. Railways are now shown to Plymstock and Catwater.

Black's Guide To Devonshire ... Fourteenth Edition
London and Edinburgh. Adam and Charles Black. 1892 (1894). E, FB.

b) Titles printed in red on reverse and 4 loose maps are inserted into a pocket in back cover.

Black's Guide To Devonshire ... Edited by A R Hope Moncrieff ... Fifteenth Edition
London. Adam and Charles Black. 1895. KB, C.

13. 1897 Single sheet with title **DEVONSHIRE.** Map overprinted **DEVON**. Lundy is inset (Aa).

Black's Guide To Devonshire ... Fifteenth Edition
London. Adam and Charles Black. 1895 (1897). E.

14. 1895 Title: **COUNTY OF DEVON FROM THE ORDNANCE SURVEY** (Ee), 480 x 490 mm. Imprints: **London. John Murray. Albemarle Street**. (CeOS) and **Copyright** (AeOS). Note **Main** and **Other Driving Roads**. Projected railway through Hatherleigh. Some names are altered or added, Milford instead of Gt and Lit Milford and Elmscott (Bb). **Budleigh Salterton** is in lower case. This map only: area is larger (Fowey to Lyme Regis); Lundy and Eddystone included in map area. The guide's sub-maps are superimposed, outlined in red with page number in top left corner. Fathom lines are erased. Colour printing.

A Handbook For Travellers In Devon ... Eleventh Edition
London. John Murray. 1895[1], 1895 (1901)[2]. BL, RGS, TQ, E[3], KB; KB.

15. 1898 Title: **DEVONSHIRE** (Ee). No imprint. Signature: **John Bartholomew & Co. Edinr.** (EeOS). Railway shown to Turnchapel. Holsworthy route corrected at Hollacombe. This state only: latitude and longitude graticule; border broken for Eddystone; key below scale bar (Dc) shows **Driving & Cycling Roads** and **Other Roads**. Lundy not inset.

The Pocket Series of Touring Maps Devonshire ... Ordnance Survey ... J Bartholomew
London. John Bartholomew & Co . (1898).[4] KB, MW, EB.

W.H.Smith & Son's Series Of Travelling Maps by J.Bartholomew No. 23 Devonshire
London. W H Smith & Son. (1898). BL, FB.

16. 1898 Imprint: **Published by A & C Black, London**. (CeOS). Signature: **John Bartholomew, Edinr**. (EeOS). Key: **Main Roads shewn thus**. Railways to Yealmpton and Northam. Holcombe added, Herdacott and St Nicholas replace Hardycot and The Ness. Lundy inset.

Black's Guide To Devonshire ... Sixteenth Edition
London. Adam and Charles Black. 1898. C, MW.

17. 1901 Railways to Bude, Lynton and Ashton to Exeter.

Black's Guide To Devonshire ... Sixteenth Edition Edited by A R Hope Moncrieff
London. Adam and Charles Black. 1898 (1901). KB.

1. Contains sectional transfers taken from a new Bartholomew map (**173**).
2. The text, title page and map(s) are identical to the 1895 edition. Spine has London, Stanford. There is an Index and Directory dated 1899 added as well as an advertising section (for 1901/1902) which is dated May 1901.
3. Exeter has two copies, one of which retains the old Walker map, though updated.
4. The maps were either issued in green covers with title: *Map Of Devonshire - Price 2/-*. with reference to Bartholomew's Half Inch map inside cover, or with Philips' blue cover. Labels pasted inside listed the title of the series and the maps covered (these vary).

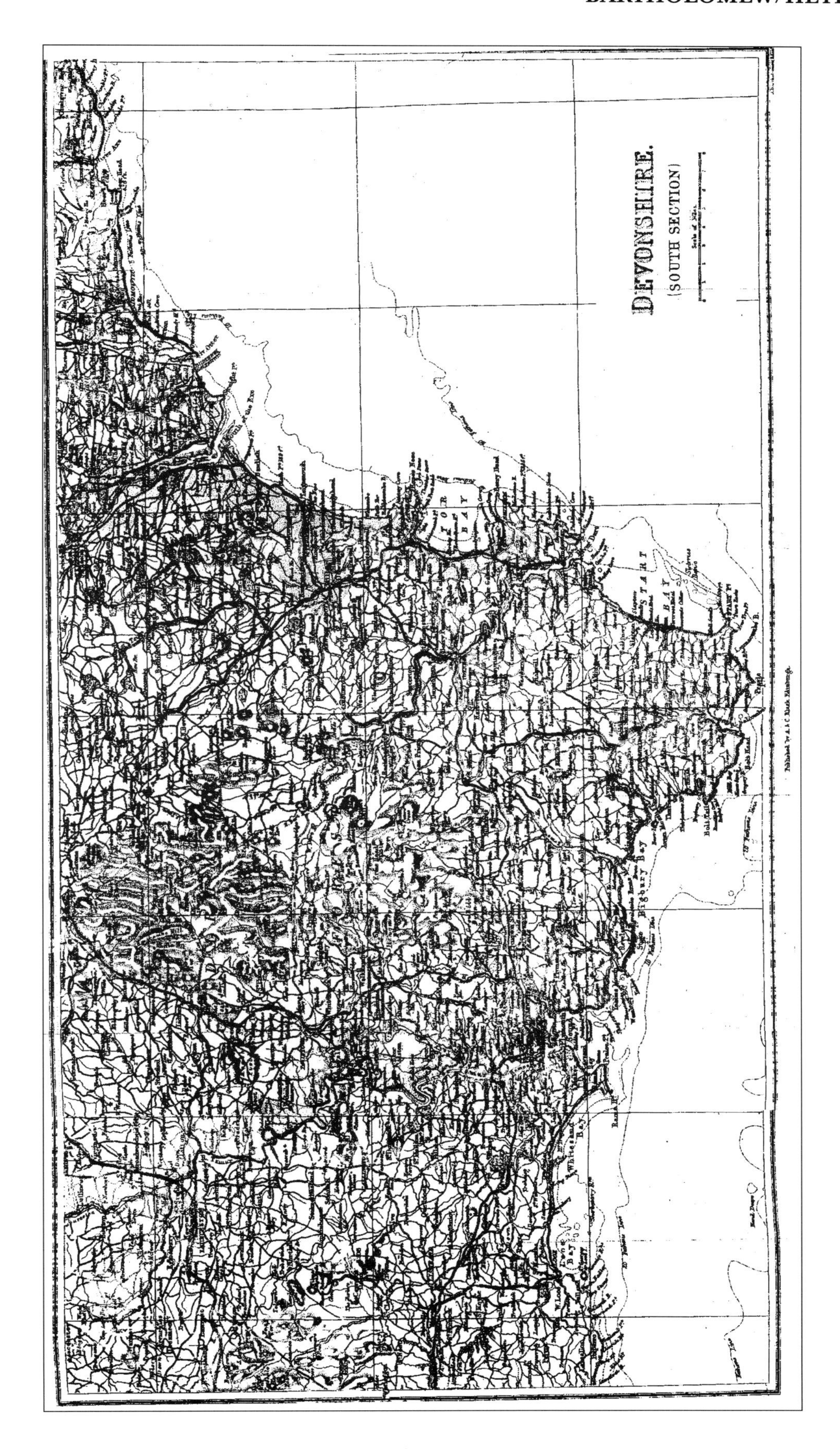

150.7 Bartholomew/Heydon *Black's Guide To Devonshire*

151

THOMAS MURBY

1874

Thomas Murby was an active publisher from 1860 until the early-1880s.[1] His first publications were books of music for schools. In 1861 he printed a set of musically arranged tales and songs for *Chambers's Library for Young People* after which he seems to have specialised in the educational sector producing books on a wide variety of subjects. Books on arithmetic, children's readers, lessons in animal physiology and even a grammar came out of his publishing house.

Murby produced a part series of educational booklets with maps *c.*1874.[2] Apparently only a few counties were published, each containing a brief description of the county taken from existing sources, together with a county map: Stafford, Norfolk, Kent, Surrey, Middlesex, Devon and a double sheet map of Yorkshire. Some of the maps in this series were engraved by W Dickes, who was known as an illustrator and publisher who experimented with the Baxter process, whereby printing is carried out using oil colours from several blocks, both wood and copper.[3]

Murby produced few maps: apart from this series he published a series of wall maps of the continents (and one map of Palestine), all produced photolithographically from reliefs (*c.*1874), and a small atlas of 16 maps, *Murby's Scholars' Atlas* (*c.*1878).

Size: 100 x 135 mm. **English Miles** (10=15 mm).

THE COUNTY OF DEVONSHIRE. Imprint: **THOS. MURBY. 32 BOUVERIE ST. FLEET ST**. (CeOS).

1. 1874 *Murby's County Geographies*
London. T Murby. (1874). BL.

1. Although the British Library has two Scripture Manuals attributed to Thomas Murby dated 1913 these are probably reissues or by another Murby.
2. David Smith; 1985; p. 147.
3. Ian Mackenzie; pp. 99 & 47.

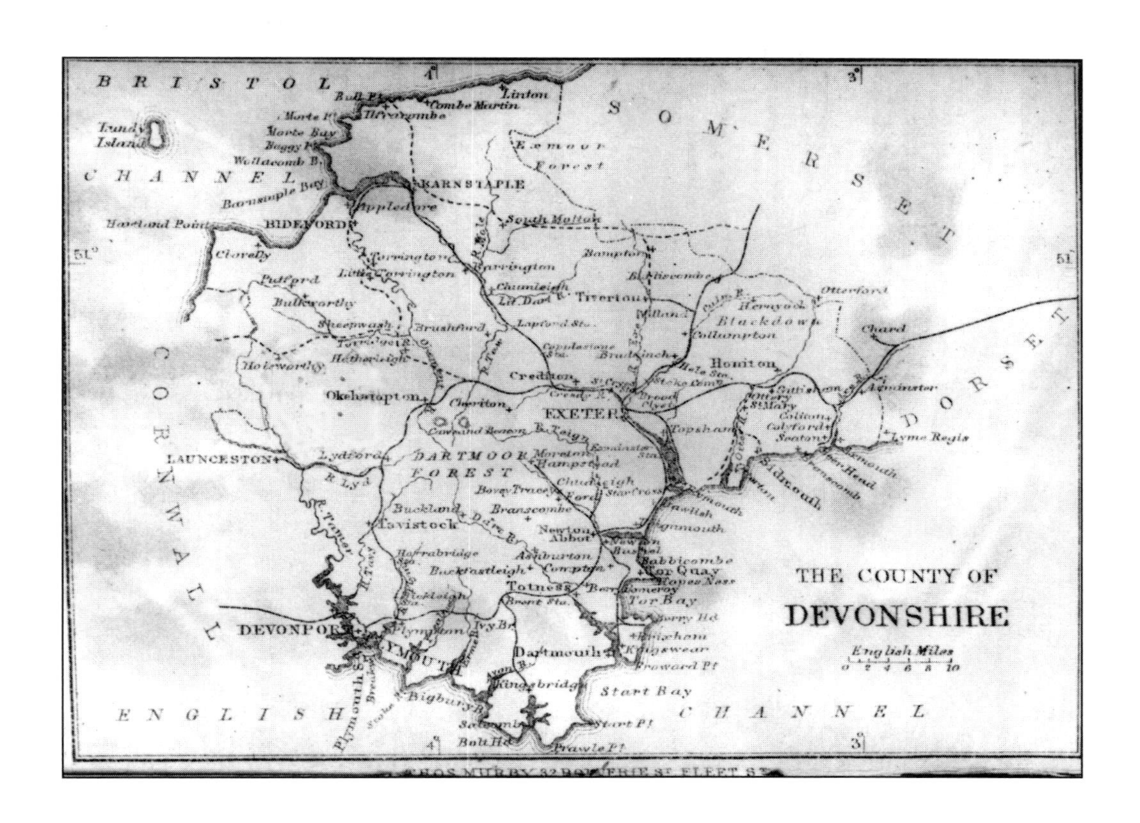

151.1 **Murby** *Murby's County Geographies*

152

WELLER/COLLINS

1875

William Collins[1] (1789-1853), first member of Britain's first Temperance Society, founded his publishing company primarily to publish the works of the great Scottish cleric Dr Charles Chalmers. His son, William II (1817-95, became a partner in 1843 and was knighted 1881) and grandson William III (1846-1906) continued the firm which still survives today. As early as 1827 William Collins was touring Devon and selling atlases but it was not until 1856 that the company produced its first atlas. They issued many books and maps for schools.[2]

Edward Weller, engraver of the successful *Weekly Dispatch* plates (**136**), engraved new plates, *c.*1875, for William Collins' set of *County Geographies* which were edited by W Lawson and produced county by county. The map was probably first used in a small school primer which sold at twopence a copy.[3] But the authors have not yet seen a Devon copy. The map appeared in a local directory published by Percy and Co., *The Devonshire Calendar,* before Collins exploited the maps in an *Atlas of England And Wales* in 1877. Then in 1878 he produced a number of regional atlases, wrongly attributed to Bartholomew. Each volume contained sixteen maps, grouped into sets, with each set bound into a paper cover printed with county plans for parts of Great Britain. Devon is on the same buff coloured sheet as Somerset with a coloured map of Cornwall on the reverse.

Size: 206 x 158 mm. **English Miles** (10=21 mm).

DEVONSHIRE. Imprint: **William Collins Sons & Co. London & Glasgow**. Signature: **Edwd Weller.** (EeOS). Plate number **32** (AeOS). Railways; Paignton, Plymouth-Launceston, Moreton Hampstead, Seaton, Torrington, Barnstaple-Taunton and to Ilfracombe. The plain border is broken at Lyme Regis. Eddystone is shown.

1. 1875 *Collins County Geographies*
London. William Collins, Sons and Co. (1875). assumed.

2. 1876 Imprint and plate number omitted leaving signature. Coloured to show Parliamentary Divisions (N, S and E). Railways added Watchet, Exmouth, Okehampton.

The First Issue: The Devonshire Calendar and Register for 1876
London. Percy and Co. 1876. E.

3. 1877 Imprint as State 1 but the signature removed leaving letter **E** (but this is not the E of Edwd!). Plate number added on reverse.

Collins Series of Atlases. Atlas Of England And Wales
London. William Collins, Sons and Co Ltd. 1877[4], (1882). C, BL, B; W, EB.

4. 1878 All imprints removed. Size: 195 x 128 mm.

The South Western Counties Atlas. Consisting of sixteen maps.....by John Bartholomew
London and Glasgow. William Collins, Sons and Co. (1878). C.

1. The details are taken from *The House of Collins*; David Keir; Collins; 1952.
2. The British Library has a large wall map of Somerset (880 x 1165 mm at a scale 10 = 195 mm).
3. R Carroll; 1996; p. 330.
4. BL and Bodleian copies have a preface dated May 1877.

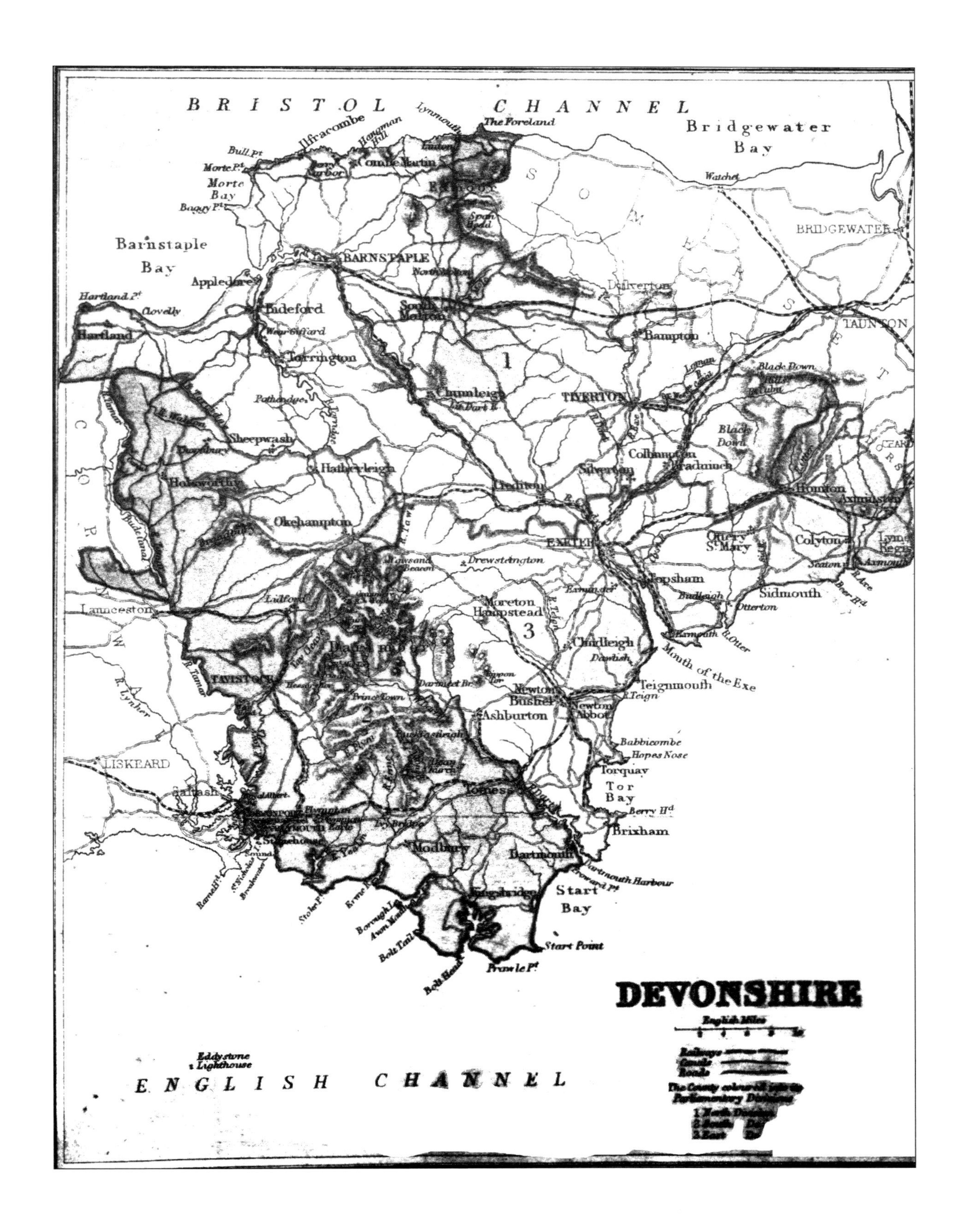

152.2 **Weller/Collins** *The Devonshire Calendar and Register*

153

BARTHOLOMEW/BLACK – Encyclopaedia Britannica

1877

The first *Encyclopaedia Britannica* appeared in three volumes in 1768-71 under the auspices of a Society of Gentlemen. Andrew Bell (1726-1809) founded the *Encyclopaedia Britannica* with Colin McFarquar. On Bell's death Archibald Constable bought the Encyclopaedia and published the 5th & 6th editions. When he went bankrupt in 1826 the Blacks bought the copyright (1828) and published the 7th edition 1830-1842. The 8th edition was published 1853-1860 and the 9th, and arguably the best, 1875–1888. Although John Bartholomew had acquired the privilege to print the *Encyclopaedia* maps in 1839 it was not until the 9th edition that county maps were included. It was very successful and in 1897 Blacks were approached by James Clarke & Co. for the rights to reprint and sell 5000 copies. In 1898 a full-page advertisement for *The Encyclopaedia Britannica, A Dictionary of Arts, Sciences, and General Literature* appeared in *The Times*. The success was astonishing with new reprinting being needed immediately. By 1899 James Clarke & Co. had bought the copyright.[1] W & A K Johnston must have taken over printing of the maps at about this time.

In 1892 *Black's Handy Atlas of England & Wales* appeared using the *Britannica* maps. The Devon map was also used in *Black's Guide To Torquay* in 1901.[2]

Size: 250 x 200 mm. **Scale of Miles** (12=33 mm).

DEVON (CaOS). **VOL VII** (AaOS). **PLATE III** (EaOS). Imprints: **ENCYCLOPAEDIA BRITANNICA, NINTH EDITION.** (CeOS). Signature: **J Bartholomew Edin^r^** (EeOS). Railways shown to date.

1. 1877 *The Encyclopaedia Britannica. ... Ninth edition. Vol VII*
Edinburgh. Adam and Charles Black. 1877. BL, NLS, GUL.

2. 1892 Title rewritten. Imprints: **Black's Handy Atlas of England & Wales** (EaOS vertical) and **Published by A & C Black, London** (CeOS). **Plate. 18** (EeOS vertical). Signatures: **John Bartholomew & Co.** (EeOS) and **The Edinburgh Geographical Institute** (AeOS). Parliamentary Division information 1885 (Ee) replaces 3 divisions. Railways added: Launceston and Okehampton to Holsworthy and Exeter-Bampton. Size now 245 x 175 mm so that map detail extends outside outer frame. Reverse has **DEVON** and **PLATE 18**.

Black's Handy Atlas Of England & Wales
London. Adam and Charles Black. 1892. BL, C, BCL, NLS, BRL.

3. 1898 Titles etc revert to state 1. Signature: **W & A K Johnston.** (EeOS). Parliamentary Division information deleted. Reverse is blank.

The Encyclopaedia Britannica ... Ninth Edition
Edinburgh. Adam and Charles Black. 1877 (1898), 1877 (1899). GL, Notts; EB.

4. 1900 **Limited** added to the imprint.

The Encyclopaedia Britannica. .... Ninth edition.[3] [Leics UL].
Edinburgh. Adam and Charles Black. 1877 (1900).

5. 1901 Imprint: **Published by A & C Black, London** (CeOS). Signature: **J Bartholomew, Edinr.** (EeOS). Britannica imprint and page number deleted. Railways to Bude, Lynton, Yealmpton, and projected Bideford to Appledore, Ashton to Exeter. Size as state 1.

Black's Guide To Torquay And The South Hams ..., Salcombe, Etc.
London. Adam and Charles Black. 1901. KB.

1. *Adam & Charles Black 1807-1957*; A & C Black; 1957; pp.6-7, 64ff.
2. Examples of the maps in an earlier lithographic state (*c.* 200 x 300 mm) have been seen; Devon possibly exists.
3. The copy reported as being in the Lincoln Cathedral collection has been sold, present whereabouts unknown.

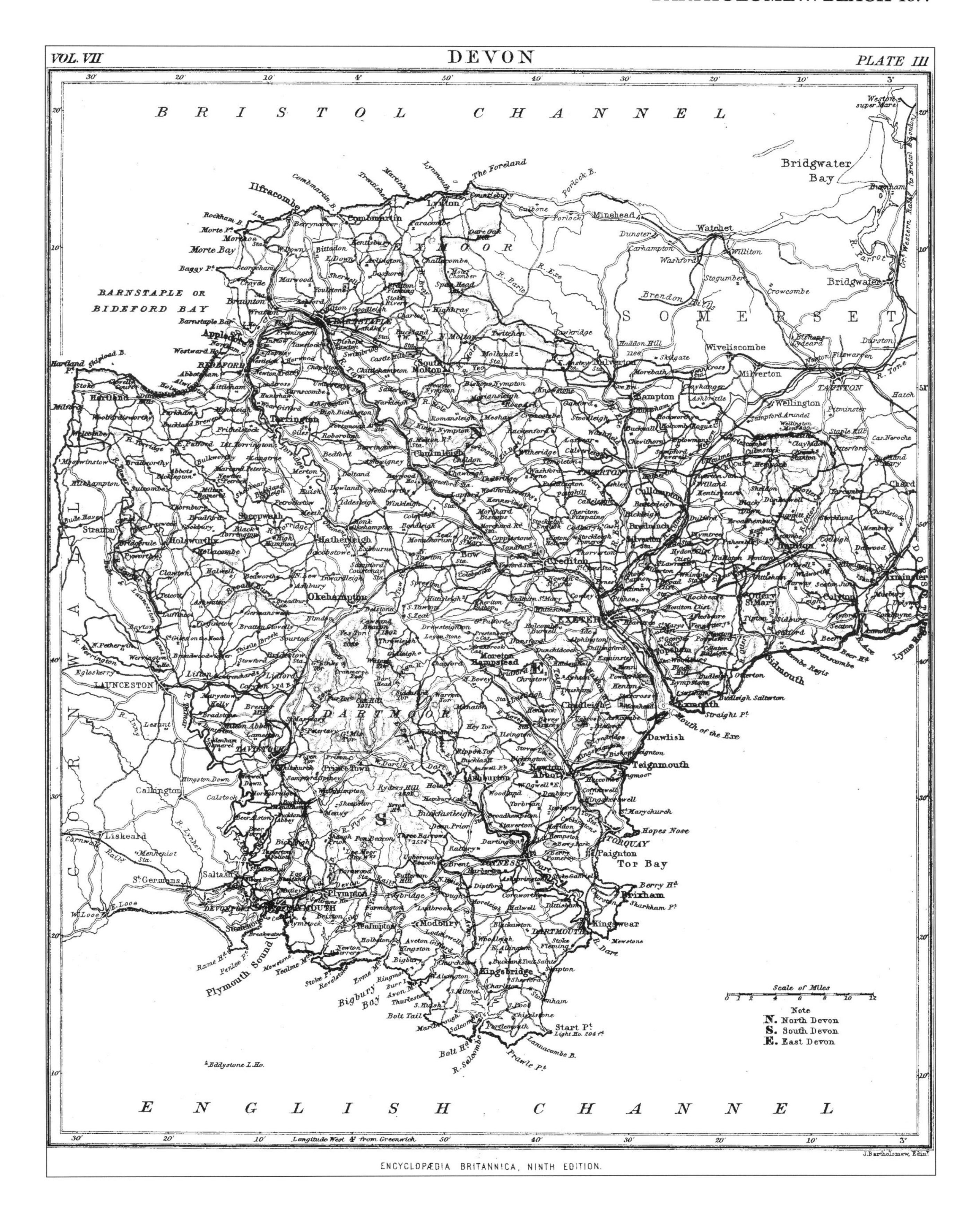

153.1 **Bartholomew/Black** *Encyclopaedia Britannica*

154

STANFORD/WORTH

1878

Richard Nicholls Worth (1837-1896) was a popular Devon journalist, geologist, author and editor of guide books. He produced two guide books covering north and south Devon in 1878/79 and these were printed and published until 1894 with maps by Edward Stanford. The south sheet map had a further printed map of the county as an inset. R N Worth is known to have spent six months researching the geology of the L&SWR Tavistock to Plymouth route in 1889 and read a paper to the Devon Association at Tavistock later that year.[1]

Worth's guides to North and South Devon were published both individually and together as *Tourist's Guide To Devonshire In Two Parts, South And North*[2]; all contained maps produced at Stanford's Geographical Establishment. Although there were several editions the number of changes made was minimal and every attempt made to keep to the original pagination.

Worth also contributed to county histories; *The Popular County Histories* series appeared *c.*1886; and a *History of Devonshire* in 1886. These were published by Elliot Stock (London, 62 Paternoster Row) and rather surprisingly there was no map of the county. Worth wrote a considerable number of works on the southwest: *Devonport, sometime Plymouth Dock* (Plymouth, 1870), *History of Plymouth from the earliest period to present time* (London, 1871), *Somerset* (London, 1881), *Dorset* (London, 1882), *History of Plymouth* (Plymouth, 1890) and he also edited many other *Tourist's Guides*, Wye, Kent, Surrey, West Riding and Channel Islands, for example.

154 - SOUTH

Size: 245 x 300 mm. **Scale of Statute Miles** (15 = 75 mm).

MAP TO ACCOMPANY THE GUIDE TO SOUTH DEVON. The title is (AcOS) but **DEVONSHIRE** is written across the county. Plain border broken for Axminster. Scale Bar (AeOS). Imprint: **London: Printed by Edward Stanford, 55 Charing Cross: S.W.** (CeOS). Signature: **Stanford's Geog^{l}. Estab^{t}.** (EeOS). There is no railway to Holsworthy or Ashton. The Eddystone is shown pictorially with the Scale Bar below (BeOS). There is a small inset map of the county (Ee): size 75 x 78 mm; **English Miles** (20=16 mm); **DEVON** is written across the county with a title **GENERAL MAP OF THE COUNTY** (AcOS).

1. 1878 *Tourist's Guide To South Devon. R N Worth*
London. Edward Stanford. 1878. BL, NLS, E.

Tourist's Guide To South Devon. R N Worth. Second Edition
London. Edward Stanford. 1880. NLS.

2. 1880 Railway now shown to Holsworthy but not to Ashton (and also on the inset map).

Tourist's Guide To South Devon. Second Edition
London. Edward Stanford. 1880. E.

Tourist's Guide To Devonshire In Two Parts, South And North
London. Edward Stanford. 1880. NLS.

1. F Booker; 1967 (1974); p. 240.
2. Spines and title pages have either *Tourist's Guide* or *Tourists' Guide.*

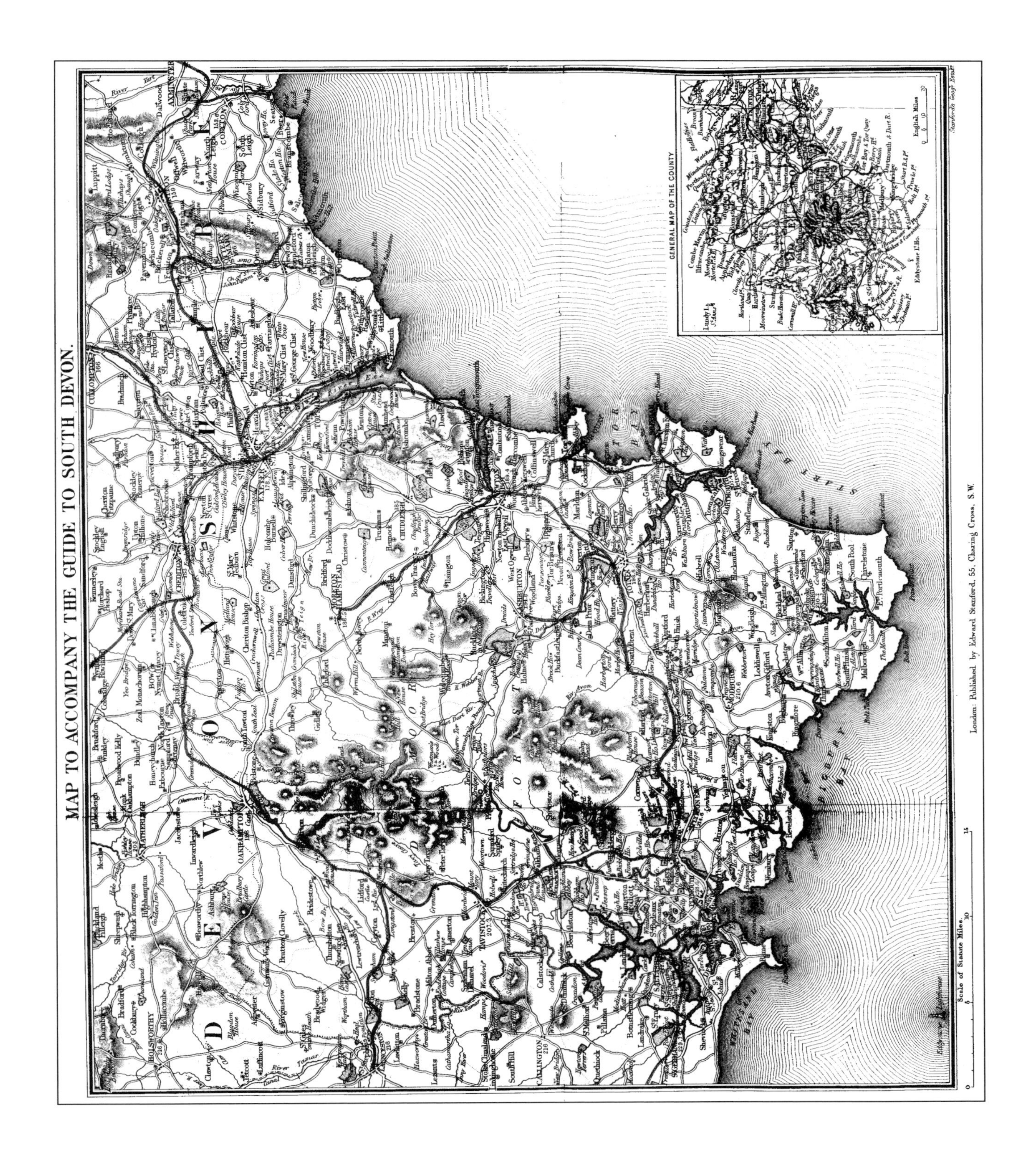

154.S.1 Stanford/Worth *Tourist's Guide To South Devon*

3. 1883 Railway now shown to both Ashton and Holsworthy and Bampton-Tiverton (and also on the inset map).

Tourist's Guide To South Devon. Third Edition
London. Edward Stanford. 1883. E, NLS.

Tourist's Guide To Devonshire In Two Parts, South And North. Second Edition
London. Edward Stanford. 1883. DEI, TQ.

Tourist's Guide To Devonshire. Third Edition
London. Edward Stanford. 1886. NLS.

4. 1886 Railway now shown from Exeter to Tiverton (but not on inset map).

Tourist's Guide To Devonshire. Third Edition
London. Edward Stanford. 1886. E.

Tourist's Guide To South Devon. Fourth Edition
London. Edward Stanford. 1886, 1886 (1887). E; KB[1].

5. 1890 The imprint is changed **London: Edward Stanford 26 & 27, Cockspur St., Charing Cross, S.W**. (as is the title page address). Railways now shown past Cadbury to Bramford Speke, Launceston to Halwell added and the L&SWR from Plymouth to Tavistoke (and also on the inset map together with the Exeter-Tiverton line).

Tourist's Guide To South Devon. Fifth Edition
London. Edward Stanford. 1890. BL, NLS, E.

6. 1894 Railway to Kingsbridge (also on key map).

Tourist's Guide To South Devon. Sixth Edition
London. Edward Stanford. 1894. BL, RGS, NLS, TQ.

1. Contains advertising section with dates for 1887.

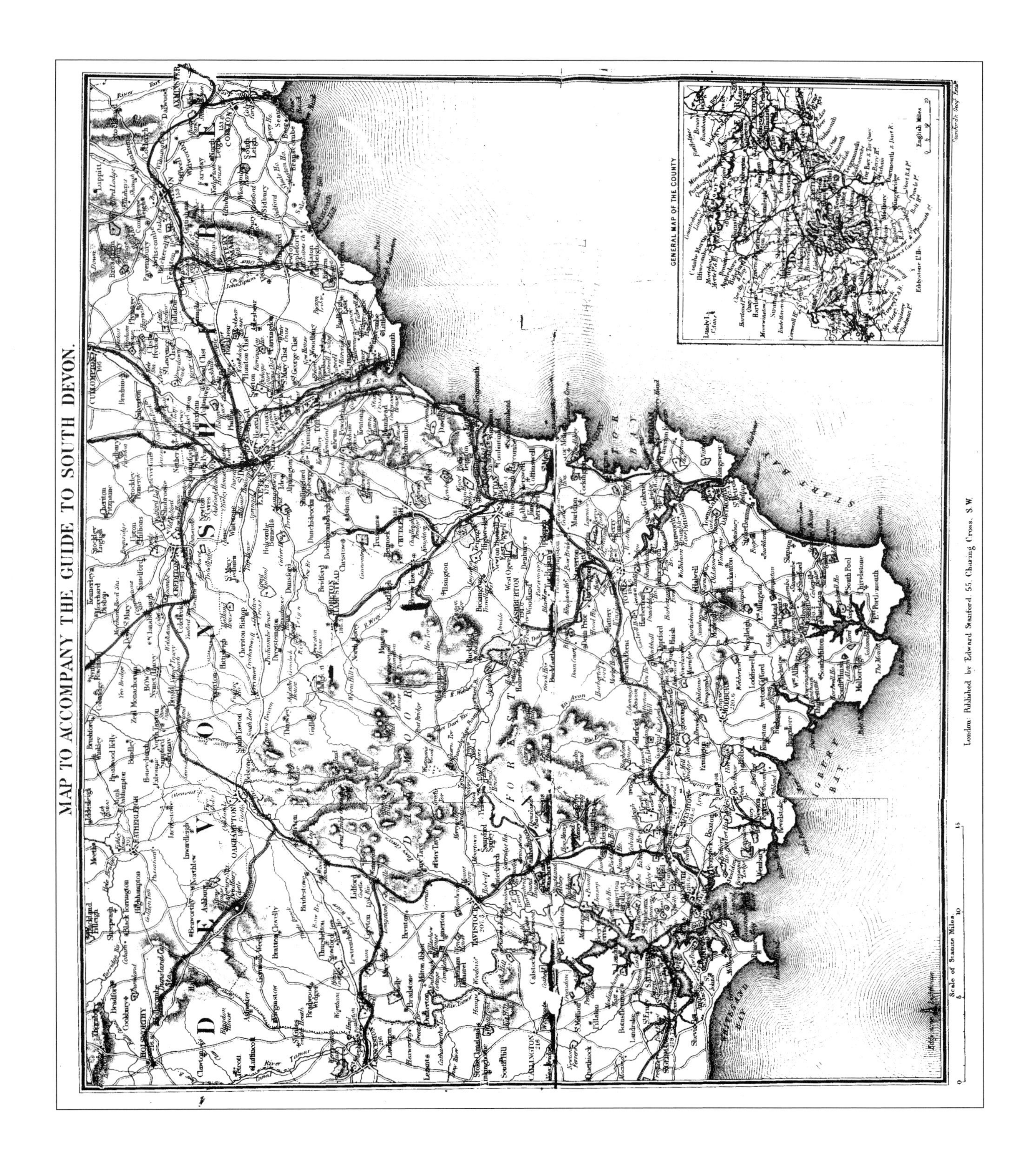

154.S.3 Stanford/Worth *Tourist's Guide To South Devon*

154 – NORTH

Size: 245 x 300 mm. **SCALE OF MILES** (10 = 50 mm).

MAP TO ACCOMPANY THE GUIDE TO NORTH DEVON. The title is (AcOS) but **DEVONSHIRE** is written across the county. Plain border broken for Exmouth. Scale Bar (AeOS). Imprint: **London: Edward Stanford, 55 Charing Cross: S.W.** (CeOS). Signature: **Stanford's Geog^{l} Estab^{t}** (EeOS).

1. 1879 *Tourist's Guide To North Devon. R N Worth*
London. Edward Stanford. 1879. BL, NLS, E, T.

Tourist's Guide To North Devon. Second Edition
London. Edward Stanford. 1880, 1880 (1882). BL, E; KB.

Tourist's Guide To Devonshire In Two Parts, South And North
London. Edward Stanford. 1880. NLS.

2.1883 This has railway from Bampton to Tiverton with 2 tracks forming a loop around the town. Line to Ashton added.

Tourist's Guide To Devonshire In Two Parts, South And North. Second Edition
London. Edward Stanford. 1883. TQ.

Tourist's Guide To North Devon. Third Edition
London. Edward Stanford. 1883. NLS, T.

3. 1883 Railway shown from Bampton to Tiverton and on to Ashley Court just south of the town, passing Tiverton to the East.

Tourist's Guide To North Devon. Third Edition
London. Edward Stanford. 1883, 1883 (1885). E; KB.

4. 1886 Railway shown from Exeter to Tiverton

Tourist's Guide To Devonshire. Third Edition
London. Edward Stanford. 1886. E, FB.

Tourist's Guide To North Devon. Fourth Edition[1]
London. Edward Stanford. 1886, 1886 (1888). T, RGS, FB; KB.

5. 1890 Imprint changed: **London: Edward Stanford, 26 & 27 Cockspur St. Charing Cross, S.W.**
Railway now shown from Launceston to Halwell.

Tourist's Guide To North Devon. Fifth Edition
London. Edward Stanford. 1890. NLS, E, NDL.

6. 1891 Railway now shown from Launceston into Cornwall.

Tourist's Guide To North Devon. Sixth Edition
London. Edward Stanford. 1890 (1891), 1894. KB; BL, RGS, NLS, E, TQ.

1. Apart from the preface this edition is the same as that in *Devonshire Third Edition.*

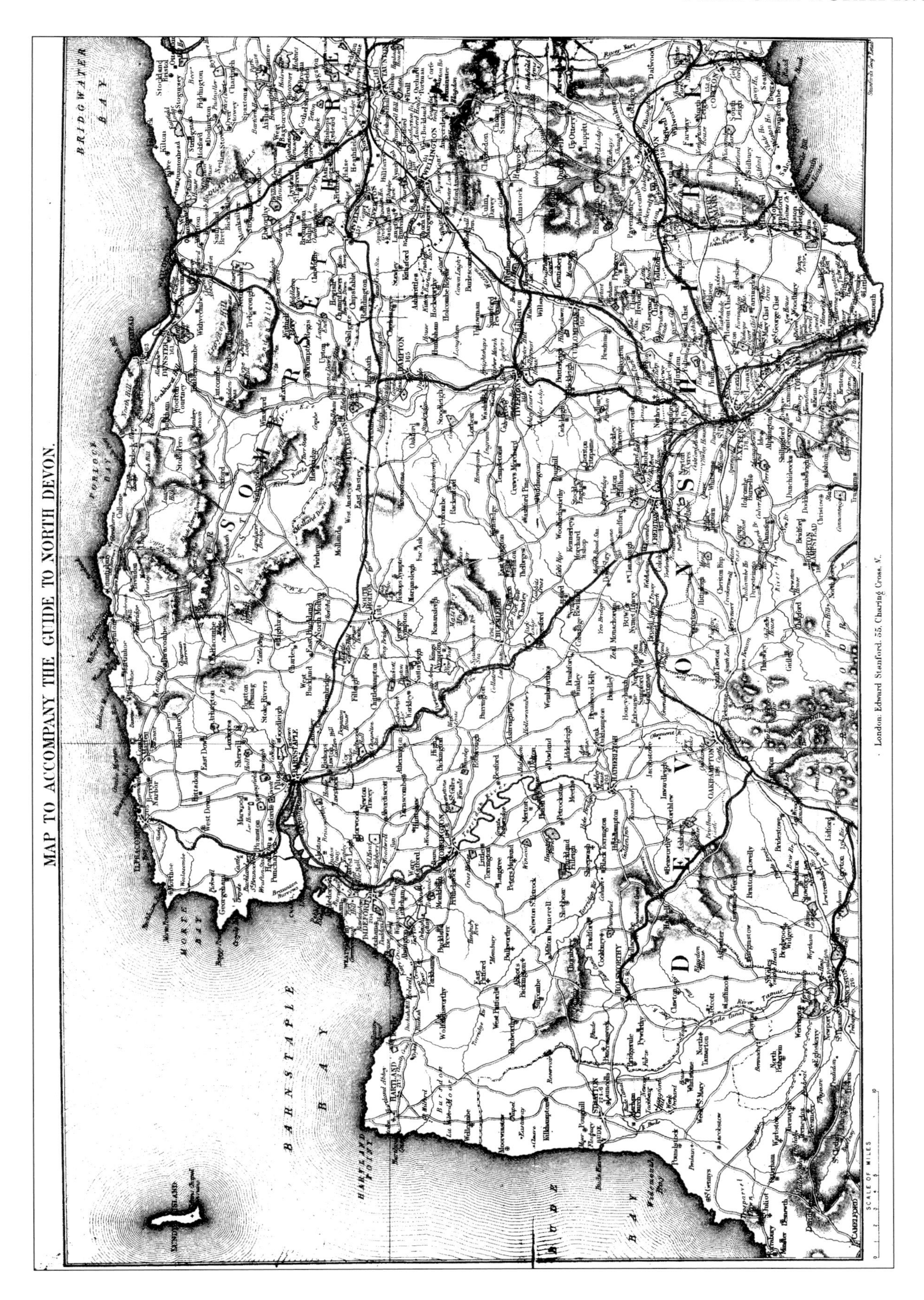

154.N.4 Stanford/Worth *Tourist's Guide To North Devon*

155

EDWARD STANFORD

1881

Stanford's Geographical Establishment produced a wide variety of maps for different publications. From 1880 Stanford issued a series of lithographic transfers for *The London Geographical Series.* These were subsequently used for *Stanford's Handy Atlas and Poll Book*. The maps are without title but the name of the county is written across the map. Many railways are shown and the heights of some hills, otherwise the maps are basic although hill hachuring is fairly complete. This transfer was apparently taken from plates of a map of the whole of England and Wales, which has not been identified.

There are code letters below certain towns (Tiverton, Barnstaple and Plymouth have B.2; Taunton has C2; Dartmouth and Torrington have B; and Exeter has C.B.2 and a C in a square). No explanation is given, though they probably refer to cathedrals and parliamentary returns.

Charlotte Maria Shaw Mason (*b.*1842) was a noted educationalist. She wrote a large number of text books for the new educational curriculum after moving to the Lake District.[1] One of the Manchester University colleges, Charlotte Mason College in Ambleside, is named after her. She died in 1923 the same year that *The Ambleside Geography Books* (spine title) - *New Edition (Revised)* appeared. This contained exactly the same maps (State 1) as the original but was now published by Kegan Paul, Trench, Trubner & Co and the Parent's Educational Union Office, London. There may have been an earlier edition as this series of books reproduced the original titles and not only did Charlotte Mason's name adorn the cover there was also a preface written by her in 1907 together with a note referring to the origin of the series.

Size: Two halves 139 x 77 mm each. **ENGLISH MILES** (20=35 mm).

No title but **DEVON** written across the county. Imprint: **Stansford's Geog^{r}. Estab^{t}. London**. (EeOS). Scale bar (AeOS). Printer's mark **R 2** (EeOS). The reverse to the west part has **p.241** and **DEVONSHIRE. I.**; that to the east has **p.244** and **DEVONSHIRE II**.

1. 1881 *The London Geographical Series ... By Charlotte M. Mason...Book III for Standard IV The Counties of England*
London. Edward Stanford. 1881. BL, C, B.

2. 1886 Titles added: **DEVON, WEST** and **DEVON, EAST** (both CaOS). Page numbers **19** and **20** (EaOS). The map is overprinted in yellow for Conservative seats and pink for the Liberals, and the names and boundaries of Parliamentary constituencies overprinted in red.

Stanford's Handy Atlas and Poll Book
London. Edward Stanford. 1886-June. BL, C.

3. 1886 The date **1886** is added below the two titles. Slate colour for the Unionists added.

Stanford's Handy Atlas and Poll Book. Second Edition
London. Edward Stanford. 1886-October. BL, BCL.

1. E Chalmondeley; *The Story of Charlotte Mason*; London; 1960.

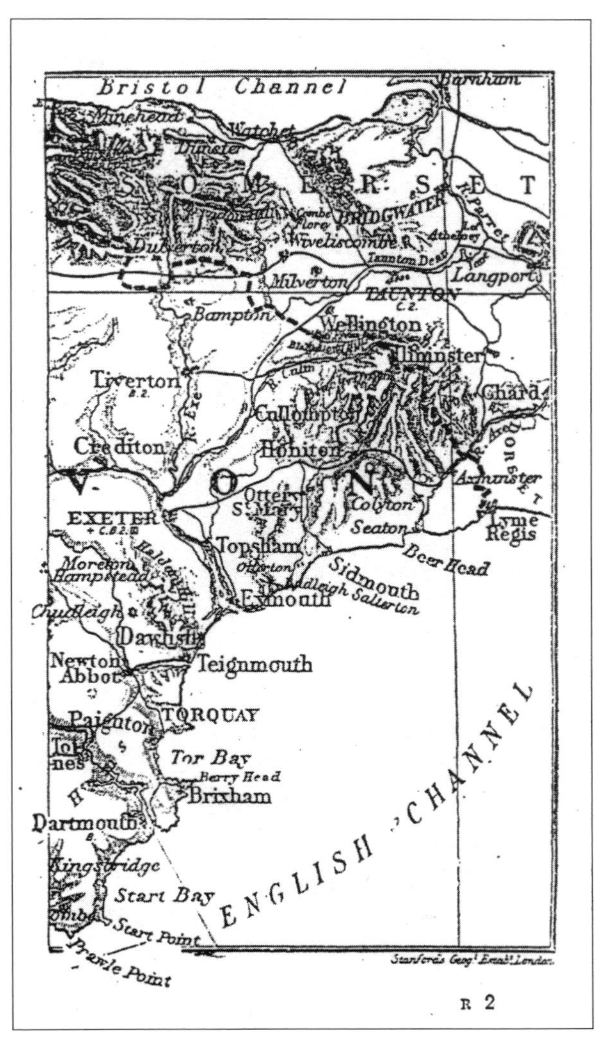

155.1 Stanford *The London Geographical Series*

156

BARTHOLOMEW/BLACK - Key Map

1882

Adam and Charles Black published the *Enyclopaedia Britannica* (**153**) but they are probably best remembered for their guide books. From *c.*1862 they produced the *Guide To Devonshire* (**142**) with the comment that: these road and railway guides *should find a corner in the portmanteau of every person about to undertake a journey of pleasure or business.* The guide was revised and reprinted (with its own pagination) in 1882 and contained two new maps by John Bartholomew. A plain index map of the county was printed on the inside cover.[1] A detailed county map (**150**) was inserted. These guides were also printed by R & R Clark of Edinburgh.

Size: 145 x 190 mm. No scale.

KEY MAP OF DEVONSHIRE (Ee). Signature: **J. Bartholomew. Edin.** (EeOS). Map sections named with page number, outlined and all over-printed in red. Railways to date. Dashed lines represent the sections of the book, dotted lines for county boundary.

1. 1882 *Black's Guide To Devonshire. Eleventh Edition*
Edinburgh: Adam and Charles Black. 1882, 1883, 1883 (1884). KB; E, P1; KB.

2. 1885 Railway now shown to Ashton passing Chudleigh.

Black's Guide To Devonshire. Twelfth Edition
Edinburgh: A & C Black. 1885, 1885(1886), 1886 (1888). MW; TQ, GUL; E, NDL; KB, Pl.

3. 1889 Signature: **John Bartholomew & Co. Edinr.** (EeOS). The red over-printing of the North Section boundary now follows the county boundary. Railways: L&SWR Tavistock to Plymouth, Launceston to Halwell (not named) and Exeter to Bampton.

Black's Guide To Devonshire. Thirteenth Edition
Edinburgh: Adam and Charles Black. 1889, 1889 (1890), 1889 (1891). FB; KB; KB.

4. 1892 Signature: **J Bartholomew. Edinr**. Map sections names and numbers strengthened; typeface altered for *SECTION* notes. North Devon now on P. 167.

Black's Guide To Devonshire. Fourteenth Edition
London and Edinburgh: Adam and Charles Black. 1892. Pl, TQ, KB.

5. 1894 Imprint: **Published by A & C Black. London**. (CeOS).

Black's Guide To Devonshire. Fourteenth Edition
London and Edinburgh. Adam and Charles Black. 1892 (1894). E, FB.

6. 1895 New title: **SKETCH MAP OF DEVONSHIRE.** Reference to SECTIONS removed leaving dotted boundaries. Railway lines thickened; Cornish lines Launceston-Camelford with projected to Wadebridge, Bodmin, Fowey, Looe; Bude and Halwell Junction added with projected Holsworthy-Bude.

Black's Guide To Devonshire. Fourteenth Edition
London and Edinburgh. Adam and Charles Black. 1895. E.

Black's Guide To Devonshire. ... Edited by A R Hope Moncrieff ... Fifteenth Edition
London and Edinburgh. Adam and Charles Black. 1895. C, E, KB.

1. The first editions had an index map including Cornwall and Dorset; this is not included here.

7. 1898 Railways: Bude, Ashton-Exeter, Launceston-Wadebridge, to Yealmpton with projected (dotted) to Modbury, to Budleigh Salterton and projected to Exmouth.

Black's Guide To Devonshire. Sixteenth Edition
London and Edinburgh. Adam and Charles Black. 1898, 1898 (1901). C, MW; KB.

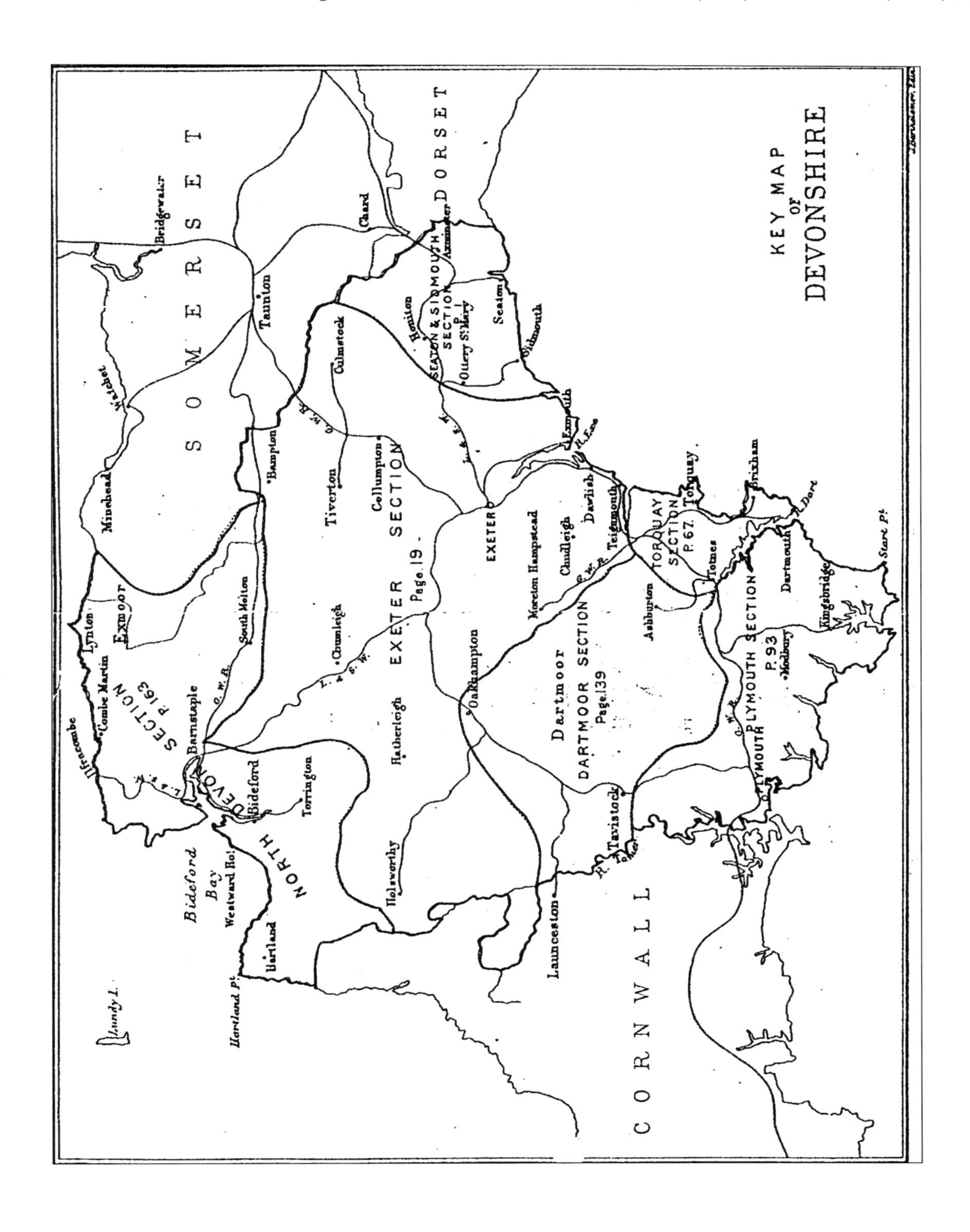

156.1 Bartholomew/Black *Black's Guide to Devonshire*

157

BARTHOLOMEW/DULAU

1882

In the 1880s the company of Dulau & Co. began publishing a series of *Thorough Guides.* The first volume was on the *English Lake District.* Subsequent volumes covered tourist areas such as Scotland, the Peak District, Wales and Ireland, as well as the watering places of the Eastern Counties. Two volumes were issued covering Devon and Cornwall: in 1882 Volume III, *North Devon and North Cornwall,* appeared; and in 1884 Volume VII, *South Devon and South Cornwall,* followed. Both volumes were written by Charles Slegg Ward, and he and M J B Baddeley were the series editors. The guides contained many sectional maps which were transfers from the earlier Bartholomew plates produced for the *Imperial Map* of A Fullarton (see **150**) and which later appeared in the *Royal Atlas.* Both north and south guide books had, in the inside front cover, the identical index map showing both counties. The map shows the main rivers, main roads and railways with, superimposed, those areas covered by each of the sectional maps included within that volume.

Mountford John Byrde Baddeley (1843-1906), a school master, earned his reputation as the compiler of these *Thorough Guide books for pedestrians.* He settled in the Lake District which he popularized as a pleasure resort. According to the advertising text in the guides: *In English topographical writing for tourists, the Thorough Guide Series is so far ahead of any other that there can scarcely be said to be a good second to it.* (*Saturday Review,* August 28th, 1886). *The Times* (August 3rd, 1887) even went so far as to compare a *Baddeley* with a *Baedeker.* A cursory glance at the contents page and page numbering of different editions seems to imply little up-dating, but a lot of new detail was added for each issue.

The sectional maps in the guides were sometimes new, and sometimes were the same as previously issued, but up-dated. From *c.*1902 Dulau used sectional transfers taken from Bartholomew's new layer-coloured maps (**174**).

The South Devon guide was re-issued in the twentieth century: the Sixth edition (revised) appeared 1902; the Seventh edition (revised), published by Thomas Nelson and Son (London, Edinburgh, Dublin and New York), appeared in 1908;[1] an eighth edition appeared in 1915 also by T Nelson with new shapes to the outline maps; the final edition of the southern guide was printed *c.*1925 and published by Ward, Lock & Co.[2] The Index map for 1908 was the same but updated with the relevant sectional map coding and a new inset map of Dorset; in 1915 the sectional maps were revised; the 1925 edition had a new index map, a close copy of the original.

The North Devon guide was reissued in 1903. When the guides were reissued in 1912 by Thomas Nelson, cycle and motor routes had been added and the imprint was *John Bartholomew & Co., Edinburgh.* The Devon guides were printed by either Strangeways & Sons or by J S Levin (1885, 1888, 1889, 1890).

1. Thomas Nelson amalgamated with John Bartholomew in 1888. See L Gardiner; 1976; p.34.
2. An edition of *South Devon and South Cornwall* published by Ward, Lock & Co. in 1925 advertises the *North Devon and North Cornwall* volume.

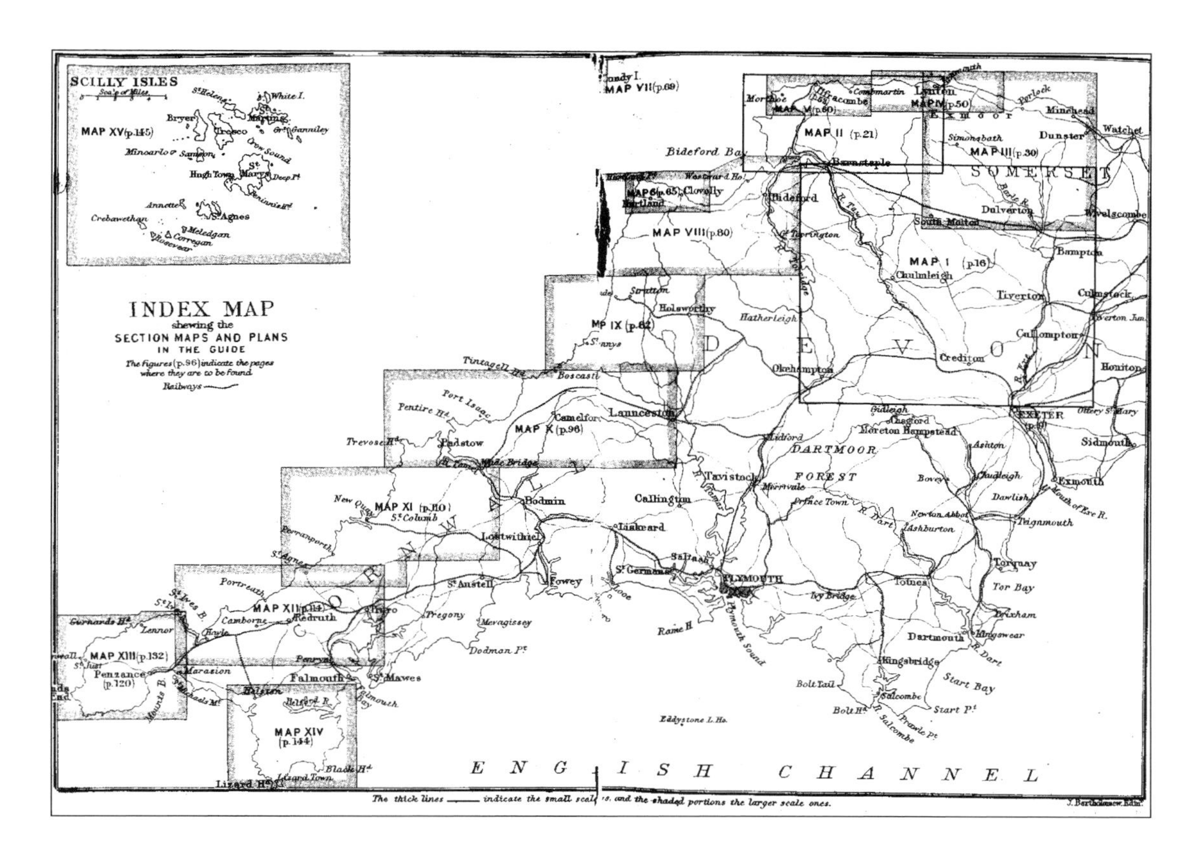

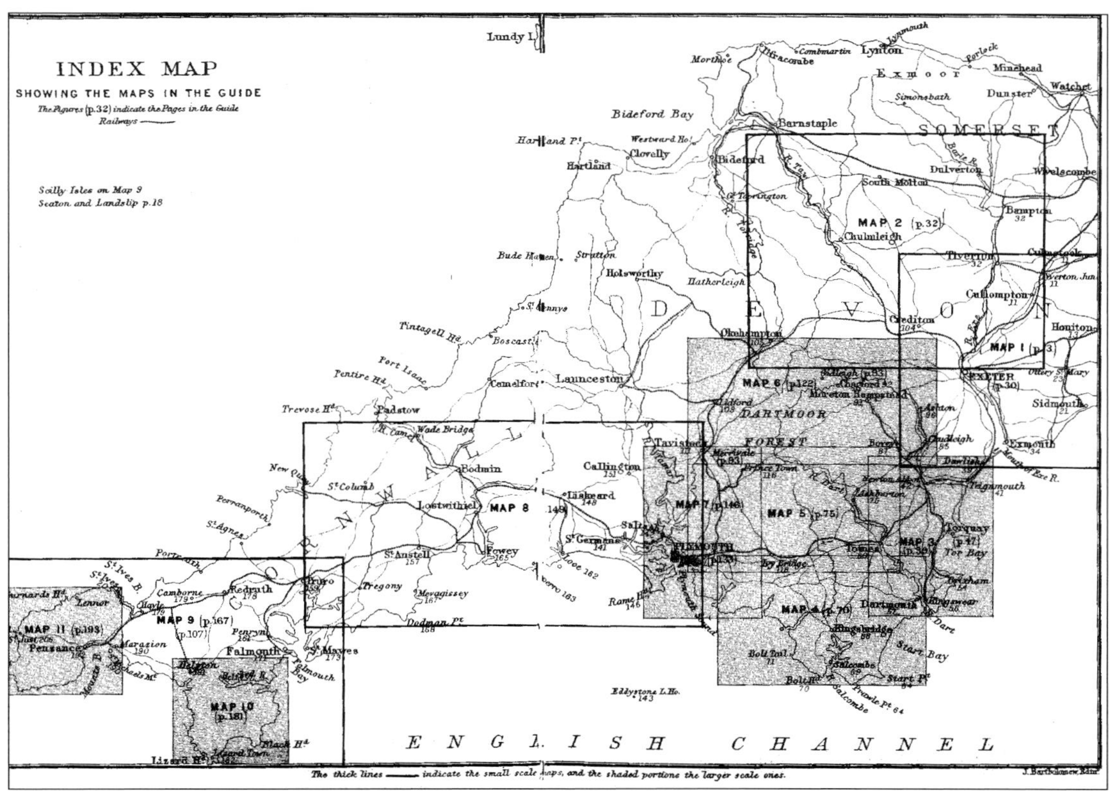

157.N.5 & 157.S.3 Bartholomew/Dulau *Thorough Guide Series*

157 – NORTH

Size: 145 x 199 mm. No scale.

INDEX MAP SHOWING THE MAPS IN THE GUIDE. Imprint: **J.Bartholomew Edinr**. Devon and Cornwall showing the areas of the maps within the northern guide by number and page. Two railway lines from St. Austell to Newquay and Redruth to Portreath. Map 8 has both Roman and Arabic numerals.

1. 1882 *Thorough Guide Series ... North Devon And North Cornwall*
London. Dulau & Co. 1882, 1883. E; KB.

2. 1884 Wording **Map 8** is removed

Thorough Guide Series ... North Devon And North Cornwall Second Edition
London. Dulau & Co. 1884. E.

Thorough Guide ... North Devon And North Cornwall Third Edition, Revised
London. Dulau & Co. 1885, 1885 (1886). RGS, T, KB; E.

3. 1888 The note below the title referring to page numbers and the page numbers on inset maps are erased. Tiverton Area is now included as Map I and maps of the Lizard and Lands End (X & XI) are added. Railway Launceston to Halwill added.

Thorough Guide ... North Devon And North Cornwall Fourth Edition, Revised
London. Dulau & Co. 1888. KB[1], E.

4. 1890 Map is colour printed with areas outlined in red. Lynton map shape changed. Lundy is keyed and arrowed to Map VII inset. Map IX now includes St Mawes. Portreath line and railway west out of St. Austell deleted. Note added (CeOS) on colouring of red lines. **Copyright** added (AeOS) and imprint altered to: **John Bartholomew & Co. Edinr**.

Thorough Guide ... North Devon And North Cornwall Fifth Edition, Revised
London. Dulau & Co. 1890. E, OrE, T.

5. 1892 Map IX is reduced (omitting St Mawes). Map II not outlined red. Place names added, eg Prawle Point, Bolt Head and St. Just. **Copyright** deleted (AeOS) and imprint reverts to **J. Bartholomew. Edinr**.

Thorough Guide ... North Devon And North Cornwall Sixth Edition, Revised
London. Dulau & Co. 1892 (1893), 1892 (1894). KB; E, KB.

6. 1896 New (lower) title: **INDEX MAP shewing the SECTION MAPS AND PLANS IN THE GUIDE**. Inset of the **SCILLY ISLES** (Aa). Some maps are reduced, especially along the Cornish coast. Map numbers and shadings are altered (eg Lundy has a separate map). Page notes return. Railway to Kingsbridge, Holsworthy-Bude. Map of Redruth added note (CeOS) on thickness of lines.

Thorough Guide ... North Devon And North Cornwall Seventh Edition, Revised
London. Dulau & Co. 1896, 1897, 1897 (1898), 1899, 1899 (1900), 1899 (1901).
BL, E; TQ, FB; T; RGS; FB, E; KB.

1. This edition was published early and a paper inserted inside the end papers (dated March 26, 1888) explains the absence of the two new maps of the Lizard and Land's End owing to delays. The text is the fourth edition and includes the ticket prices for 1887. Printed by J S Levin.

157 - SOUTH

Size: 145 x 200 mm. No Scale.

INDEX MAP SHOWING THE MAPS IN THE GUIDE; Imprint: **J.Bartholomew Edinr.** Sub-note referring to area scales (CeOS). Devon and Cornwall showing the areas of the maps within the southern guide by number and page.

1. 1884 *Thorough Guide Series No. VII ... South Devon And South Cornwall*
London. Dulau & Co. 1884. E.

Thorough Guide Series South Devon and South Cornwall Second Edition, Revised
London. Dulau & Co. 1885, 1885 (1887). RGS; E.

2. 1889 Imprint: **John Bartholomew & Co. Edinr.** Exeter now Map 1 and Chulmleigh Map 1A and Maps IX and X inside Map 8 (sic) added.

Thorough Guide Series South Devon and South Cornwall Third Edition, Revised
London. Dulau & Co. 1889, 1889 (1890). RGS, E, Pl; KB.

3. 1891 Imprint: **J. Bartholomew Edinr**. All maps are re-numbered without Latin numerals. Many coastal names are added and most towns etc have page numbers added. A note is added below the title **Scilly Isles on Map 9** and **Seaton and Landslip p.18.** (Ab).

Thorough Guide Series South Devon and South Cornwall Fourth Edition, Revised
London. Dulau & Co. 1892 (1891), 1892, 1892 (1893). KB[1]; E, RGS, Pl; FB, KB.

4. 1895 Title now (Ac) with inset map for southern Dorset (Aa). **Copyright** added (AeOS). The maps, especially in Cornwall, are all altered to Latin numerals. Place page numbers are erased. Railways to Kingsbridge, L&SWR Plymouth-Tavistock and Launceston to Wadebridge added.

Thorough Guide Series South Devon and South Cornwall Fifth Edition, Revised
London. Dulau & Co. 1895, 1895 (1897). RGS, E; KB.

1. Although the title page is dated 1892 several railway adverts are dated 1891 indicating either an earlier printing to cope with demand or leftover adverts being used.

158

ANON – J.P.

1883

Following the success of the puzzle pictures in *The Children's Friend* (**147**) the Wesleyan Conference introduced a further series in 1880. These were called *Biographical Puzzles* and of the fourteen produced four have been found with maps: Lincoln, which was part of a puzzle on John Wesley, Yorkshire (John Wycliff), Leicestershire and Derbyshire. No copy of Devon is known.

Then in January of 1883 they issued a new series of county puzzles in their magazine called *Early Days.* This publication was first produced in 1824 as *The Child's Magazine and Scholar's Companion* and was re-titled in 1846. It became the *Kiddie's Magazine* in 1917 until it closed in 1958.

In the new series No. I was Bedford and the set continued monthly being completed with Yorkshire No. XL in 1886. Devon would have appeared in September, 1883. *Early Days* was issued by the Wesleyan Conference Office, 2 Castle Street, City Road and sold at 66 Paternoster Road, London.

The very small map shows only three towns, Barnstaple, Exeter and Plymouth with their respective rivers, the Taw, Exe and Tamar and the page (144) is headed *PUZZLE-PICTURES No. IX – DEVONSHIRE.* The pictogram puzzle describes the county. The reverse has the answers to the previous county Derbyshire.

Like the previous puzzle series produced by Seeley & Co. in 1865 the pages, maps and illustrations were formed on one wood block but unlike the previous set some were signed, by the otherwise unknown, J.P., including in this series Devon.

Size: 25 x 33 mm. Scale is irrelevant.

The map is (Bb) on a page 186 x 133 mm entitled **PUZZLE-PICTURES. No.IX - DEVONSHIRE HIEROGLYPHICAL** (Ca). With the page number **144** (Aa). There is no title to the map. Signature **J.P.** (Ae). The map and illustrations form part of the text, or puzzle; *Devon in the S.W. [pen] [inn] SULA of ENGLAND........*

1. 1883 *Early Days*
London. Early Days Office. B.

158.1 **Anon – J.P.** *Early Days*

159

BRYER/KELLY

1883

In 1883 Kelly & Co. introduced a new map for their *Directory of Devonshire and Cornwall* to replace the earlier Becker map of 1856 (**132**). This map was again printed by the Cheffins company at their Steam Printing Works but the engraver employed was F Bryer. By this time Kelly's address was Gt Queen Street, London, where they had moved to *c.*1870. Although many copies of the directories exist, unfortunately most have lost their maps. The Bryer map had a long life. It was still being used in the *Directory of Devonshire 1930,* but now with advertising [Buchanan Antiques (Bampfylde House, Exeter) above and below]. After 1939 the county directory was replaced by directories of town and local neighbourhoods.

Size: 482 x 510 mm. **Scale of Miles** (8=50 mm).

KELLY'S MAP OF DEVON. Imprints: **Cheffins Steam Printing Works; 6 Castle St. London, E.C.** (AeOS), **London: Kelly's Directory Office 51 Gt. Queen Street** (CeOS) and **Engraved by F. Bryer, 19 Craven St. Strand, London** (EeOS). Plain border and no detail outside the county boundary.

1. 1883 *Kelly's Directory of Devonshire and Cornwall with maps Engraved Expressly for the Work*
London. Kelly and Co. 1883. E.

2. 1889 Title is now moved (Ae) and is on 3 lines (not two). Only Kelly's imprint remains. Town plan insets are added: **PLAN OF EXETER** (107 x 126 mm) (Ea); and **PLAN OF PLYMOUTH** (125 x 150 mm) (Ee). Both with their own scale bars: 2 inches to 1 mile and 2 ½ inches to 1 mile. Railways: Exeter to Bampton, Holsworthy-Bude, Launceston-Looe-Liskeard-Padstow, Lee Moor, Tavistock West with junction to Calstock, Plymouth branch and Pomphlet (Yealmpton line).

Kelly's Directory of Devonshire and Cornwall
London. Kelly and Co. 1889. E, NDL.

3. 1893 Imprint changed: **London: Kelly's Directories Limited 182 to 184 High Holborn WC** (CeOS). Railway to Kingsbridge and projected Yealmpton-Modbury and to Turnchapel. The Plymouth inset is updated with railway link and housing.

Kelly's Directory of Devonshire and Cornwall
London. Kelly Directories. 1893. Truro.

4. 1897 Railways to Turnchapel, Yealmpton and Modbury, Lynton, Westward Ho! and Northam, Budleigh Salterton. Some Cornwall lines are erased and Bude and Calstock links now shown pecked. Projected lines: Torrington to Okehampton, Ashton to Chagford and to Exeter. The word 'Barracks' is erased from Princetown.

Kelly's Directory of Devonshire and Cornwall
London. Kelly Directories. 1897. E.

Kelly's Directory of Devonshire[1]
London. Kelly Directories. 1897. FB

5. 1900 Railway Ashton to Exeter. Both inset plans have further street changes. The Parliamentary Districts are listed above the Plymouth inset with a note that 'the figures on the map denote the elevation above the sea level'.

Kelly's Directory of Devonshire and Cornwall
London. Kelly Directories. (1900). Pl.

1. Although the directories were usually titled Devonshire and Cornwall the counties were paginated separately and sometimes so bound.

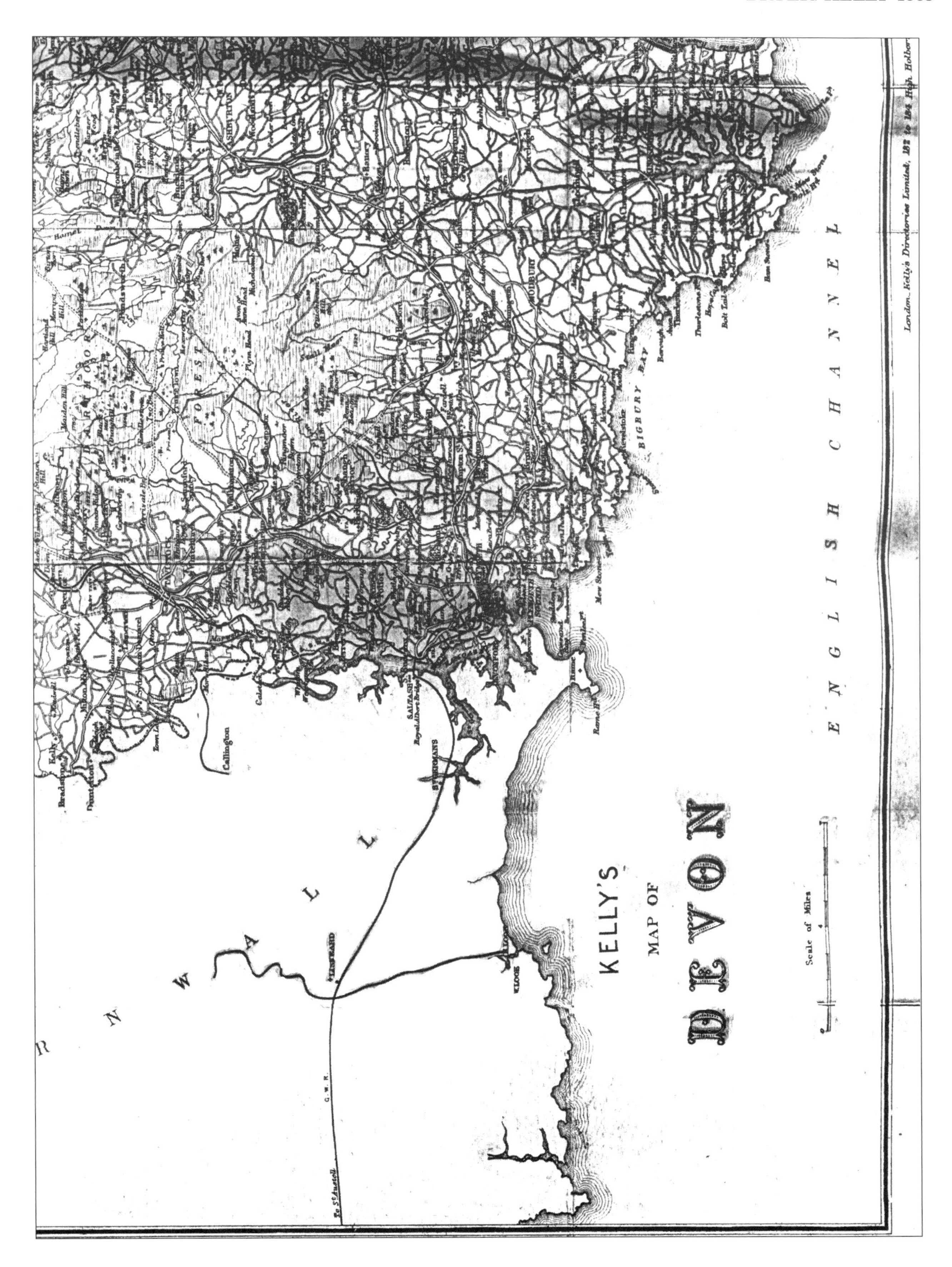

159.1 Bryer/Kelly *Kelly's Directory*

160

WESTERN MORNING NEWS

1885

As a service to its readers the *Western Morning News* newspaper published a map in 1885 detailing the new electoral districts as affected by the Redistribution of Seats Act of the same year. This map was coloured to show changes and had a Reference Key (Ee): Parliamentary Divisions are Red; Petty Sessional Areas are Black; and Boroughs are shaded.

There were (at least) two states of the map. The railway system as shown on both states is a little confusing. Although the line from Newton Abbot is shown to Paignton, which opened for traffic on 1st August 1859, neither the extension to Kingswear nor to Brixham, which were completed in 1864 and 1868 respectively, are included. In addition, the wrong route is shown to Ilfracombe and the Exeter to Bampton line via Tiverton is not shown although this would certainly have been near completion at the time of printing. The copy in the Westcountry Studies Library also has the line to Kingsbridge, although projected since the 1870s, which would not be completed until the 1890s.

Size: 540 x 737 mm. No Scale.

DEVON & CORNWALL ELECTORAL DISTRICTS, ACCORDING TO THE REDISTRIBUTION ACT OF 1885. Printed expressly for AND ISSUED EXCLUSIVELY BY THE "WESTERN MORNING NEWS" C°.L^D. PRICE THREEPENCE. Plain border broken for North Foreland, Axminster and Lands End. Inset of the **SCILLY ISLANDS** (Ac). The railway is shown to Kingsbridge.

1. 1885 Loose sheet.

Devon & Cornwall Electoral Districts According To The Redistribution Act Of 1885
Plymouth. Western Morning News. 1885. (E).

2. 1885 Railways are shown to Holsworthy and from Launceston to Halwill, but the line to Kingsbridge is omitted.

Devon & Cornwall Electoral Districts...
Plymouth. Western Morning News. 1885. (DEI).

160.1 Western Morning News – Devon and Cornwall Electoral Districts

161

EDWARD STANFORD

1885

In 1885 *Stanford's Parliamentary County Atlas* appeared with maps that were originally transfers from *Stanford's Library Map of England and Wales.* Although most county maps for *Stanford's Parliamentary County Atlas* are on one page, Devon is on two sheets. The map is overprinted in yellow, mauve and green to show the Parliamentary divisions and the Parliamentary boroughs. The names of surrounding counties have been twisted to fit the space available. The maps are detailed with coloured divisions, roads and railways, heights but no hills and are followed by 8 pages of statistics, boundaries etc.

A single transfer of the whole county was used in an ornithological work by D'Urban and Rev. Murray Mathew, vicar of Buckland Dinham in Somerset. An *excellent and enthusiastic ornothologist* Murray contributed many notes to the pages of the *Zoologist.* William Stewart Mitchell D'Urban FLS was an Exeter resident, living at 22 Lower Terrace in Mount Radford. He was curator of the Albert Memorial Museum, Exeter. The work has two other maps: one shows bird migration routes of the British Isles and is signed *Mintern Bros. Lith.*; and the second shows the migration routes of Europe. Both maps have D'Urban's signature in the title implying that he drew them both. J G Keulemans drew the 4 chromolithographic plates of birds. The printers were Taylor & Francis of Red Lion Court in Fleet Street. The book appeared only one year after W. Pidsley's volume on Devon birds (**168**)

Size: 177 x 238 mm. **SCALE OF MILES** (10=33 mm in border).

DEVON (SOUTH) (AcOS) and plate numbers: **V.19a**. (EaOS) and above this **33.** Imprint: **London: Edward Stanford, 55 Charing Cross** (CeOS). Signature: **Stanford's Geographical Estabt** (EeOS).
and
DEVON (NORTH) (AcOS) and plate numbers: **V.19b** (EaOS) and above this **34**. Imprint and signature as above. The maps overlap considerably. The South sheet includes Tiverton and the North sheet has Topsham. Railways are shown up to date (Tiverton-Bampton 1884 but not Tiverton-Exeter 1885).

1. 1885 *Stanford's Parliamentary County Atlas And Handbook Of England And Wales*
London. Edward Stanford. 1885. Hull, BL, RGS, W, NLS, B.

2. 1892 New title: **COUNTY OF DEVON. to illustrate BIRDS OF DEVON BY MESSRS D'URBAN & MATHEW (Ee). SCALE OF ENGLISH MILES** 20=64.5 mm (Ee). Imprint erased and signature amended: **Stanford's Geogl Estabt.** One transfer of the *Library Map* combining both of the above onto one sheet: size 253 x 276 mm. This covers a slightly larger area than that above: towns and villages added east and west, eg Pinney Bay and Humble Rock or Fowey and Tintagel. Names removed, eg BRISTOL CHANNEL, CORNWALL and SOMERSET.

The Birds Of Devon By W S M D'Urban and the Rev. Murray Mathew
London. R H Porter. 1892. NLS, TQ, KB.

The Birds Of Devon By W S M D'Urban and the Rev. M A Mathew Second Edition
London. R H Porter. 1895. FB.

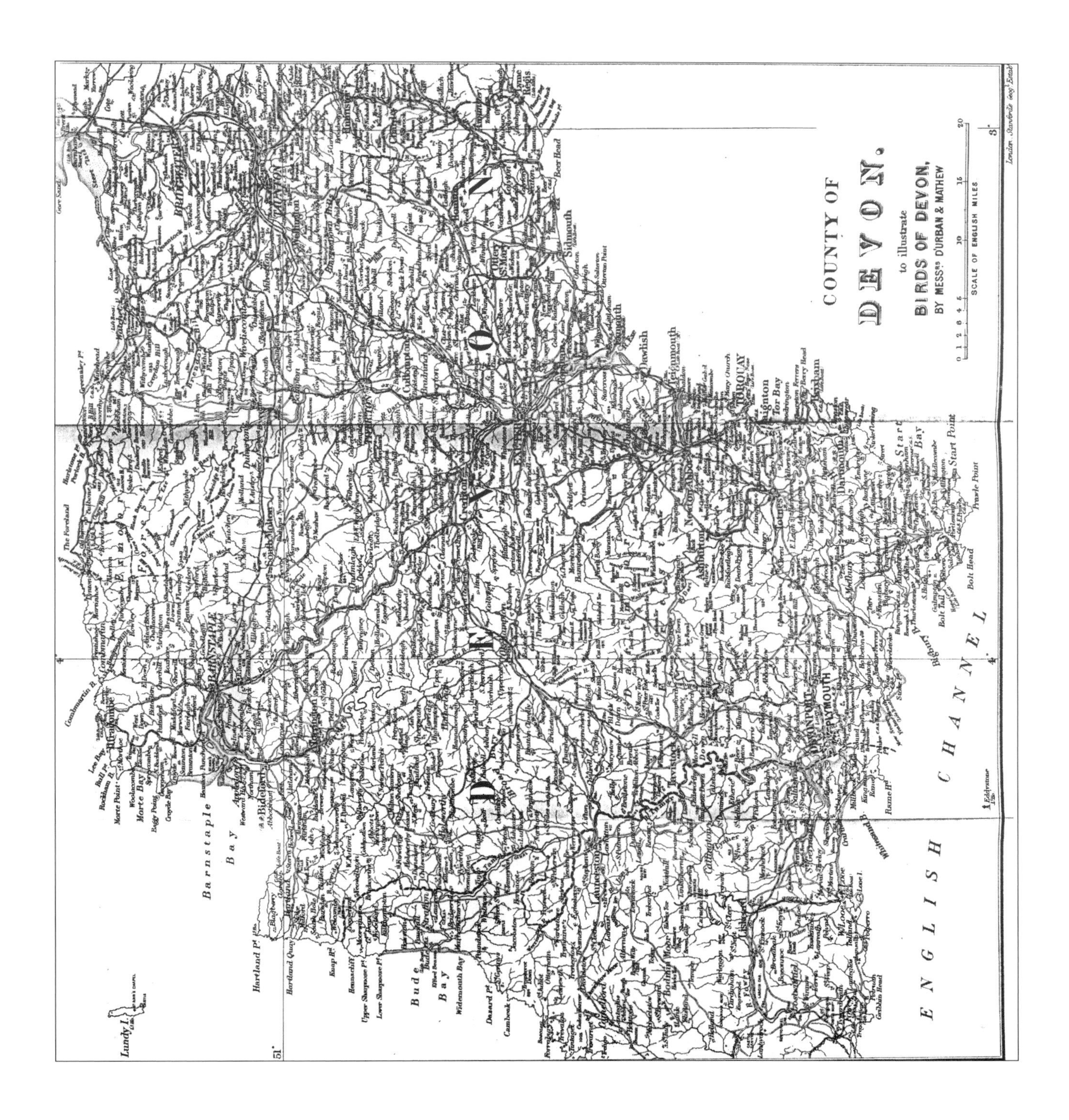

161.2 Stanford *The Birds Of Devon* by D'Urban and Mathew

162

OWEN JONES/ORDNANCE SURVEY

1885

Throughout the 1800s the Ordnance Survey surveyed many western industrial areas and a large number of towns at various scales. However, it was not until 1875 that "the Battle of the Scales" ended when Cornwall was mapped at the new standard scale of 1:250,000. Probably because of its industrial importance and tin-mining it was given priority. The rest of the West Country was surveyed between 1882 and 1889, with Devon having the dubious pleasure of being the last of the counties to be re-surveyed at 1:250,000 (but at 6 inch to the mile for uncultivated areas). The survey was completed in 1888-9.

Nevertheless, during the last twenty years of the nineteenth century various maps of the county were produced at the offices of the Ordnance Survey. The following transfer was probably taken from plates of a larger map of England and Wales: the 15-sheet quarter-inch map begun in 1859 but only published in that form in 1891. The map, *Devonshire: New Divisions Of County,* shows the Parliamentary and Petty Sessional divisions and boroughs with red, blue and purple overprinting and Owen Jones' signature in a lozenge shaped panel. The areas outside the county have been removed. Most main roads are shown but not all railways; and although almost exactly the same as two later maps from the Ordnance Survey (**164** and **181**) it is not from the same plates.

Size: 470 x 470 mm in T-shaped frame. **Scale - Four Miles to an Inch** (4 + 12 = 100 mm).

DEVONSHIRE, NEW DIVISIONS OF THE COUNTY. Signature: **R Owen Jones Lt Colonel R E.** (Ed). Imprint: **Zincographed at the Ordnance Survey Office, Southampton. 1885.** (CeOS). Railways to Plymouth, Exeter to Torrington, through Oakhampton, to Tavistock and Lifton, Exmouth, Tiverton, Ashburton and Moreton Hampstead. The line to Dartmouth has only been completed as far as Goodrington (it was actually completed in 1864).

1. 1885 *Redistribution of Seats Act, 1885. County of Devon. Return to an address ... 7th July 1885*
London. Henry Hansard & Son. 1885. GUL.

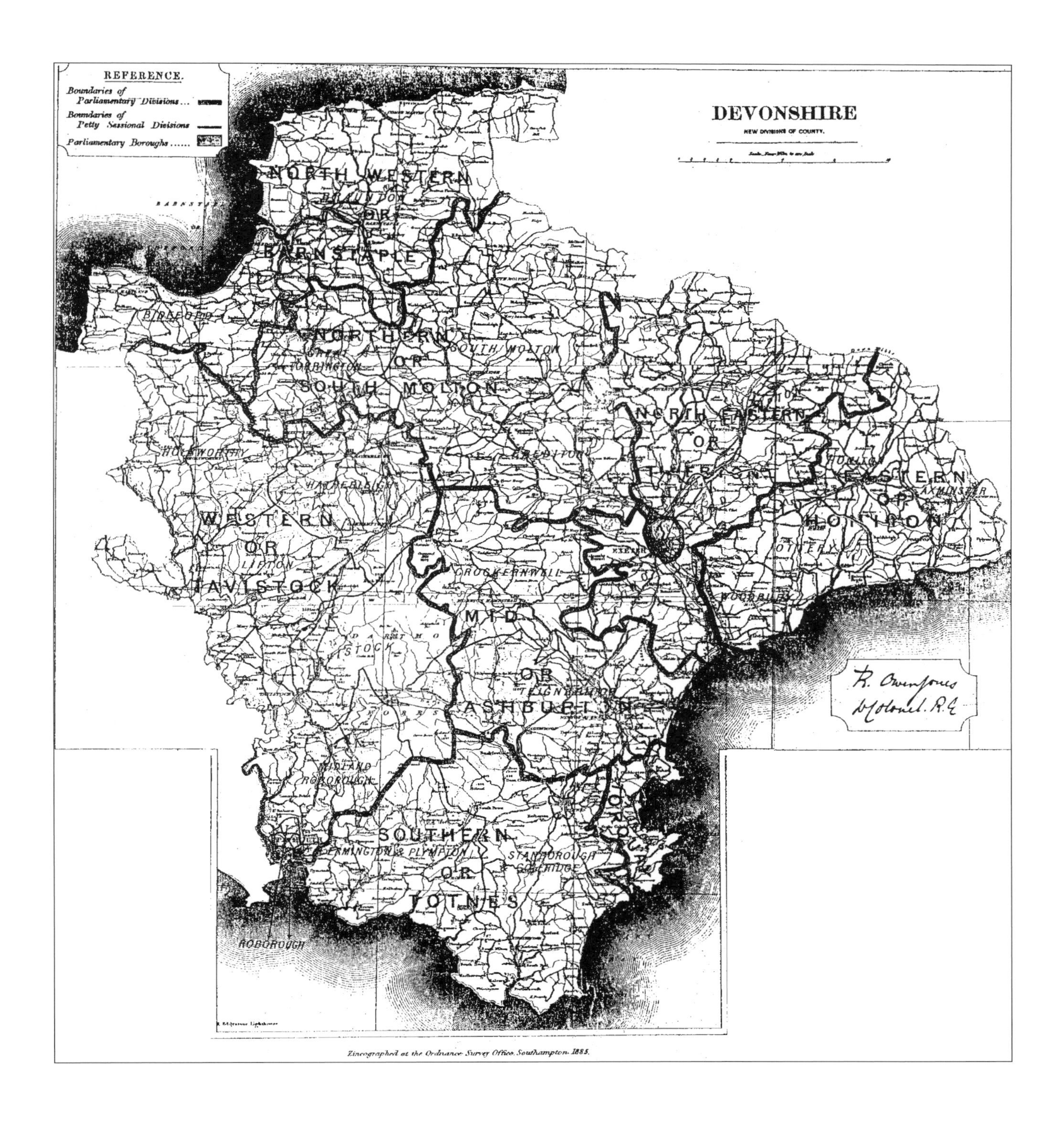

162.1 Jones/OS Produced for the *Redistribution of Seats Act*

163

G W BACON & CO.

1885

Although G W Bacon published a large number of county maps, the majority were transfers taken from plates that he had bought up from auctions rather than maps that he had specially engraved. However, in c.1885 a small pocket tourist guide to England and Wales appeared which included sectional maps of the country. Although the maps were not intended to be county maps, Devon is, in fact on one sheet with a small part of Cornwall and slightly larger areas of Somerset and Dorset.

The *Hand Book Of England And Wales* (cover title) was a complete set of county descriptions averaging 10 pages to a county, with each county section being paginated separately. The Preface was written by G W Bacon, who probably composed the whole guide. The guide was printed by C F Hodgson and Son, Gough Square, Fleet Street. At the back of the handbook there was a complete map of England and Wales in sections. Devon was included on sheet 11 (of 13) and covered the area from Polperro to Langton Herring and Bude to Shepton Mallet.

The text has two references to 1884 and the railway network, based on sheet 11 (state 1) is consistent with a date between 1883, when the railways reached Ashton (1882) and the Princetown line was taken over by the GWR and reopened, and 1885 when the Exeter route to Tiverton was commenced. Although the second known edition, state 2 below, has end papers dated 1876 by hand, this is doubtful as the line from Launceton to Halwill was not completed until 1885 and the extension from Exeter to Tiverton and on to Bampton was completed in two stages in 1885 and 1886. The Ashton extension is erroneous: the line was connected to Exeter in the early 20th century.

Size: 145 mm x 170 mm. No scale.

Sh.11 (EaOS) **Adjoining Sheet 7, 10, 12** in margins. No imprint or signature. Railways to Holsworthy (with projected extension southwest into Cornwall), Ashton (with projection to just west of Crediton) and Princetown.

1. 1885 *Tourists' Guide And Book Of England And Wales*
London. G W Bacon & Co. (1885). EB.

2. 1886 Railways added: Launceston to Halwill and Exeter to Tiverton and Bampton. Additions, eg Brean in Barnstaple Bay and Greenaley Point near Minehead; Prawle Point and Sherborne across border; LYME BAY named.

Tourists' Guide And Hand Book To England And Wales
London. G W Bacon & Co. (1886). KB.

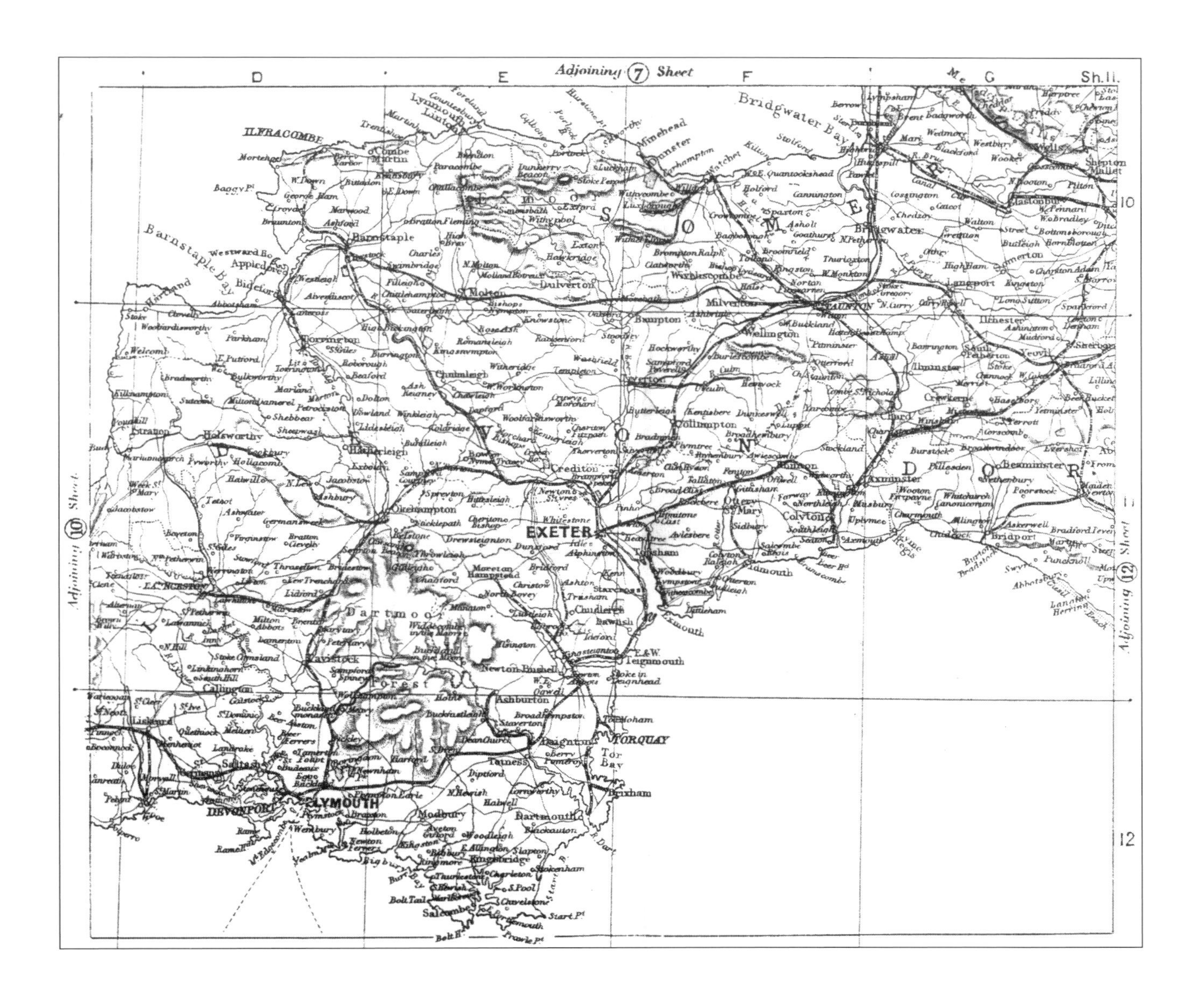

163.1 Bacon & Co. *Handbook Of England And Wales*

164

ORDNANCE SURVEY

1888

Although the Ordnance Survey had published a map in 1885 (**162**) to show the New Divisions of the County they did not use this when they published a Diagram of the Sanitary Districts in 1888. The new map had the same coastline, as might be expected, but the interior detail present in the earlier map was not shown. Instead the detail was limited to showing the extent of the Sanitary Districts together with symbols for the parish towns. Although the map has no roads surprisingly it shows more railways. It does also have larger scale plans of Exeter and Plymouth and more references.

The map is superimposed by the sheetlines and reference numbers of the 1-inch and 6-inch Ordnance Survey maps and the sheetlines of the 25-inch maps. This map was sold individually as loose sheets and also as part of a complete set of the *Ordnance Survey of Devonshire on the Scale of Inch to a Mile.* Sizes refer to printed area, the maps were without border.

Size: 530 x 630 mm. **Scale of this Diagram 4 Miles to 1 Inch**.

DIAGRAM of the ORDNANCE SURVEY of DEVONSHIRE Shewing CIVIL PARISHES.. Map has inset plans of both Exeter (Ee) and Plymouth (Ce). There is a table of References. Railways to Plymouth, Exeter to Torrington, through Okehampton, both lines to Tavistock and Lifton, L&SWR through Holsworthy, Exmouth, Seaton, Sidmouth, Hemyock, Taunton to Barnstaple, Ilfracombe, Ashton, Exeter to Morebath (Bampton), Tiverton, Ashburton and Moreton Hampstead. The line to Dartmouth has been completed with branch to Brixham. There is an erroneous line from Teign Grace directly north to Crediton. Price 2d.

1. 1888 *Diagram Of The Ordnance Survey Of Devonshire Shewing Civil Parishes*
Southampton. Ordnance Survey Office. 1888. BL.

2. 1888 **DIAGRAM of the Sanitary Districts in DEVONSHIRE Shewing also CIVIL PARISHES. REFERENCE TO COLOURS** with coloured U S (Urban Sanitary) Districts. Lundy Island is shown as part of Bideford Union. Teign Grace to Crediton line deleted. Price 6d.

Diagram Of The Sanitary Districts In Devonshire
Southampton. Ordnance Survey Office. 1888. BL.

3. 1891 **INDEX to the ORDNANCE SURVEY of DEVONSHIRE Shewing CIVIL PARISHES.** Imprint: **Photozincographed and Published at the Ordnance Survey Office, Southampton, 1891.** (CeOS). **Price 2d. All rights of reproduction reserved.** Scale is **Index** not Diagram. The map has the sheet lines of both the *One Inch Map (New Series)* and the *6 Inches to a Mile* as shown in the **References** table. Inset of Plymouth now (Ce) and References (Ee) with inset of Exeter above (ie raised).

..

Index to the Ordnance Survey of Devonshire
Southampton. Ordnance Survey Office . 1891. KB.

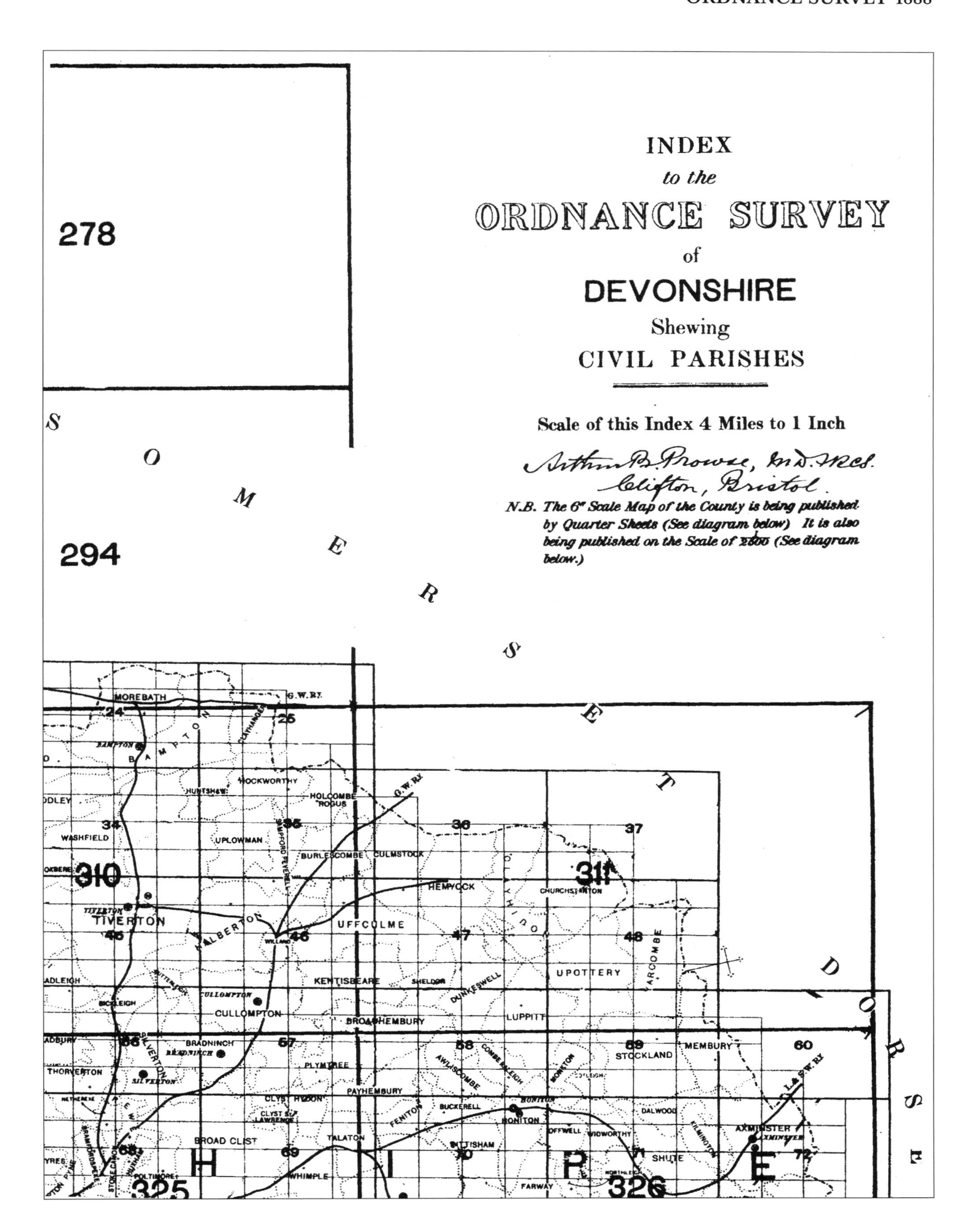

164.3 Index– Ordnance Survey

165

CASSELL & CO./L&SWR

1888

Amongst many railway publications and timetables published by the various operators were some which included county maps. In 1888 (and 1889) Cassells published *The Official Guide to the London & South Western Railway.*

This guide included 2 panorama views showing the L&SWR lines London to Ilfracombe and Brighton to Padstow. One of the maps included depicted the major part of Devonshire, *Route Map. - IV Exeter To Plymouth And Ilfracombe*, and showed the L&SWR routes from Exeter to Ilfracombe and Devonport with branches in heavy lines (and the GWR route to Plymouth and its branches in thin lines). The Dartmoor Loop is shown as complete but the L&SWR line to Tavistock from Plymouth would not be completed until 1890. Lines connect Barnstaple and Ilfracombe with Lynmouth and Lynton, presumably this was to signify the coach connections.

The map shows only the parts of Devon west of Exeter. It omits Hartland and the areas east of the Exe but it does show the extent of the company's involvement together with the connections to the coastal resorts both north and south.

There are vignettes of Devonshire countryside, eg Exeter Cathedral, Clovelly, Lynton, Hound Tor, Bickleigh and Plymouth. The text on the reverse refers to Holsworthy and is numbered p. 132.

Size: 123 x 80 mm. No scale.

ROUTE MAP. - IV EXETER TO PLYMOUTH AND ILFRACOMBE. (CeOS). Railways are shown from Exeter to Exmouth, to Torrington, Ilfracombe, Okehampton, Devonport, Launceston, Holsworthy and Bude.

1. 1888 *The Official Guide to the London & South Western Railway*
London. Cassell and Co. Ltd. 1888, 1889. RGS; RGS.

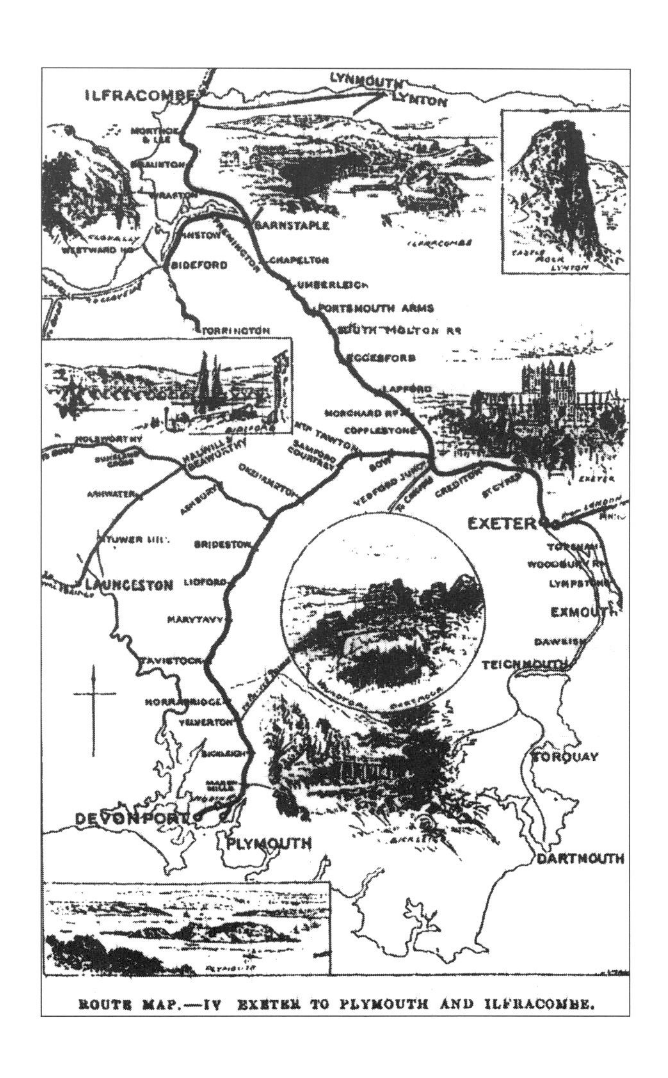

ROUTE MAP.—IV EXETER TO PLYMOUTH AND ILFRACOMBE.

165.1 Cassell & Co. *The Official Guide to the London & South Western Railway*

166

SWISS & CO. (Cary)

1890

A number of maps of Devon can be found in works on hunting. The Walkers probably produced the first series, issued in *Hobson's Fox-Hunting Atlas* and later as *Walker's Fox-Hunting Atlas* (see **116**). Edward Weller (see **136**) also engraved at least one Devon hunting map: a *Map of the Stag-Hunting Country in the Counties of Devon & Somerset* showing W. Somerset and N. Devon emphasising the hilly border regions.

Swiss and Co., a firm operating from Devonport, seem to have specialised in hunting maps.[1] Alfred H Swiss was registered as printer at 112 Fore Street in *White's Directory* of 1878. Swiss & Co. - printers to H M Stationery Office, printers, stationers (retail), booksellers, insurance agents, map publishers and relief stampers - were still registered at that address in 1930.

Gall & Inglis were successful publishers who had bought part of the stock of George Frederick Cruchley when it came up for auction in 1877 (see appendix II). Some of the plates sold included those of John Cary that Cruchley had acquired over thirty years previously. Among them were the plates to *Cary's New Map of England and Wales with part of Scotland* of 1794; a complete revision appearing in 1832 as *Cary's Improved Map of England and Wales with A Considerable Portion Of Scotland.* This was published on 65 sheets and sold in book form and as a box of folding maps. Devon is covered on sheets 2, 3, 9, 10 and 16. The series was sold first by Cary, later by Cruchley as the *Reduced Ordnance Map of England & Wales* - advertised as *Half the scale and half the price* - and also sometimes sold with a George Philip & Son label from their Educational and Geographical Depot at 32, Fleet Street. There was minimal up-dating; railways were added and new conurbations, eg Torquay, were extensively revised but the basic map and outlying areas were almost the same as they had been in 1794.

The maps described below, the fox-hunting maps of Devon produced *c.*1890, are transfers from these plates with added railways and hunt information.[2] Swiss and Gall & Inglis must have worked closely together as the various imprints show. The maps are usually linen-backed, folded and bound into a small booklet (sometimes with a hunting scene depicted on the cover, signed by Imogen Collier), which includes the names of the various hunts, their sizes and principal officers. The hunt meeting places are coloured, circled and numbered. The map has a two mile vertical numbered grid and main roads and railways are coloured.

Another map by a local publisher that utilised the same plates was L Seeley in his *Map Of The Environs Of Torquay From The Ordnance Survey.* Torquay booksellers, stationers and printers, L Seeley & Son had premises at *Lawrence Place, Torquay, Opposite the Tree*! The map, 490 x 615 mm, covers south Devon but misses the east edge and has railways consistent with a date of *c.*1890. The map has an imprint: Seeley's Library, Torquay. This version must pre-date Swiss as it still shows marine information such as rocks and coastal depths later erased.

Beatrix Cresswell wrote at least two books for the *Homeland Handbook* series. One of these, *Dartmoor and its Surroundings* (St. Bride's Press, London, probably second edition) included two maps - of North and South Dartmoor - which were transfers from the same plates.

1. R Carroll (1996; p. 411) notes a book on a Lincolnshire hunt, *The History of the Belvoir Hunt* by T F Dale, which included a lithograph fox-hunting map by Gall & Inglis *By the Courtesy of Mr A H Swiss, Devonport* (London, 1899).
2. The British Library (Maps 4.a.40) has a copy of a post 1900 edition of this map. It covers the area from Lannsalloes and St Garis to Budleigh Salterton and Watchet in the east with an inset map of the Continuation East Of Watchet. The whole of the counties north to south are shown. Later railways such as the route to Appledore, Exmouth to Budleigh and Christow to Exeter are depicted. Although an early single-sheet map has not been seen, from the various advertisements in the separate booklets it would seem that this map existed in at least two states *HUNTING MAP OF DEVONSHIRE. Size, 40 ins, x 36 ins, comprising both the North Devon and South Devon maps enumerated above.* The same advertisement is in a N Devon copy but with changes to the meets.

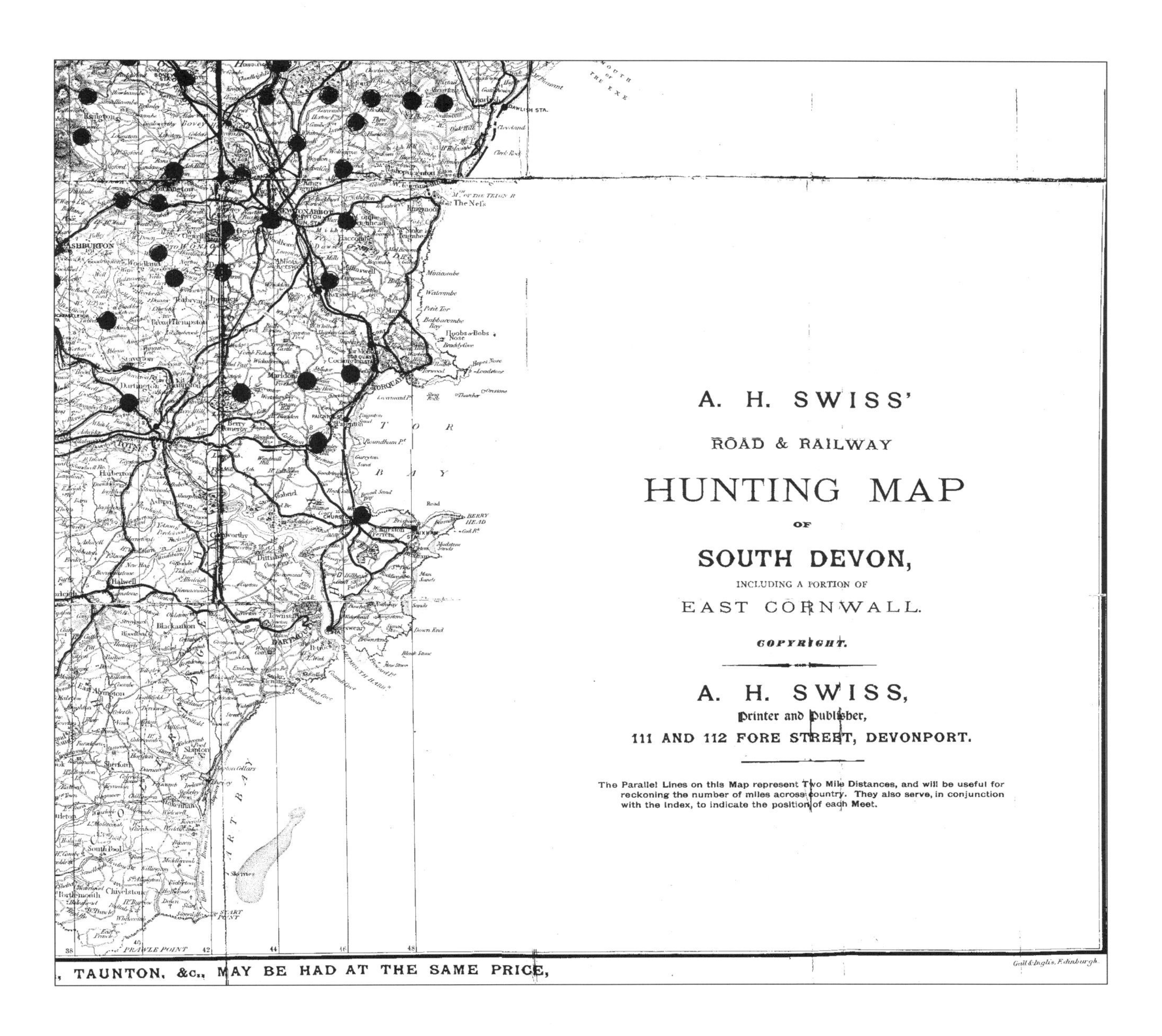

166.2 Swiss & Co. *Road and Railway Hunting Map* – South sheet

Size: 976 x 753 mm. **SCALE HALF AN INCH TO A MILE** (5=65mm).

SWISS & CO.'S NO. 1 HUNTING MAP. Signature: **SWISS & CO. Publishers. Fore Street, DEVONPORT. AND MAY BE HAD OF ALL PRINCIPAL BOOK-SELLERS** (within the title panel Aa). Imprint: **Gall & Inglis, Edinburgh** (CeOS).

1. 1890 *Swiss & Co.'s No. 1 Hunting Map The East Cornwall, Devonshire & West Somerset District* (cover title)
Devonport. Swiss & Co. (1890). Assumed.

2. 1891 Now on two sheets: South Sheet 620 x 872 mm. North Sheet 520 x 872 mm. Each has a **Scale of Miles** (15=50 mm). (Scale 1M=3.3 mm.). Both maps have imprint: **COPYRIGHT. A.H. SWISS, Printer and Publisher, 111 And 112 FORE STREET, DEVONPORT.** Signature: **Gall & Inglis, Edinburgh** (EeOS). Notes outside the border: **THE NUMBER OR MILES BETWEEN ANY TWO PLACES CAN EASILY BE ASCERTAINED BY REFERENCE TO THE ABOVE SCALE OF DISTANCES** (DeOS) and **A CORRESPONDING MAP, UNIFORM WITH THIS, EXTENDING TO ..., &C.. MAY BE HAD AT THE SAME PRICE** (CeOS). Railways to Sidmouth, Hemyock, Ashton and Launceston to Halwill Junc^{n}. The cover has a variant of the title: 'No.2 [or 3] HUNTING MAP showing the Meets of Hounds [with the names and number of the hunts], the price 3/6 and the publisher's name.

A. H. Swiss' Road & Railway Hunting Map Of South Devon including a portion of East Cornwall
Devonport. A H Swiss. (1891[1]). FB.

A. H. Swiss' Road & Railway Hunting Map Of North Devon including a portion of West Somerset
Devonport. A H Swiss. (1891[2]). TB.

3. 1895 Changes to meeting places, some circles are crossed out, repositioned, or omitted. Changes to North map include the addition of *The Exmoor Foxhounds* with red squares.

A. H. Swiss' Road & Railway Hunting Map Of South Devon ... (Pl).
Devonport. A H Swiss & Co. (1895).

A. H. Swiss' Road & Railway Hunting Map Of North Devon ... EB, FB.
Devonport. A H Swiss & Co. (1895[3]).

1. The cover drawing is so dated.
2. In T Burgess's copy of North Devon - an advertisement omits some hunts and adds one, suggesting a later state.
3. An advertisement lists: *THE PLYMOUTH DISTRICT HUNTING MAP.* Size, 26 ins x 25 ins with 1 mile circles from Plymouth, and showing the meets of foxhounds.

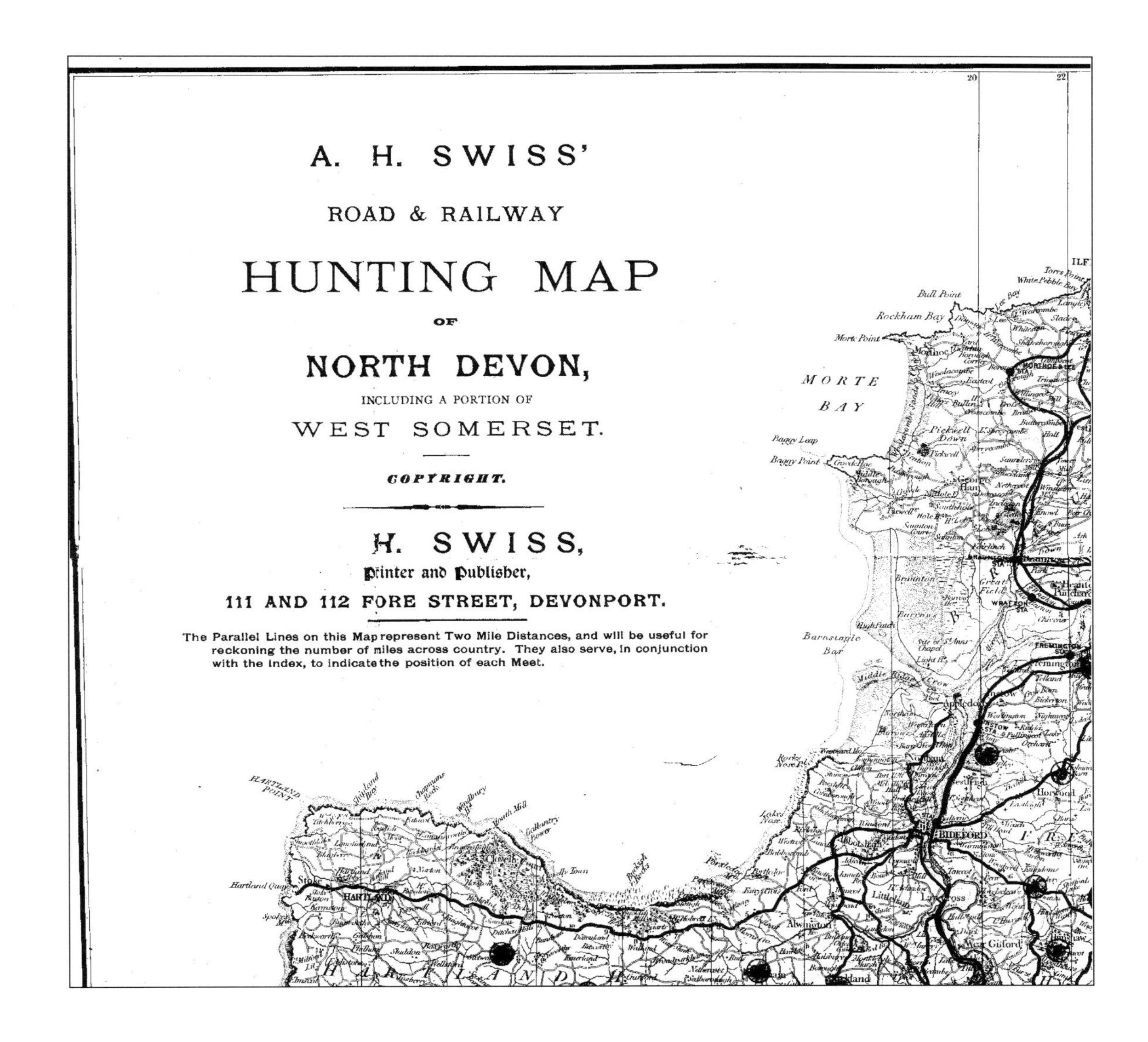

166.3 Swiss & Co. *Road and Railway Hunting Map* – North sheet

167

JOHN JAQUES & SON
1890

The firm of John Jaques and Son created a number of indoor and outdoor games throughout the nineteenth century, among them a wide variety of card games. The firm was founded in 1795 becoming John Jaques and Son in 1861, and later adopted the simpler J Jaques and Son. John Jaques II won a place in sporting history when he introduced Croquet into England at the Great Exhibition of 1851 and today the company continues to lead the market in specialist sporting equipment as Jaques And Co. Ltd.

In about 1890 they produced a charming set of cards based on counties, possibly for children. Each county was represented twice: once as an outline map of the county with towns and cities marked with a small circle; the second, matching card, portrayed a caricature. The object would have been to turn up two matching cards and make a pair; a form of memory. Each character card also had either a rhyme or a limerick. Although Edward Lear did not invent the limerick he did help to make it popular with his books *A Book of Nonsense* (pre-1862) and *More Nonsense* (1872). Devon has a four-line limerick about a lady carrying Devon cream and her cat, but the Somerset rhyme is highly amusing:

A Zomerzet varmer awoke with a pain,
Said he, "is it my stomach, or is it my brain?
Whichever it is, there is summit amiss,
But a gallon o' zider will zettle all this."

The cards were coloured. Devon has an orange coloured compass. The old lady has a red cloak, green dress and ribbon, orange hat and basket, brown staff, and pink face and hands.
Another set of children's cards produced about the same time was the pack published by Raphael Tuck. Containing maps of the regions of Great Britain, Devon is shown with Cornwall, Somerset and Dorset. These were reward cards given to good pupils.

Size of card: 58 x 77 mm. Scale is irrelevant.

Imprint on box: **Skits - A Game of the Shires - Published by JAQUES & SON. 102 HATTON GARDEN, LONDON.**

Card one - **DEVONSHIRE** (Ce) and plain map. Compass (Ea). Number 39 (Aa).

Card two - Map of Devon has been overdrawn with a little old lady carrying a basket and a cat and holding a stick. There is a verse above and below the map.

1. 1890 *Skits - A Game of the Shires*
London. J Jaques & Son. (1890). EB[1].

1. The authors are grateful to Eugene Burden for supplying information on these cards and permission to use the illustration. See also his letter to *The Map Collector*, Issue 56.

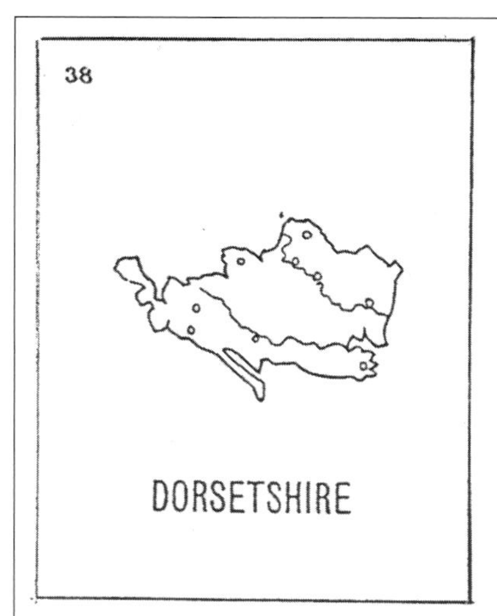

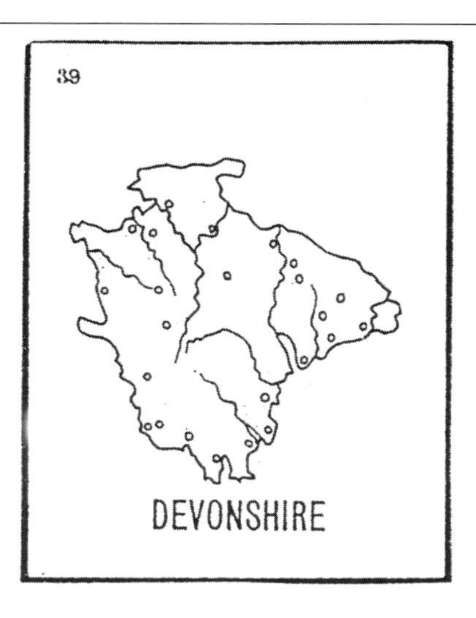

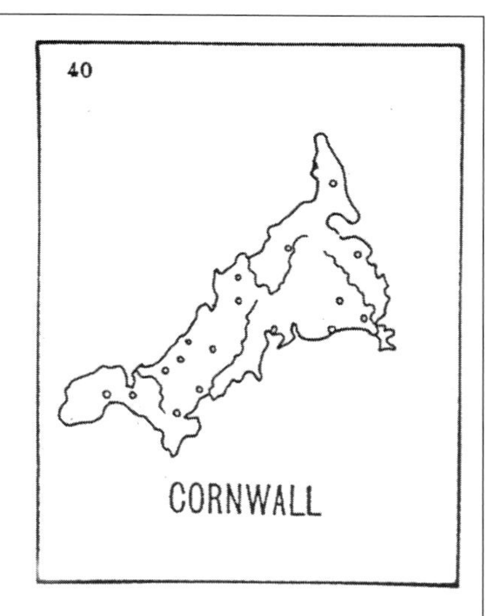

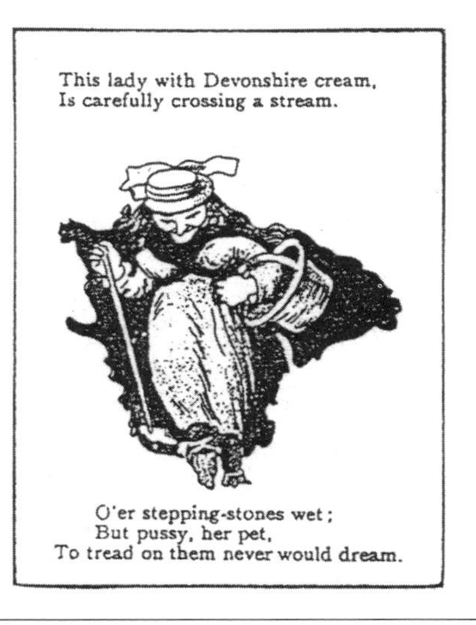

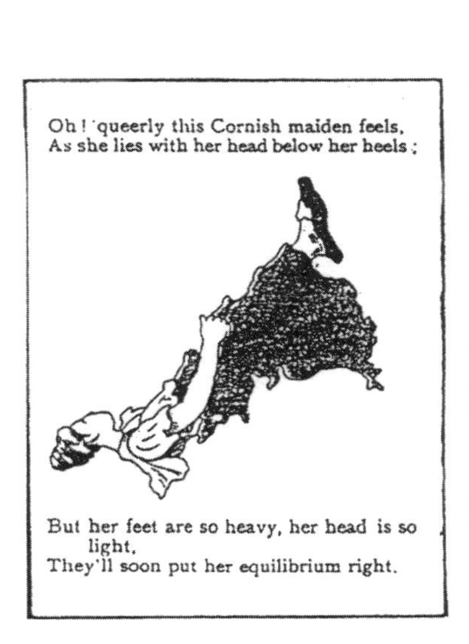

167.1 Jaques & Son *Skits – A Game of the Shires*

168

W J SOUTHWOOD

1891

Two Victorian maps of Devon are to be found in books on birds: one was issued by R H Porter for D'Urban and Mathew (see Stanford **161**); the other appeared in *The Birds of Devonshire* by William Pidsley. This was issued in 1891 and contained an outline map of the county. The map shows principal towns and the rivers. Both Eddystone lighthouse and Lundy have been included and the map area extends from Bude to Lyme Regis and north to Malmshead.

The book was edited by H A Macpherson MA and included an introduction by him with a short memoir on the late John Gatcombe, a local naturalist. The large but very simple map accompanying the work was signed W J Southwood, but nothing is known about the engraver. Two publishers are noted on the title page: W W Gibbings had premises at 18 Bury St. WC. London; and James G Commin was a Devon bookseller at 230, High St., Exeter.[1]

In his preface (dated November, 1890) William Pidsley wrote: The exciting cause of this handbook must be looked for in the omission of other Devonshire Naturalists to provide a book of reference on the Ornithology of our County; an omission that may perhaps be accounted for, by the seriousness of the undertaking. However, Pidsley must have been in contact with the Reverend Mathew (who he thanks in the preface) and D'Urban, who he quotes frequently, and he must have beeen aware of their work (**161**) which was to be published in 1892. The latter book is far superior to Pidsley's: there are more illustrations; the text on the birds includes ornithological notes; while Pidsley's is limited to bird sightings and slayings (also included by D'Urban and Mathew).[2]

Size: 460 x 510 mm. **SCALE** (12 =74 mm) **Miles.**

DEVON. Printed for W^m E H PIDSLEY'S BIRDS OF DEVONSHIRE (Dd) with imprint: **W J SOUTHWOOD, LITHO. EXETER** (Ae).

1. 1891	*The Birds of Devonshire by William E H Pidsley ...* London: W W Gibbings. Exeter: J G Commin. 1891.	NLS, E, KB.

1. James G Commin was still advertising his services in 1930: *new and second-hand bookseller - scarce, out of print books sought for free of charge* - as well as offering binding of any sort and *valuations for fire or probate. Kelly's Directory*, p. 997.
2. D'Urban and Mathew wrote on page xiii of their work: *Although a work on the Birds of Devonshire has already been issued by Mr W E H Pidsley, it did not seem so thoroughly to exhaust the subject as to exclude another which should deal more fully with the numerous points of interest in the County Ornis.*

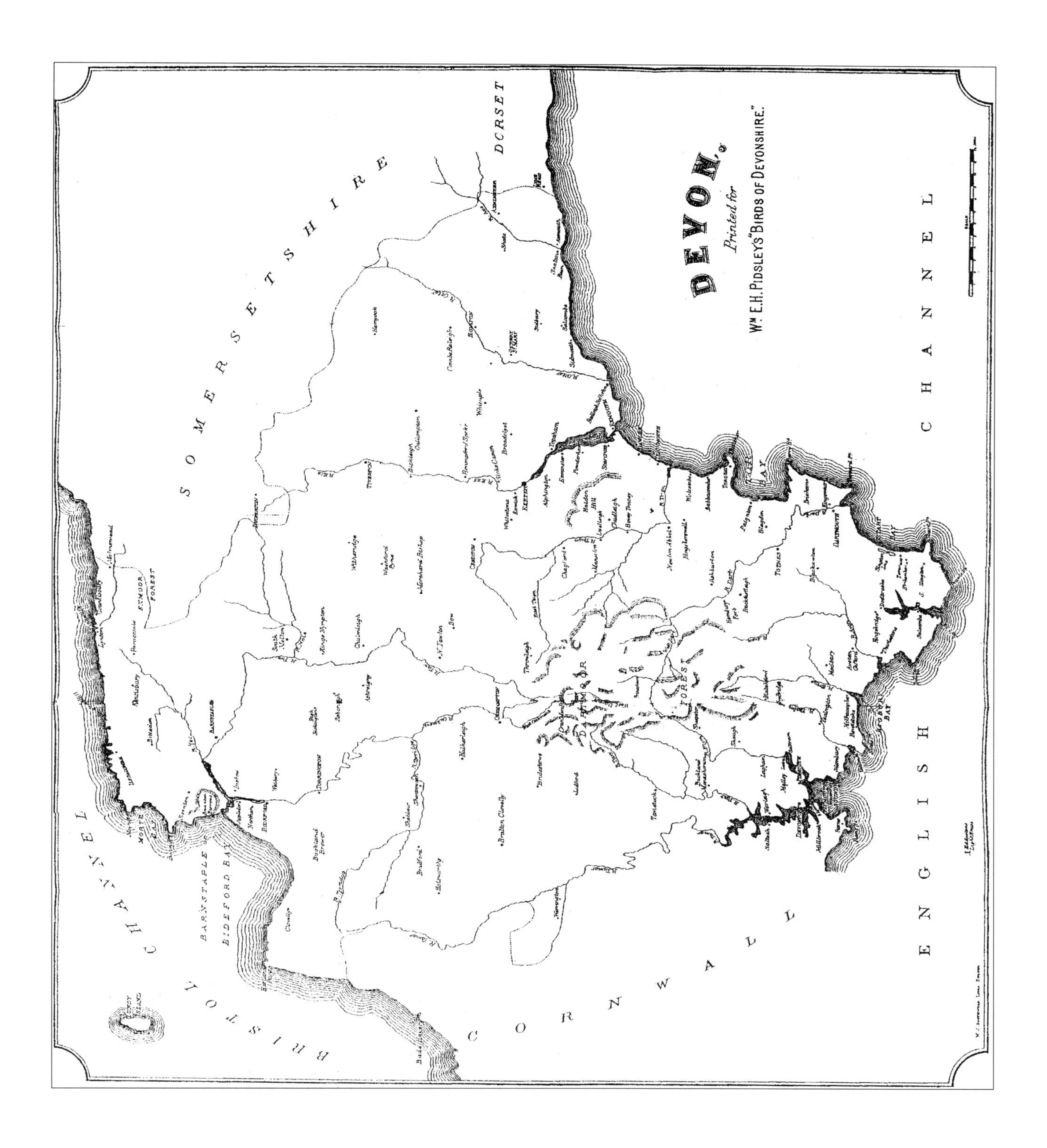

168.1 Southwood *The Birds of Devonshire* by William Pidsley

169

ORDNANCE SURVEY

1892

A large number of maps of Devon, based on, and produced by, the Ordnance Survey were becoming available between 1885 and 1900. These were mostly taken from the updating of Devon being executed about the same time. As seen (**162** and **164**) some of these were used to show the new divisions or the sanitary districts etc but one was simply a single sheet map of the county with the additional listings of characteristics and parishes. This was followed by a second map, revised from the first, as an index to the various sheet numbers of the six inch scale maps available. The maps showed the names and/or reference letters and boundaries of the Parliamentary Divisions, towns and parishes. It was printed with and without a border. This map can be differentiated from the earlier (1891, **164.2**) map, which was superimposed by the sheetlines and reference numbers of the 1-inch and 6-inch Ordnance Survey maps and the sheetlines of the 25-inch maps, by the lack of detail in the earlier map. The later map has a fair amount of road detail and is also larger: Foreland to Bolt Tail measures 600 mm compared to 450 mm. The maps were all sold as single sheets.

Size: 650 x 670 mm. Scale (2+8=85 mm) **Three Miles to One Inch**.

ORDNANCE SURVEY OF DEVONSHIRE (Da). Imprint**: Photozincographed at the Ordnance Survey Office, Southampton, 1892.** (CeOS). Also **Price of this index sixpence**. There are enlarged sketch maps of Plymouth (Ae) and Exeter (Ee). Table of **CHARACTERISTICS** below title and a right hand table has the **AREAS OF PARISHES** making the complete sheet 650 x 1000 mm. Railways to Plymouth, Exeter to Torrington, through Okehampton, both lines to Tavistock and Lifton, L&SWR through Holsworthy, Exmouth, Seaton, Sidmouth, Dartmouth, Brixham, Hemyock, Taunton to Barnstaple, Ilfracombe, Ashton, Exeter to Morebath (Bampton), Tiverton, Ashburton and Moreton Hampstead.

1. 1892 *Ordnance Survey Of Devonshire*
Southampton. Ordnance Survey. 1892. BL.

2. 1892 Border added surrounding map and table of parishes. New prices, coloured and uncoloured.

Ordnance Survey Of Devonshire
Southampton. Ordnance Survey. 1892. BL.

3. 1892 New title: **INDEX to the SIX INCH SCALE of the ORDNANCE SURVEY of DEVONSHIRE**. Price: **Price Two Shillings and Sixpence uncoloured** and **Three Shillings coloured** (both CeOS). The individual sheet lines superimposed on the map.

Index to the Six Inch Scale
Southampton. Ordnance Survey. 1892. BL.

4. 1900 New Title: **ENGLAND & WALES. DIAGRAM OF DEVONSHIRE Shewing Unions, Sanitary Districts, Boroughs and Civil Parishes; AND THE SHEET LINES OF THE ORDNANCE SURVEY MAPS ON THE SCALE OF 25.344 INCHES TO 1 MILE (1/2500) Surveyed in 1855-88.** Parish areas removed, size: 650 x 740 mm. New date: **Photozincographed ..., 1900.** Also **Price 3s** (both Be). New note: **N.B. The Boundaries on this Diagram are revised up to date (26-10-99)** (De). Table of **CHARACTERISTICS** now left of title.

Diagram Of Devonshire
Southampton. Ordnance Survey. 1900. RGS.

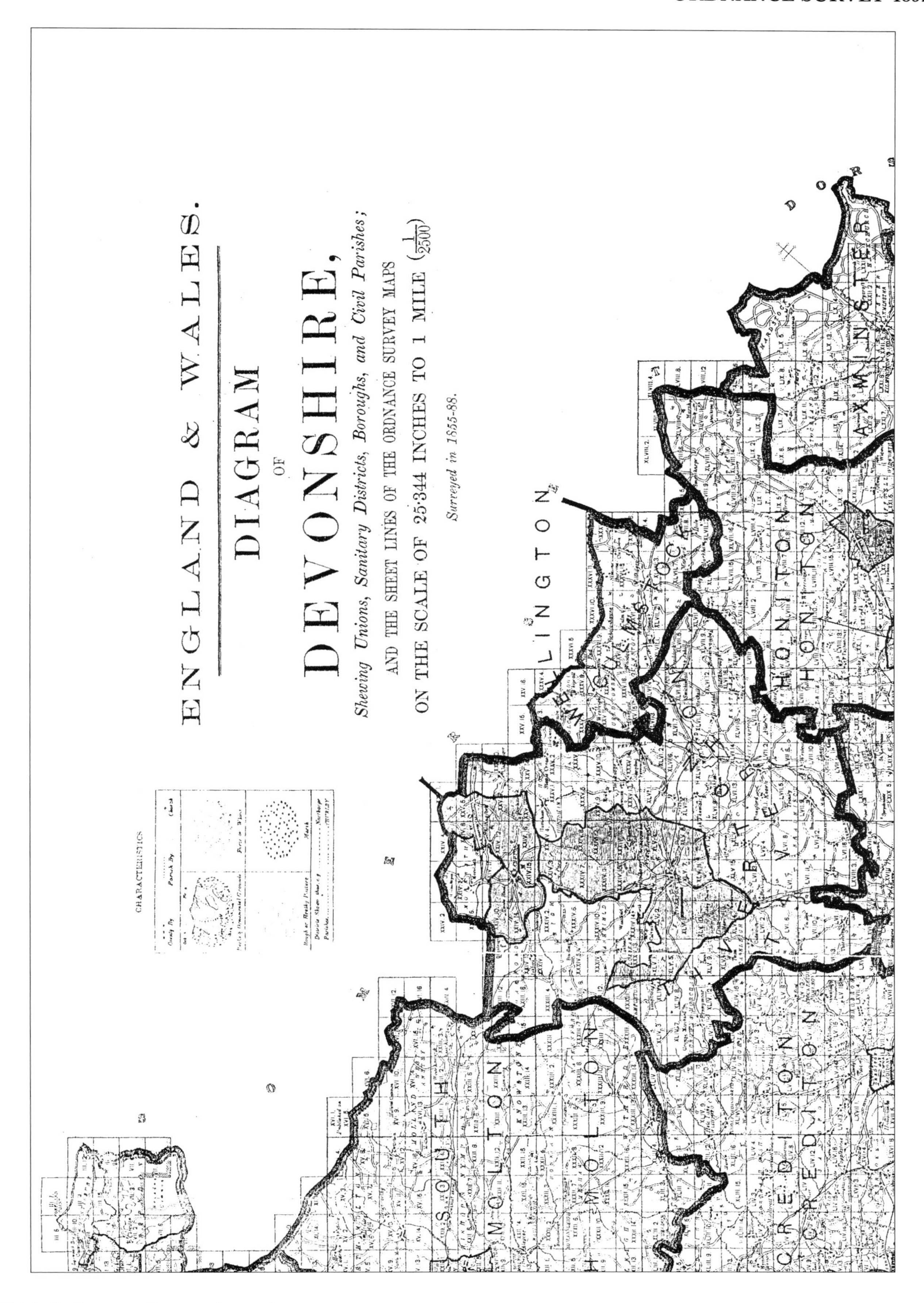

169.4 Ordnance Survey loose sheet

170

JOHN LLOYD WARDEN PAGE

1893

The Victorian age brought out the urge for people to wander and the freedom to do so. This was accompanied by a stream of guidebooks, some like the much earlier William Camden eager to list and describe the antiquities and histories, some in the style of a modern guide book suggesting walks and places to visit. Among these was a charming work *The Rivers Of Devon.* Taking the river as the starting point the author, John Lloyd Warden Page, describes the countryside to be found along the banks of most of the rivers of Devon. Possibly as a companion volume he wrote *The Coasts Of Devon* which appeared in 1895 (**172**).

Warden Page was a local historian and geographer who had already produced books such as *An Exploration of Dartmoor and its Antiquities, An Exploration of Exmoor and the Hill Country of West Somerset* (London. Seeley & Co. 1890, 1895) as well as *Okehampton: its Castle.* To complete his river book a map of the rivers of Devon was placed opposite the last page (p. 348). The map, a photo-lithograph, was printed by James Akerman of London and drawn by John Lloyd Warden Page himself. The book was published by Seeley & Co. of Essex Street in the Strand and printed by Billing & Sons of Guildford.

The map, as the title suggests, shows all the main rivers and most of the important tributaries. It also shows major roads and the railways with both the Great Western and London and South Western lines labelled. It delineates Dartmoor giving the heights of most of the better known tors (and some other hills such as Dunkery Beacon on the Somerset border). Interestingly not all canals are shown; while the Bude Canal, used until 1891, can be identified, the canals from Morwellham to Tavistock and Bideford to Great Torrington are missing, possibly because they were no longer important.

Size: 245 x 225 mm. **Eight Miles to the Inch.**

MAP OF THE RIVERS OF DEVON in a square frame (Ea) which also contains the explanation to symbols used for towns, villages, principal roads and elevations scale. Signature: **J Ll Warden Page fecit 1893** (Ae). Imprint: **Photo-lithographed & Printed by James Akerman. 6 Queen Square. WC.** (EeOS).

1. 1893 *The Rivers Of Devon from Source to Sea with Some Account of the Towns and Villages on Their Banks* London. Seeley and Co. Ltd. 1893. BL, BRL, TQ, T, E.

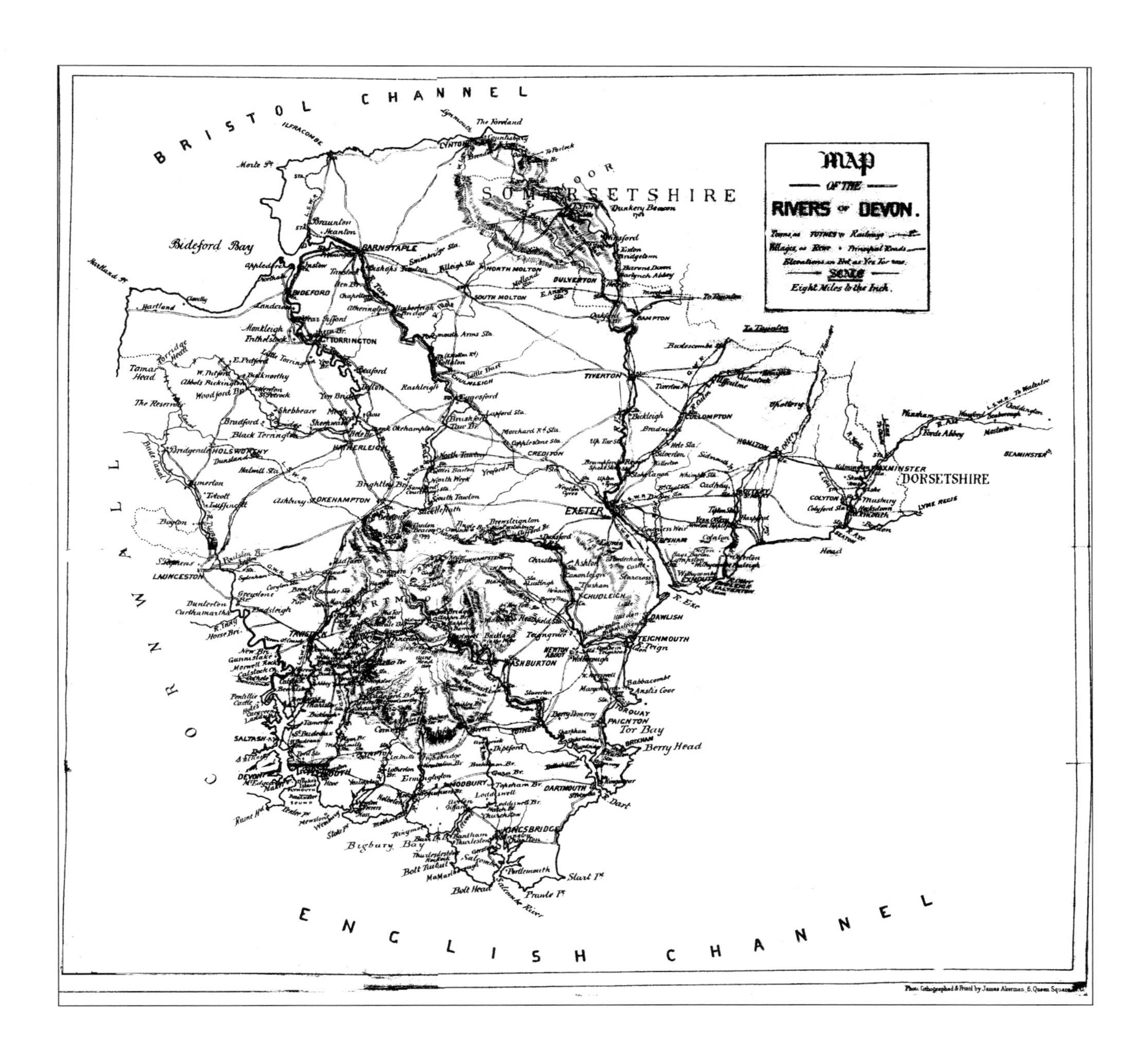

170.1 **Page** *The Rivers Of Devon*

171

FRANCIS SIDNEY WELLER

1894

Francis Sidney Weller was the son of Edward Weller whose name is associated with many maps prepared in the middle of the nineteenth century. Edward Weller's maps engraved for the *Weekly Dispatch* were very successful (see **136**). He also engraved the maps for McLeod's *Physical Atlas* (**140**) and produced some maps for *Philips' Atlas of the Counties of England* (**141**).

Francis inherited his father's interest in geography and in 1894 an atlas appeared with his signature on the maps; this was Mackenzie's *Comprehensive Gazetteer of England and Wales.* The plates from which the transfers were taken appear never to have been used for direct intaglio printing. The maps have either the county name or "Plan of ..." - there were 14 city plans including one of Plymouth, Stonehouse and Devonport. The maps, 39 English county maps and 10 Welsh counties, were produced between 1893 and 1895[1] and include large areas of adjoining counties. Most have a key to railways, roads and canals. There is overprinted colour to identify Parliamentary divisions. The text refers to the census returns 1881and agricultural figures for 1893.

Size: 216 x 278 mm. **Scale of Miles** (20=55 mm).

DEVONSHIRE (Ee) with scale below and Key to railways, roads and canals. Signature: **F S Weller F R G S** (EeOS). Imprint: **WILLIAM MACKENZIE, LONDON, EDINBURGH & GLASGOW** (CeOS). Eddystone lighthouse appears in the graduated border. **Longitude West of Greenwich** is also in border (Ce). Though the railway is shown to Turnchapel and Plymstock the line was not completed until 1897.

1. 1894 *The Comprehensive Gazetteer of England and Wales. Edited by J H F Brabner, F R G S*
London, Edinburgh and Dublin. William Mackenzie. (1894). BL.

1. Smith 1985 (p. 145) mentions that variants were used in other publications as early as 1884 (!) and as late as 1906 exhibiting appropriate changes such as the marking of the pre-1885 Parliamentary divisions and the revision of railway information. No earlier copy of Devon has so far been discovered.

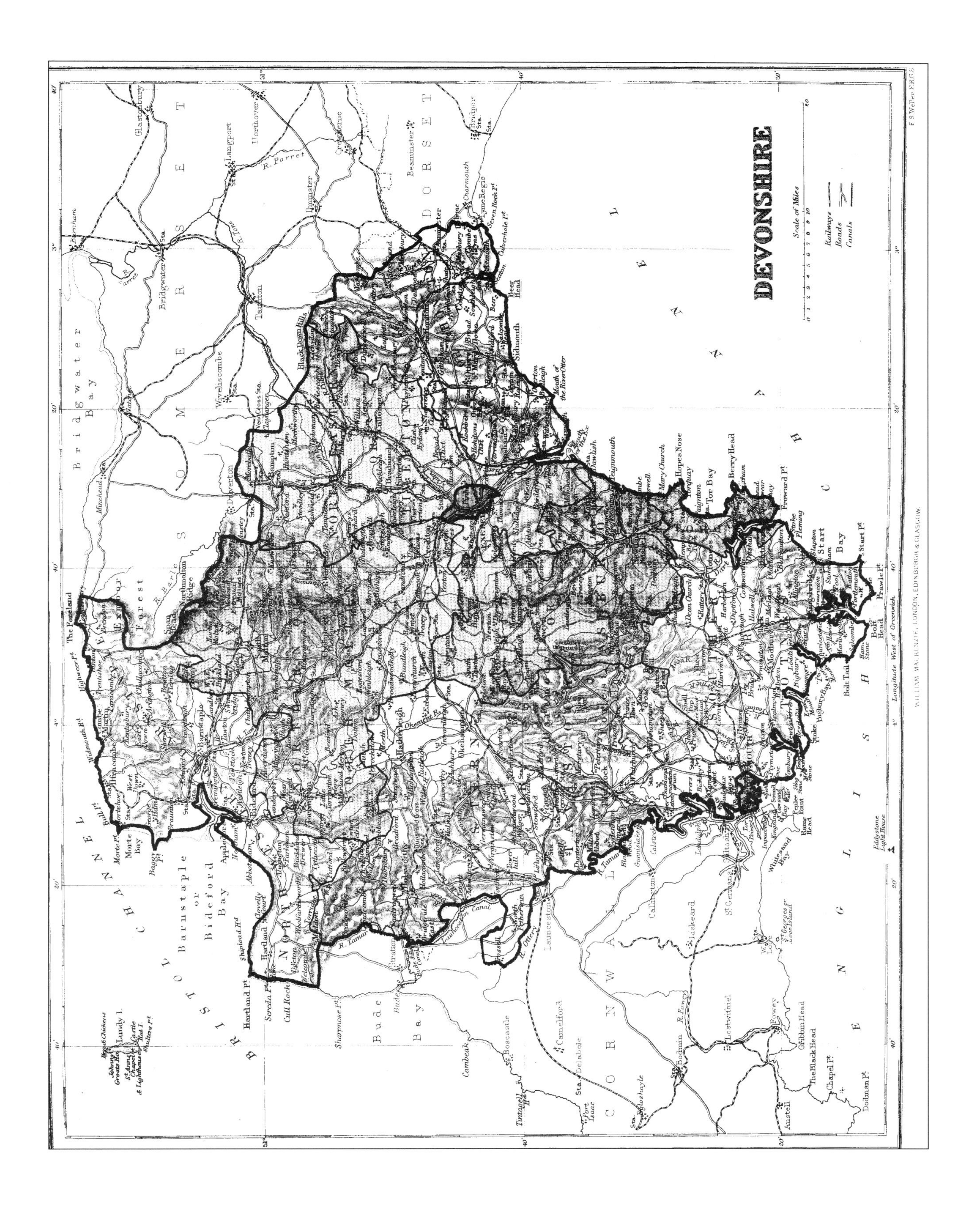

171.1 Weller *The Comprehensive Gazetteer of England and Wales*

172

JOHN LLOYD WARDEN PAGE

1895

Warden Page, a west country historian and geographer, had already produced a number of books on Devon, including *The Rivers Of Devon* in 1893 (**170**). Possibly as a companion volume he wrote *The Coasts Of Devon* which appeared in 1895. To complete his book a map of the coast of Devon was bound in after the last page (p. 437). The map was drawn by Page and is signed by him. A large number of views are included in his book, some by Page, and others signed by Alexander Ansted. The book was printed and published by Horace Cox who had premises at Windsor House in Bream's Building in east London.[1]

The map, as the title suggests, shows all the coast, rivers and most of the important tributaries. It also shows the railway with the Great Western and London and South Western lines labelled, only major roads are shown and Dartmoor is sketchily drawn but the heights of most of the better known tors (and some other hills such as Dunkery Beacon on the Somerset border) are shown. The map apart from the scale is very similar to that in *The Rivers Of Devon.*

Size 345 x 325 mm. **4 Miles = 1 Inch.**

MAP OF THE COASTS OF DEVON (Ea) and explanation to roads, railways and elevations.
Signature **J Ll Warden Page 1895** (Ae). Inset map **LUNDY ISL^{D}.** (Aa).

1. 1895 *The Coasts Of Devon And Lundy Island*
London. Horace Cox. 1895. BL, BRL, MCL, T, E.

1. Cox & Barnett were registered at 6 Bream's Buildings in Chancery Lane 1816-1823 according to Todd (1972).

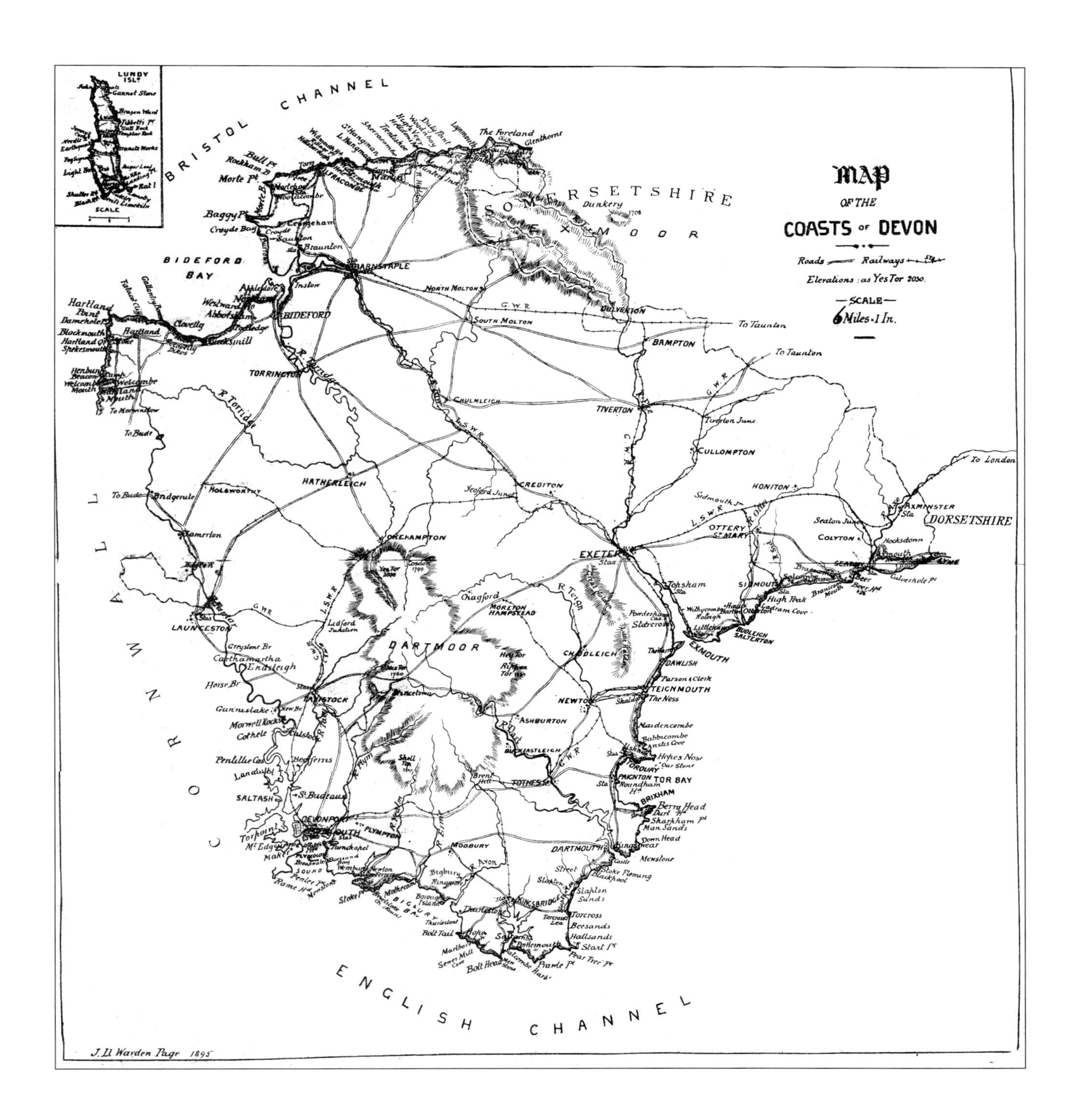

172.1 **Page** *The Coasts Of Devon*

173

FIRKS/EVERSON

1895

This chromolithograph map, designed by Charles Everson, was probably designed as a wall map, perhaps for schools. The huge size and bright colours would have made it very attractive to children (or as a poster in today's Tourist Information Office!). As well as a number of boats out to sea, there are four large colourful corner vignettes: of miners digging turf on the moor, possibly for use in the tin smelting works behind them (Aa); of farm workers unloading a haywain after the haymaking (Ea); flower sellers (Ae); and a fisherman on shore with basket of fish and crab pot (Ee). The map shows the Great Western and London and South Western Railways (which are colour-coded) and both the Bude and Torrington (to Halwill Junction) projected lines are shown. Also depicted are canals, seaports and watering places, bishoprics, Parliamentary and Municipal Boroughs, forts and even battlefields. There is an inset map of the Scilly Isles.

G Firks and Son was registered at 1 East Street in 1878 as a *printer, book-seller, stationer and music-seller* and as *stationers* at 41 Old Town Street in a directory of 1930.

Westward Ho!, a new town, was created by a property developer exploiting the name of Kingsley's novel, published in 1855. The hotel was followed by the church in 1870 and then by terraced houses. The first county map to show the town was Philip's Geography map of 1872 (**149**). The United Services College was opened in 1874 to become the scene for Kipling's *Stalky & Co.* Helped by publicity, and perhaps by this map, the town had become a popular resort by the end of the century, with the golf club its sole redeeming feature. The railway line was opened in 1901.

Size: 1375 x 1197 mm. **Scale of English Miles** (10=105 mm).

G FIRKS & SON`S "WESTWARD HO!" MAP OF DEVON & CORNWALL DESIGNED BY CHARLES EVERSON (Ca). Imprint: **PUBLISHED BY G.FIRKS & SON, WEST OF ENGLAND EDUCATIONAL DEPOT, 10A OLD TOWN STREET, & 1, EAST STREET, PLYMOUTH.** (CeOS). A reference key (Ce) and degrees of latitude and longitude are shown in the border.

1. 1895 *G Firks & Son`s "Westward Ho!" Map of Devon and Cornwall*
Plymouth. G Firks & Son. (1895). BL, B, NLS.

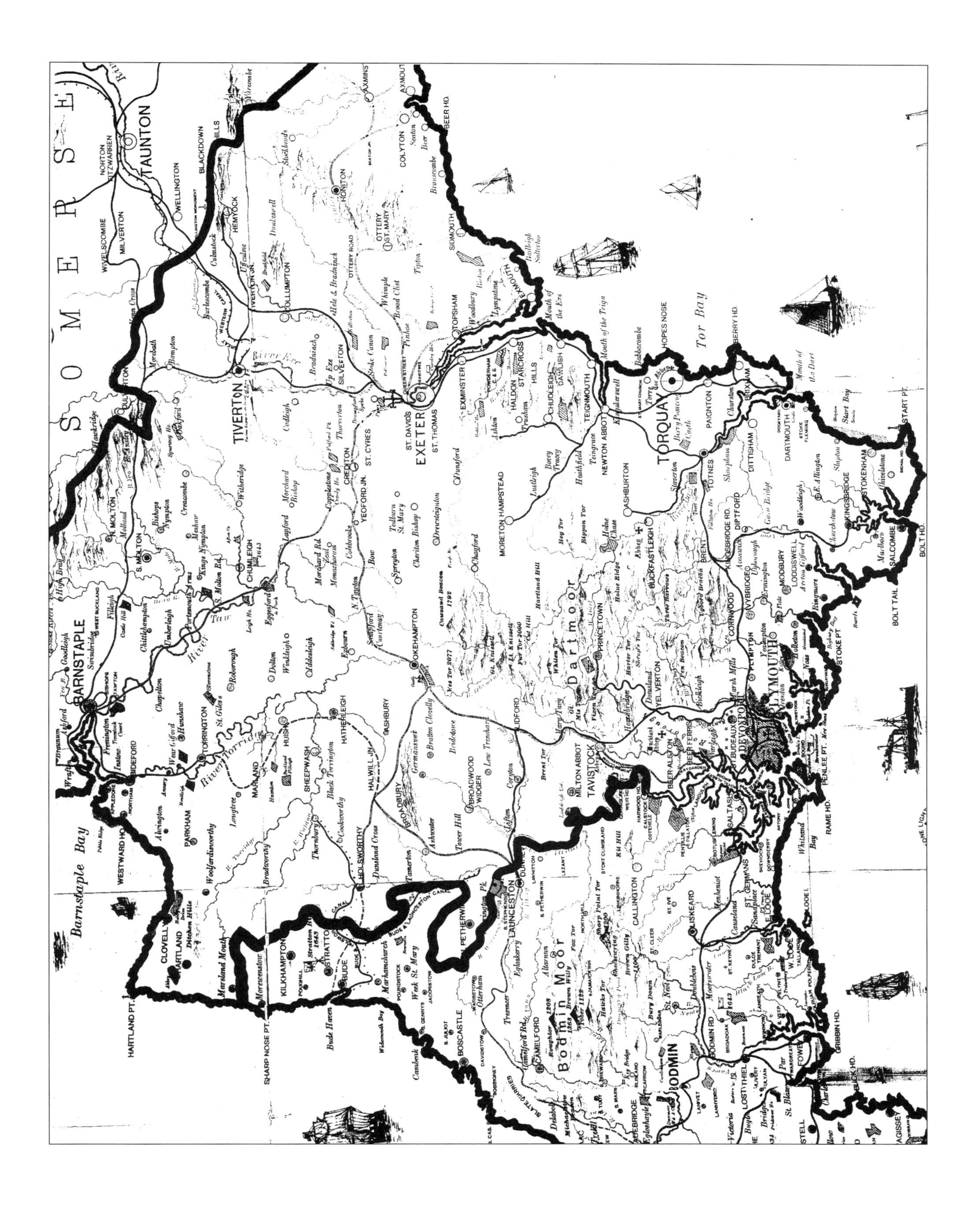

173.1 Firks & Son *"Westward Ho!" Map of Devon and Cornwall Detail*

174

JOHN BARTHOLOMEW

1895

Towards the end of the century John Bartholomew & Co., now managed by John George (1860–1980), produced a series of half-inch to the mile travelling maps *for Tourists & Cyclists.* These new maps were larger than the county maps produced and used in the later Black's and Murray's guides (**150**) and were completely overprinted in colour depicting accurately the heights of the landscape. It was, in fact, the first systematic commercial application of layer colouring to maps in Britain. A key to the height of land and the scale of colours was added for ease of use. There were 37 sheets covering England and Wales. Devon was produced on two sheets covering North and South Devon, and for clarity and detail they are excellent examples of late nineteenth century map production.

The maps had a long and varied life. The first sheets with colour contouring (South Devon and Surrey) appeared in 1895 and, according to the Bartholomew Printing Records held at the National Library of Scotland, North Devon and Sussex followed in 1896. The South Devon map was first published by W H Smith & Son in 1895 (as map 24 in their series of Travelling Maps) and North Devon as map 26. The plates were also used to produce sectional maps for Murray's 11th edition of the *Handbook for Travellers in Devonshire* in 1895-1901 (**150.14**). Sectional maps of North and South Devon were included although a north sheet map did not include the Lynton railway which was shown on the first recorded (and seen) North Devon sheet. The Dulau *Thorough Guides* also used sectional transfers from about the same date.

From 1896 the map series was numbered: North Devon and South Devon became 35 (1898) and 36 (1899) respectively.[1] From 1901 the maps were printed with the Cyclists' Touring Club symbol. Bartholomew published and sold these maps, either on paper or mounted on cloth, under his own imprint from The Geographical Institute, first in Park Road and from 1911, Duncan St., Edinburgh. Early in the next century G W Bacon was selling maps produced from the same plates but without the layered colour.[2] Other Mapsellers and Geographical Booksellers such as Sifton Praed and Co. Ltd, of 67 St James's Street London, were offering the the maps as boxed sets. The North and South sheets, with new titles and numbering were still on sale in the 1950s.[3]

174 – SOUTH

Scales 2 Miles to an Inch (5=65 mm).

Size: 480 x 705 mm both folding to 195 x 105 mm.

SOUTH DEVON (CaOS). Imprints: **NEW REDUCED ORDNANCE SURVEY OF ENGLAND AND WALES. - Scale, 2 miles to an inch**. (AaOS) and **By JOHN BARTHOLOMEW, FRGS** (EaOS). Signatures: **THE EDINBURGH GEOGRAPHICAL INSTITUTE.** and **JOHN BARTHOLOMEW FRGS** (EeOS). Railway is shown to Plymstock. Coloured and contoured with a coloured Height Bar (Ee) and note **Eddystone Lighthouse 14 Miles SW of Plymouth** (BeOS). Road key is: Main Driving Roads and Other Driving Roads.

1. 1895 *New Reduced Ordnance Survey South Devon – by John Bartholomew*
W H Smith & Son's Series of Travelling Maps
London. W H Smith & Son. (1895). EB.

1. We are grateful to Eugene Burden for information concerning this series. This section is based on his correspondence and also Tim Nicholson's letter to the *IMCoS Journal*, Summer 1999, Issue 77, pp. 62-63.
2. *Bacon's New Half-Inch Maps Cycling and Motoring Devon* was a map of the county covering the area of both the previous maps folding to pocket size. A larger example folding into boards 240 x 190 mm (unfolding to 1150 x 730 mm) was *Bacon's New Library Map Of Devonshire And Part of Somerset.* These both the same plates as before but the layout was substantially different. In both, parts of east and west Devon were included in inset maps and the *New Library Map* had an *Alphabetical Index-Gazetteer* together with a *Table of Distances By Road* below. These maps did not utilise the layer printing (ie not hypsometrically tinted) and were either uncoloured or with standard colour printing.
3. South Devon was renamed Dartmoor (sheet 2) and North Devon became Exmoor (sheet 3).

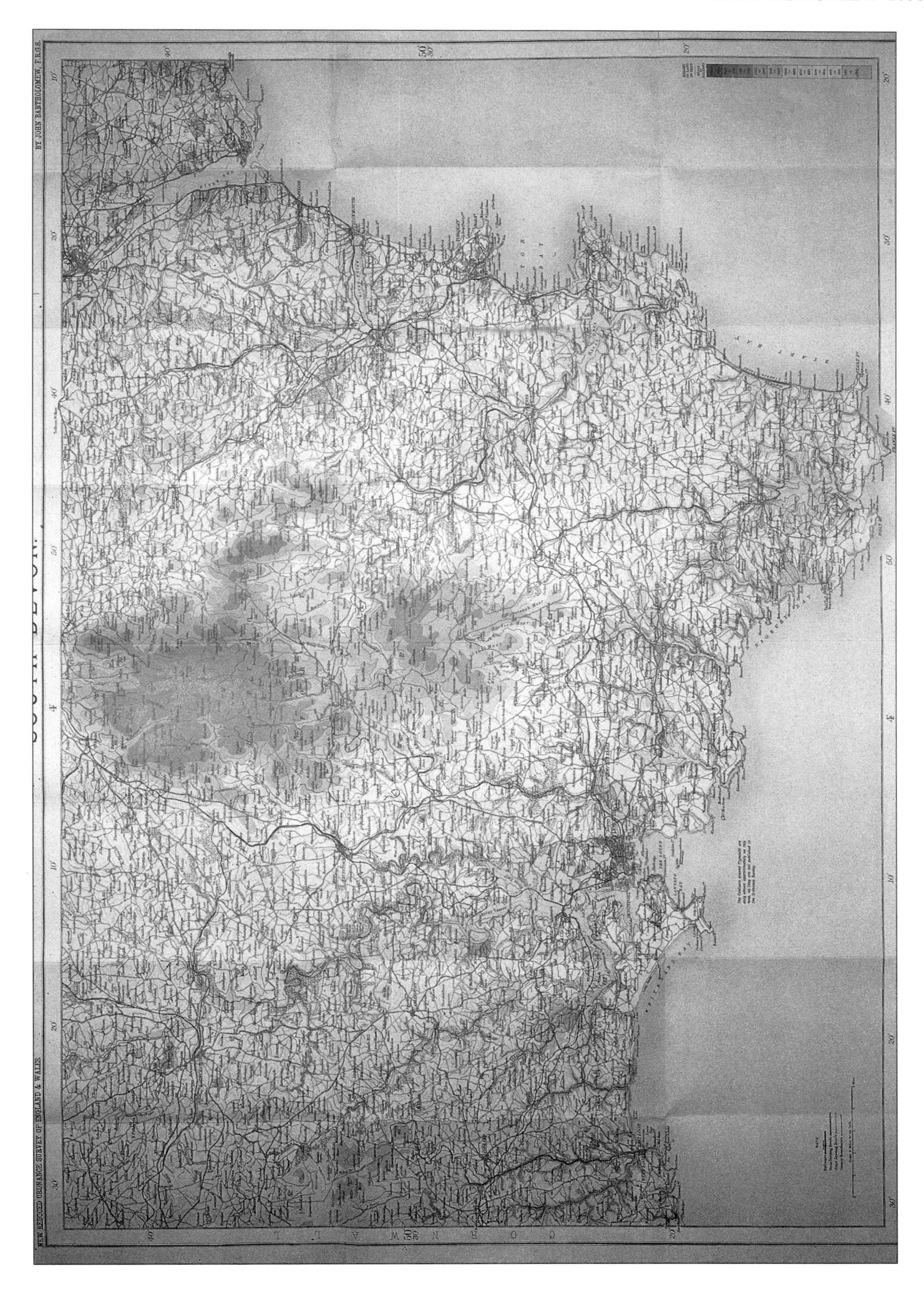

174.S.1 Bartholomew *South Devon*

2. 1896 Key to roads is now: Driving and Cycling Roads and Other Roads. Note: **For continuation Northward see NORTH DEVON** (Ca). **Copyright** added (EeOS). Railways: Ashton-Exeter complete with line to Chagford (wrong route) with stations, eg Fingle Br., and Lynton under construction.

Bartholomew's Reduced Ordnance Survey - South Devon
Edinburgh. John Bartholomew & Co. (1896). TN, EB.

3. 1898 Number added: **SHEET 36** (AaOS). Imprints: **BARTHOLOMEW'S REDUCED ORDNANCE SURVEY OF ENGLAND & WALES. SCALE 2 MILES TO AN INCH** (CaOS) and title: **SOUTH DEVON** (EaOS). Railway to Turnchapel and Yealmpton as well as Budleigh Salterton. Mineral railway downgraded NE of Liskeard. Line to Chagford deleted. No note: North Devon ... . Layer colours has low tide note. Wider to the east, eg Sidmouth with slight loss west (West Looe just present). Index to Adjoining Set (Ae).

Bartholomew's Reduced Ordnance Survey - South Devon
Edinburgh. John Bartholomew & Co. (1898).[1] BL, RGS, C, NLS.

4. 1901 Scale bar drawn around border. Road key is now: Driving & Cycling Routes in **EXPLANATORY NOTE** in box. New layer colours (low tide removed). Index to Adjoining Set (Ae) removed. Railway Budleigh to Exmouth added.

Bartholomew's Reduced Ordnance Survey - South Devon
Edinburgh. John Bartholomew & Co. (1901). TN, EB.

174 – NORTH

Scales 2 Miles to an Inch (5=65 mm).

Size: 480 x 730 mm - folding to 195 x 105 mm.

NORTH DEVON (CaOS). Imprints: **NEW REDUCED ORDNANCE SURVEY OF ENGLAND AND WALES. - Scale, 2 miles to an inch**. (AaOS) and **By JOHN BARTHOLOMEW, F.R.G.S.** (EaOS). Signatures: **THE EDINBURGH GEOGRAPHICAL INSTITUTE** (AeOS) and **COPYRIGHT-JOHN BARTHOLOMEW F.R.G.S.** (EeOS). Note: **For continuation Southward see SOUTH DEVON** (Ce). Covers the area from Hartland and Bude Bay to Watchet (enters inner border) and south to Oakhampton with both borders broken for Exeter. Railway shown to Holsworthy with projected line to Bude; railway to Lynton under construction; Hemyock line inadvertantly omitted. Height colour bar (Aa). Latitude and longitude with 5 mile scale bar (CeOS).

1. 1896 *New Reduced Ordnance Survey - North Devon*
Edinburgh. John Bartholomew & Co. (1896).[2] BL, RGS, B, NLS.

2. 1899 Number added: **SHEET 35** (AaOS). New title: **BARTHOLOMEW'S REDUCED ORDNANCE SURVEY OF ENGLAND & WALES. SCALE 2 MILES TO AN INCH** (CaOS). **NORTH DEVON** (EaOS). Railway completed to Bude and Lynton.

Bartholomew's Reduced Ordnance Survey - North Devon
Edinburgh. John Bartholomew & Co. (1899). NLS, EB.

1. Known in various editions; eg with or without wheel logo of the CTC (Cyclists' Touring Club) and with and without price on the cover.
2. Murray's Handbook to Devon, 11th Edition, 1895, has a sectional transfer of North Devon with no line to Lynton. This might imply that the North sheet was finished but not on general sale in 1895 (see illustration opposite and see also 150.14).

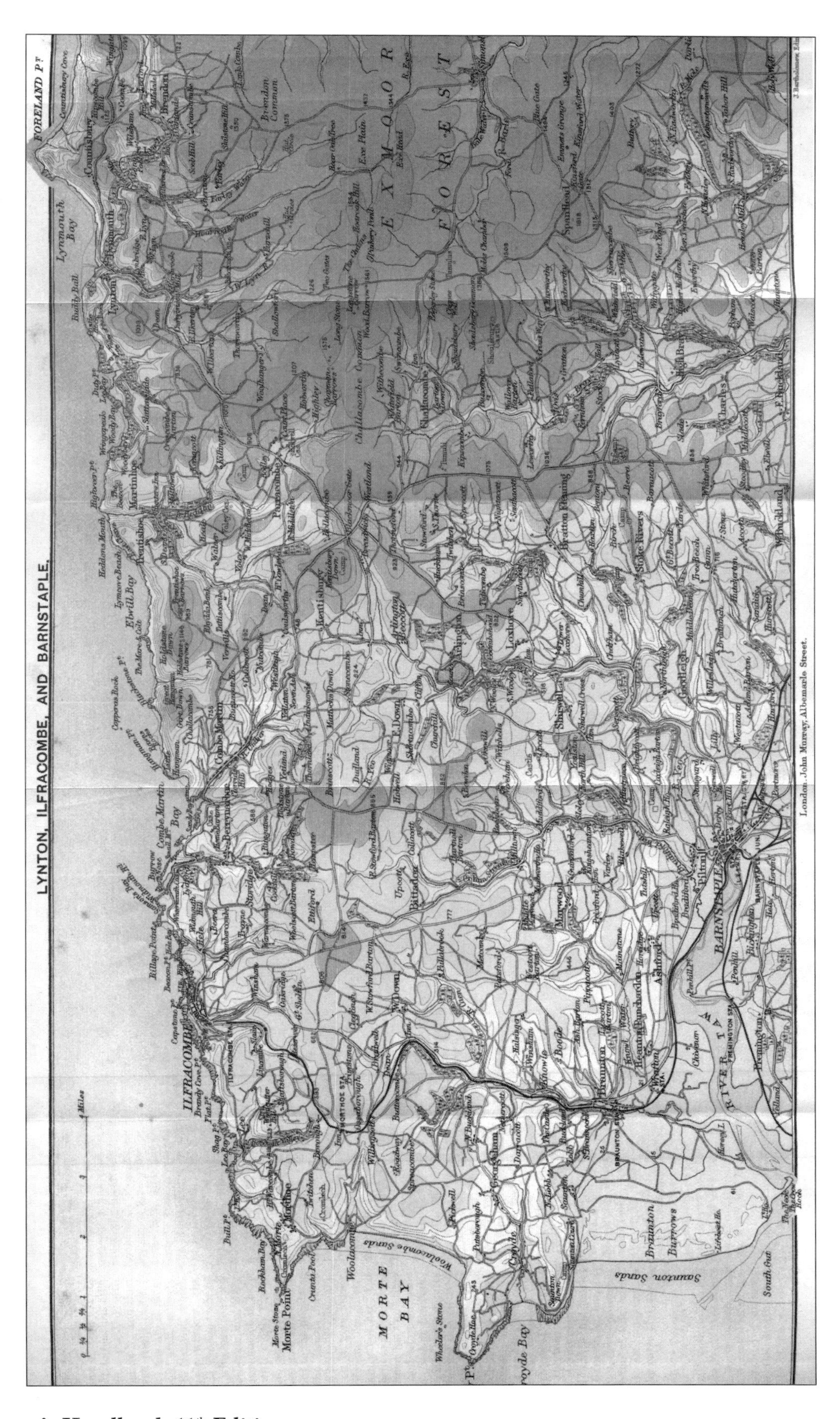

174.N Murray's Handbook 11th Edition

175

BARTHOLOMEW/PATTISONS

1897

Bartholomew took transfers from the *Imperial Map* to provide sectional maps for guide books (eg Dulau and Co's *Thorough Guides* series, **157**). From March 1899 to May 1900 transfers of the *Imperial Map* were again taken to produce a set of regional maps. These were published in 20 parts to produce *The Royal Atlas of England and Wales* (Edited by J Bartholomew and published by George Newnes Limited). South Devon was sheet XXXI with North Devon on sheets XXV and XXVI. Included in the Atlas was a second overview map of England and Wales on six sheets with a map of the West Country (Plate 12). Transfers from this map were used to produce a cyclist's map specially prepared for the Pattisons whisky company. The cyclist would not have gained a tremendous amount of useful information from this small scale map; but it did give Pattisons the opportunity to advertise their products. Roads and railways are shown but, although a note below the maps refers to a key for cycling roads, it only shows the major roads. Around the map there are vignette scenes from the Pattisons` distillery and two of their products. The map is vertical compared to the frame which has *Pattisons Limited* across the top and *Scotch Whisky Distillers, Leith, Edinburgh & London* below (ie the map has to be turned to be read correctly).

The reverse is titled *Pattisons` Scotch Whisky is invaluable to all Travellers & Sportsmen Abroad or at home who go in for Cycling - Golfing - Curling - Hunting - Coaching - Yachting - Shooting - Fishing* and there are delightful little engravings of each of the sports listed and the addresses of the Pattisons company. The maps were used to advertise the company floated by Robert and Walter Pattison in 1896 and 15 maps of England and Wales and 8 of Scotland were produced. The brothers went bankrupt in 1898 and were imprisoned for embezzlement in 1901.[1]

The plates of both these England and Wales maps were extensively used. *Darlington's Devon and Cornwall* guides (Llangollen. Darlington & Co. and London. Simpkin, Marshall & Co. Ltd. from *c.*1908) used a transfer of Cornwall from the smaller map and transfers of the larger map for an urban map of Torquay, Dartmouth and Exmouth as well as one of Plymouth and Tavistock. Sectional transfers from the smaller map were taken for *The Touring Atlas of the British Isles* (London. George Newnes.) and were still being used as late as 1948 in the *British Isles Pocket Atlas* (13th Edition: John Bartholomew & Son Ltd. Edinburgh.).

Abel Heywood's series of *Penny Guide Books* included transfers from the *Royal Atlas.* A map covering most of North Devon can be found in *A Guide to Ilfracombe and Clovelly* published by Abel Heywood and Son in Manchester in 1900.[2] Heywood, a bookseller and publisher of radical literature, who kept prices low through the inclusion of advertisements and *vile printing,* began his series in 1866 with Buxton and Kent appearing in 1872.[3] Earlier guides of Devon may have been produced.

Size of map area 225 x 170 mm. Scale (15=37 mm) **Miles.**
Size of sheet 330 x 240 mm.

DEVONSHIRE (CaOS) with imprints **The Edinburgh Geographical Institute** (AeOS) and **Copyright - John Bartholomew & Co.** (EeOS) with Scale (BeOS) and **Cycling Roads shewn thus** (DeOS). Railways to Holsworthy but not the continuation to Bude, to Sidmouth but not Budleigh Salterton and a thin line, probably representing the planned line from Barnstaple to Lynton which was opened in 1898. The map folds into a linen waterproofed cover measuring only 85 x 60 mm.

1. 1897 *Pattisons' Cyclists' Road Map of Devonshire issued by Pattisons Limited Scotch Whisky Distillers Leith Edinburgh London Manchester & Glasgow*
Edinburgh. Banks & Co. (1897). KB, MW[4].

1. T Nicholson; A Scotch Mystery; *IMCoS Journal*; Issue 79, Winter 1999.
2. *Map Of The Environments Of Ilfracombe.* Imprint: Abel Heywood & Son. 56 & 58 Oldham Street, Manchester. (Copy at MCL 942.35L).
3. John Vaughan; 1974; p. 89 and footnote.
4. Covers have different colour, otherwise they are the same. Other counties, eg Norfolk, have a different text on the cover and there are different scenes around and on the reverse. See also Tim Nicholson.

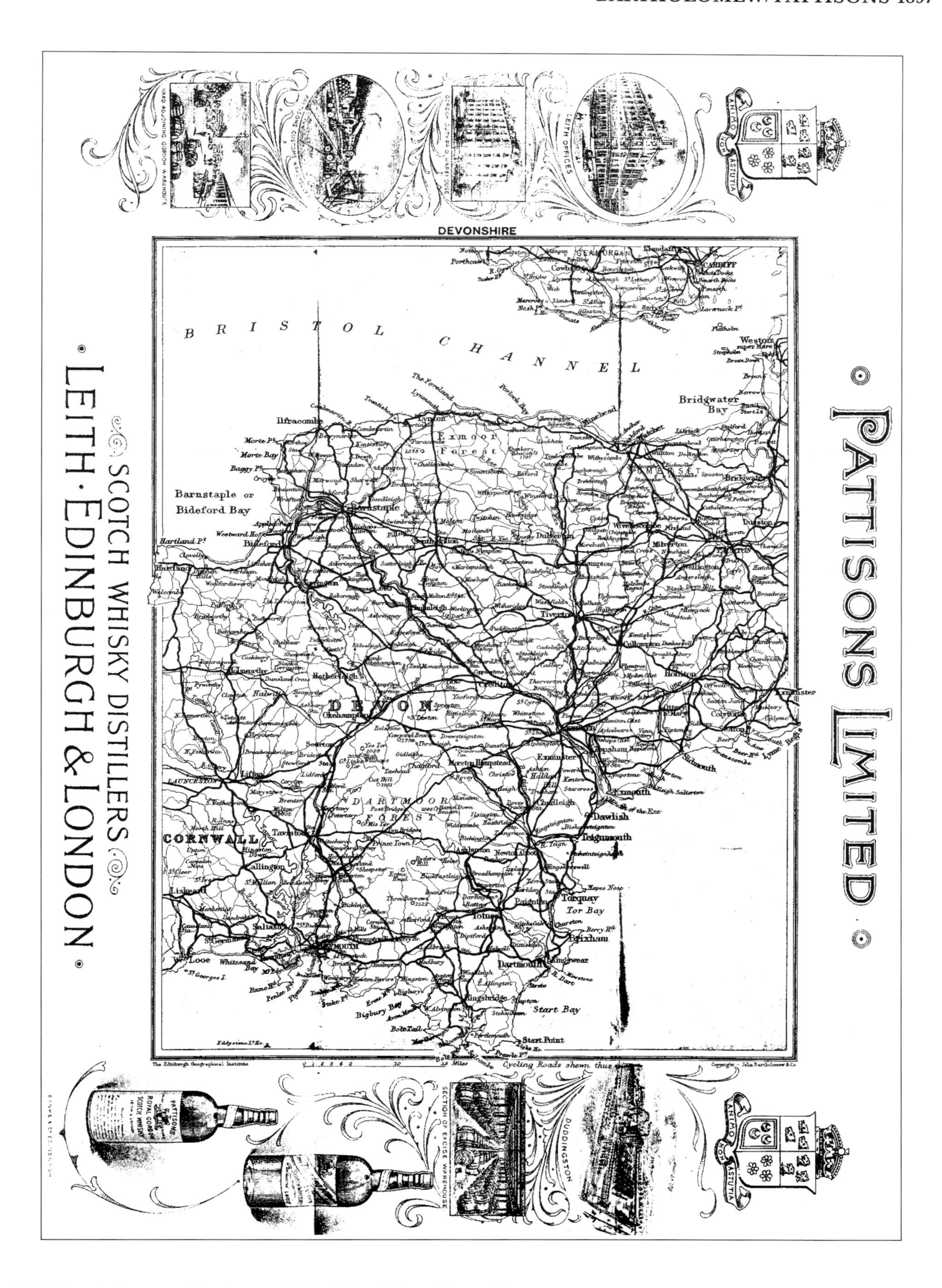

175.1 Bartholomew/Pattisons *Cyclists' Road Map of Devonshire*

176

CASSELL & CO.

1897

Cassell's ventures in cartography began with the purchase of the Weekly Dispatch plates (**136**). Cassell also started a series of county maps, *c.*1872, which was never completed, as well as producing a gazetteer with maps taken from plates produced by W & A K Johnston: *Cassell's Gazetteer of Great Britain and Ireland* - containing 25 regional maps of England and Wales, 16 of Scotland, and 8 of Ireland, 1893-8. This also included maps of cities and islands. One further venture was an attempt to produce an English version of the standard world atlas of the day, *Andrees Allgemeiner Handatlas.* This, issued as the *Universal Atlas,* was not successful and yet later, revised, it became *The Times Atlas* (1895-96).[1]

One of the more unusual maps of Devon is to be found in Cassell's series *The Rivers of Great Britain.* This was issued 1891-1897 and contained maps within the text. *The Rivers of Great Britain* was the general title for a series of three volumes: *The Royal River: The Thames from Source to Sea* (Volume I, appeared in 1891); *Rivers of the East Coast* (which was Volume II Part I, 1892); and the map of Devon was in *The Rivers of the South and West Coasts* (Volume II Part II). This final volume of the series did not appear until 1897. All three volumes (in one volume) were re-issued in 1902. The map of Devon was not amended.

Although the first two volumes in the series also contained maps, none of these were county maps. Usually the attention was on rivers and hence the maps were area and not county maps, for example the Rivers of Lancashire and Lakeland (in Volume III). The third volume covered an area from Kent and Surrey along the south coast and then the whole of the west coast to Ayrshire and the Clyde but also included rivers such as the Avon, Wye and Usk. Only this third volume has county maps and only Devon and Cornwall have an individual map – the others show two counties together for example Hants with Dorset and Kent with Surrey.

Each section of the guide was written by a different author. The writer of the section on Devon was W W Hutchings, who also wrote *London Town, Past and Prsent* in 1909. The chapter included on page 26 the small map: the map of Devon is very simple: the sea is shaded, railways are shown together with towns and rivers and while hills have been carefully hachured to show their situation, roads have not been included.
W W Hutchings also wrote *London Town, Past and Present,* 1909.

Size: 104 x 102 mm. **English Miles** (15=19 mm).

THE RIVERS OF DEVON (CeOS) with no imprints or signature.

1. 1897 *The Rivers of Great Britain. The Rivers of the South and West Coasts*
London, Paris & Melbourne. Cassell & Co. 1897. BL, KB.

1. Nowell-Smith; 1958; p. 168.

CASSELL & Co. 1897

roll of Devonshire "worthies" is only less illustrious for its poets than for its heroes. Perchance the explanation of what almost looks like a conspiracy of silence is that the streams, full of allurement as they may be, are not rich in associations of the poetic sort. Of legend they have their share, but for the most part it is legend uncouth and grotesque, such as may not easily be shaped into verse. Their appeal, in truth, is more to the painter than to the poet. For him they have provided innumerable "bits" of the most seductive description ; and neither against him nor against the angler—the artist among sportsmen—for whom also bountiful provision has been made, can neglect of opportunity be charged.

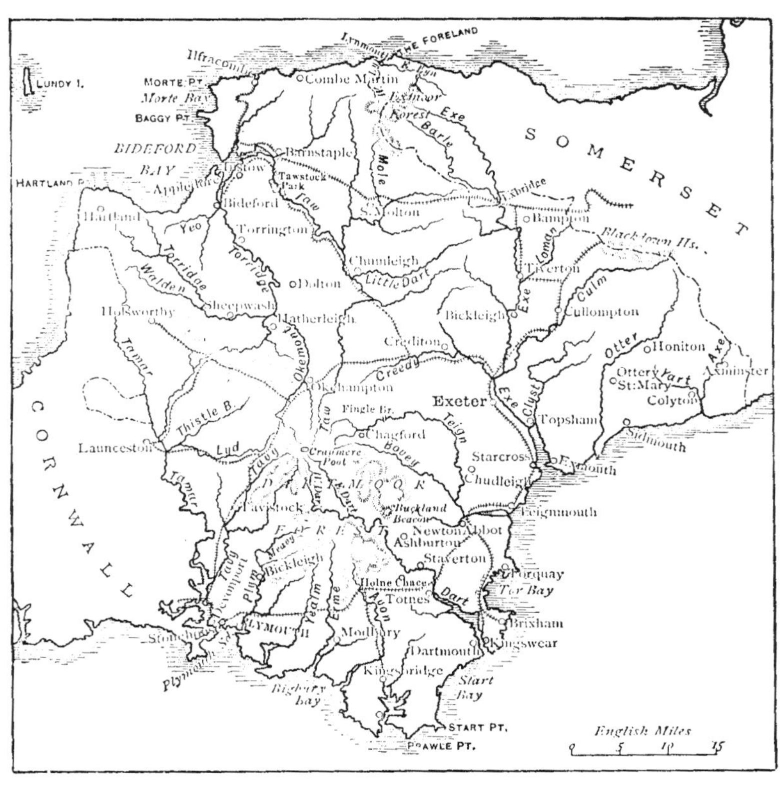

THE RIVERS OF DEVON.

It is in the royal "forests" of Exmoor and Dartmoor that nearly all the chief rivers of Devon take their rise. Of these moorland tracts, the one extending into the extreme north of the county from Somersetshire, the other forming, so to speak, its backbone, Dartmoor is considerably the larger ; and in High Willhayse and in the better known Yes Tor, its highest points, it touches an altitude of just over 2,000 feet, overtopping Dunkery Beacon, the monarch of Exmoor, by some 370 feet. Between the two moors there is a general resemblance, less, however, of contour than of tone, for while Exmoor swells into great billowy tops, the Dartmoor plateau breaks up into rugged "tors"—crags of granite that have shaken off their scanty raiment and now rise bare and gaunt above the general level. Both, as many a huntsman knows to his cost, are beset with treacherous bogs, out of which trickle streams innumerable, some, like the Wear Water, the chief headstream of the East Lyn, soon to lose their identity, others to bear to the end of their course names which the English emigrant has delighted to reproduce in the distant lands that he has colonised. Not strange is it that with loneliness such as theirs, Exmoor and

176.1 Cassell *The Rivers of Great Britain*

177

JAMES JERVIS
1897

There had been plans for an extension to the Torrington line almost before the line was completed from Bideford (1872) but nothing happened. *Philips' Atlas of the Counties* shows a projected route from Torrington, via Hatherleigh and passing north of Jacobstow to meet the L&SWR route at Sampford Courteney. The same route is found on Bartholomew's map printed by W H Smith in *c*.1875 (**150**). However, although the completion of the north Dartmoor loop meant that there was no compelling reason to connect Torrington with either Okehampton or with Halwill, certain groups still envisaged a more direct connection with the south west of the county.

Various attempts to get the line built were made and one suggested route was printed and published about 1897. The map probably drawn, but not engraved, by J Jervis ,a civil engineer, covers all but the extreme eastern portion of the county, without rivers or roads. A railway map, it shows all the then current lines with the exception of the Sidmouth and Budleigh Salterton (1897) routes. The emphasis is on a proposed line from Torrington directly to Okehampton. The line of the suggested route is basically the same as on Philips' map but passes south of Jacobstow to meet the main line at Okehampton. But this line was never built.

Torrington can, however, claim to have had one of the last railways built in Britain when a line was finally completed in 1925. However, that line ran from Torrington to Halwell Junction via Hatherleigh and not directly to Okehampton.

The four 'authorised' lines included in Jervis' map are: Barnstaple to Lynmouth (completed 1898); Bideford to Westward Ho! (1901); Holsworthy to Bude (1898); Doddiscomsleigh (Ashton) to Exeter (1903).

Size: 245 x 202 mm. No Scale.

No Title but the key (Ae) reads **TORRINGTON & OKEHAMPTON RAILWAY** and there are two further legends: **OTHER RAILWAYS** and_**AUTHORIZED.** Signature: **Jam^{S}. T. Jervis, M. Inst C.E. Engineer, 9, Victoria Street. Westminster** (Ee). Note: **C 4.5e** (Ae)

1. 1897 No Title but *Torrington & Okehampton Railway*
Loose sheet. Publication uncertain. (E).

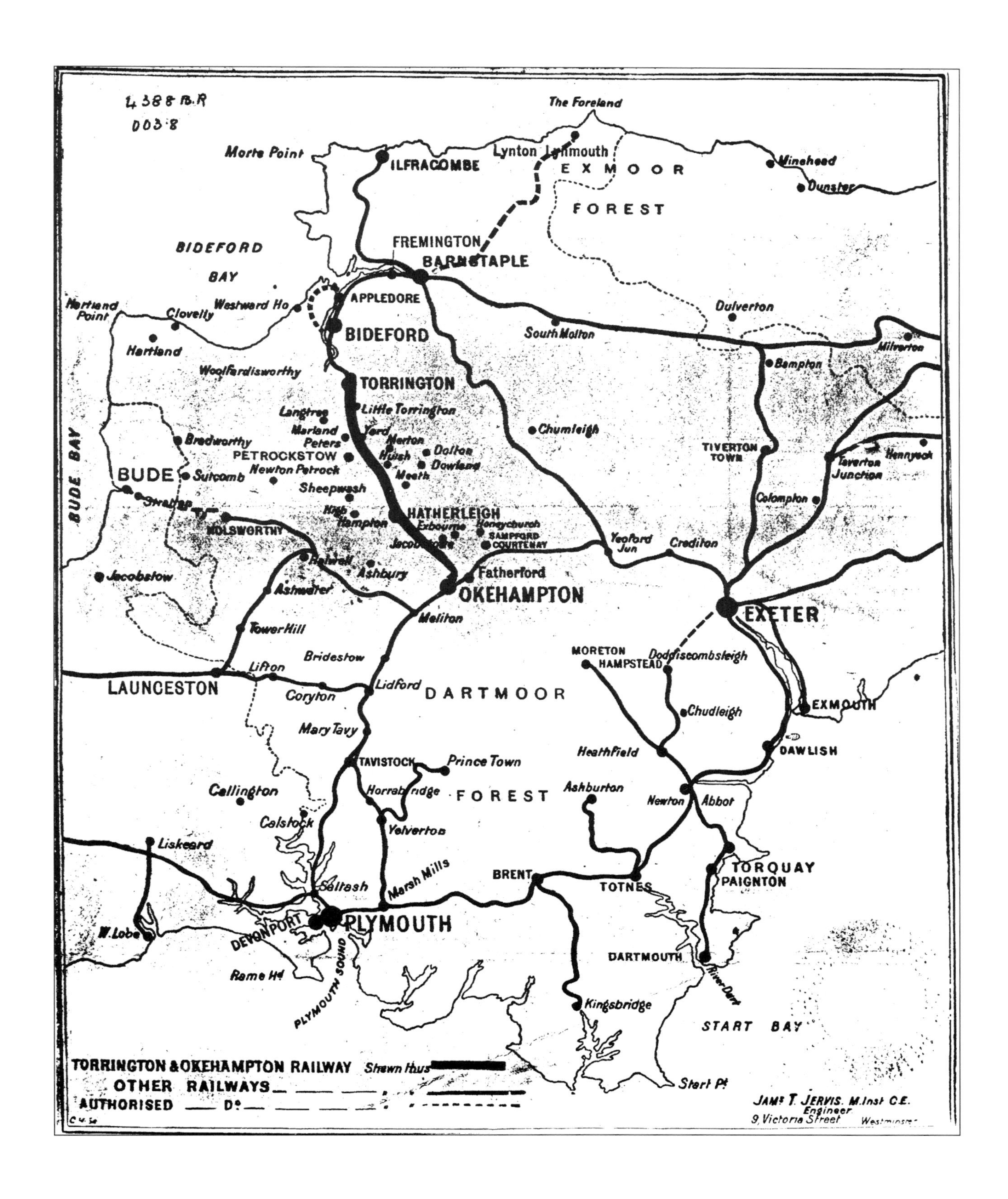

177.1 Jervis Loose sheet with *Torrington & Okehampton Railway*

178

CLARK/NORWAY

1897

Arthur Hamilton Norway (1859-1938) wrote a number of travel books based largely on his own experiences while walking through the countryside. He not only described the country and the towns but also wrote about the people and their history.

Highways and Byways in Devon and Cornwall was illustrated by Joseph Pennell[1] (1858-1926) and Hugh Thomson (1860-1920). Joseph Pennell was an American who lived and worked extensively in Europe although born and educated in Philadelphia. He was book illustrator, etcher and lithographer (and first president of the Senefelder Club[2]) and wrote a bibliography of James Whistler.[3] He wrote a series of lectures for the Slade School which were collected and published as *The Illustration of Books* and intended as a manual for the use of students (pub. T Fisher Unwin, 1896). This described the printmaking processes for the reproduction of artwork by both traditional and the newer photo-mechanical methods.

Highways and Byways in Devon and Cornwall appeared in 1897 published by Macmillan. *Devon and Cornwall* was soon reprinted (twice in 1898) and other *Highways and Byways* followed, including *Yorkshire* by Norway (1899), *Donegal and Antrim* by Stephen Gwynn, *North Wales* by A G Bradley and *East Anglia* by William A Dutt. These were all illustrated by Thomson and/or Pennell. One guide of Europe (*Highways and Byways in Normandy*) was written by Percy Dearmer. Although all of these appeared under the Macmillan label, one further guide written by Norway, *Naples, Past and Present,* was published by Methuen & Co. in 1901. Norway also wrote other works on the West Country including a *History of the Post Office Packet Service (1793-1815).* The book contained a simple map of Devon and Cornwall, showing the author's tour from Lyme Regis around the coast to Lynton.

The map is very simple and has no title. Neither roads nor railways are shown and only the principal towns and rivers that lay along the author's route are included. There is a simple north compass and a note of the Author's Route (the route being printed in red). The printers were R & R Clark who were the main printers for Black's Devon guides.

Size: 182 x 250 mm. **Scale of Miles** (24=45 mm) **Miles**.

The map has no title but there is an imprint: **R & R Clark, Limited, Printers, Edinburgh** (AeOS).

1. 1897	*Highways and Byways in Devon and Cornwall* London. Macmillan & Co. Ltd. 1897.	BL, E, FB.
	Highways and Byways in Devon and Cornwall London. Macmillan & Co. Ltd. 1898[3], 1900.	E, OrU; KB.

1. Sets of 20 lithographs printed by Joseph Pennell for and with the title *Highways and Byways in Devon and Cornwall* were sold. The British Library (10352.m.16) has a set.
2. The club, named after the inventor of lithography, Aloys Senefelder, was founded in 1910 to instruct artists in all stages of the work and to provide a common press. Among its early productions was a series of posters for the Underground Railway.
3. Was printed twice - in January and September.

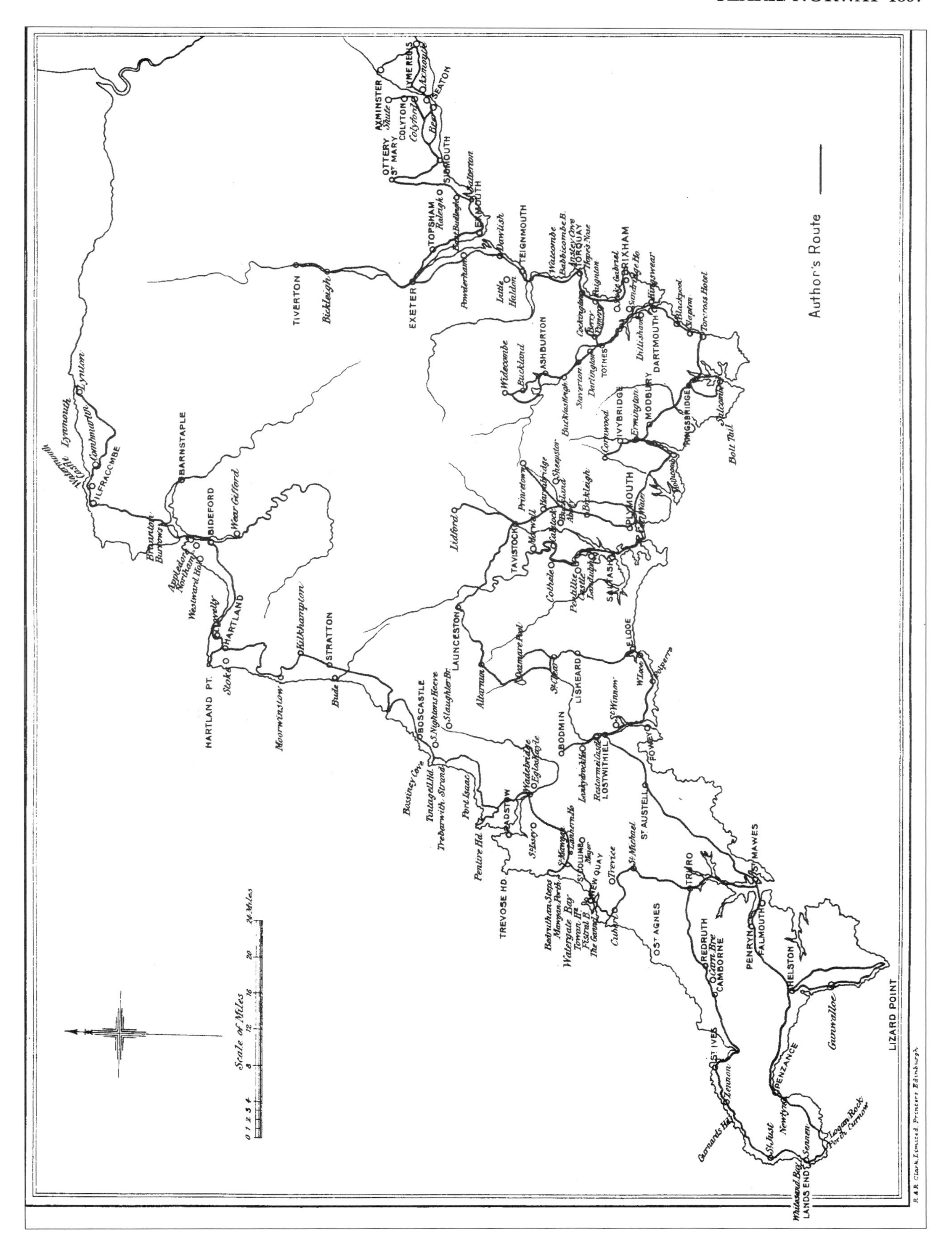

178.1 Clark/Norway *Highways and Byways in Devon and Cornwall*

179

ERNEST GEORGE RAVENSTEIN

1899

For many years the Philips family firm had been selling folding maps based on their original County Atlas maps of 1862. However, these were becoming increasingly out of date by the end of the nineteenth century and Philips engaged E G Ravenstein to engrave a new plate of England and Wales, *Philips' Topographical Map Of England And Wales* on 33 sheets, which was completed *c.*1895 and used to produce county and local area maps. This was advertised as three miles to the inch and contained useful cycling information such as dangerous hills, distances to London and distances between road junctions.

Ernst Georg Ravenstein was born 30th December 1834 and moved to England where he carried out work for the Topographical Department of the War Office and the Royal Geographical Society as well as for companies such as H G Collins, Stanfords, Blacks and Philips. He edited a number of geographical works for A K Johnston (the younger) from the 1880s. He anglicised his name to Ernest George. Ravenstein died 13 March 1913.

This map was produced at the unusual scale of 1:200,000, or 3.15 statute miles to the inch; (nearly) all larger maps of Devon up to this time had been produced at some scale relative to the inch. The county map of Devon covers an area from Dodman Point in Cornwall to Blackford just east of Highbridge and includes Lyme Regis and Lundy Island (although Prawle Point breaks the border). The *Philips' Topographical Map Of England And Wales* is advertised on the inside cover with the area covering Devon on three sheets. The Devon cycling map did not exactly coincide with the area covered by the three sheets but was a specially taken transfer. The railways to Lynton, Bude and Yealmpton are included (all 1898): although the line from Ashton to Exeter is complete, the Exmouth connection is not completed to Budleigh (both 1903) and there is no line to Northam and Appledore (1901). Copies were made on both plain paper and on Pergamoid paper, ie a specially treated semi-waterproof paper.

In the twentieth century, further transfers were taken and exploited by local traders. Varnan, Mitchell & Co. Ltd, of the *Weekly News* Offices, (at the rear of 9 High Street) in Ilfracombe were selling their *Tourist And Cycling Map Of North Devon District,* a folding map in yellow covers, for 3d about 1908. Their map covers the coast line from Hartland Point to West Porlock, and south as far as Bude and Morchard Bishop (270 mm x 405 mm).
Philips' Ten-Sheet *Road Map of England & Wales* (at a scale of 3 miles to 1 inch) was also based on the Ravenstein map. Sheet 1 (size 720 x 870 mm) of the series covered everything west of Uffculme and Whimple.

Size: 600 mm x 700 mm. **Scale: 1:200,000 or 3.15 Statute Miles to 1 inch** (10 = 80 mm).

Philips' Cycling Map Of The County Of Devon With Parts Of Somerset & Cornwall (Ee) with scale and **REFERENCES**. Signature: **E G Ravenstein Dir** (AeOS) and **G Philip & Son, 32 Fleet Street, London.**(EeOS). Imprint: **Philips' Topographical Map Of England & Wales, Cycling Edition** (AaOS).

1. 1899 *Philips' Series Of District Maps – County Of Devon* (cover title)
London & Liverpool. George Philip & Son, Ltd. (1899). EB.

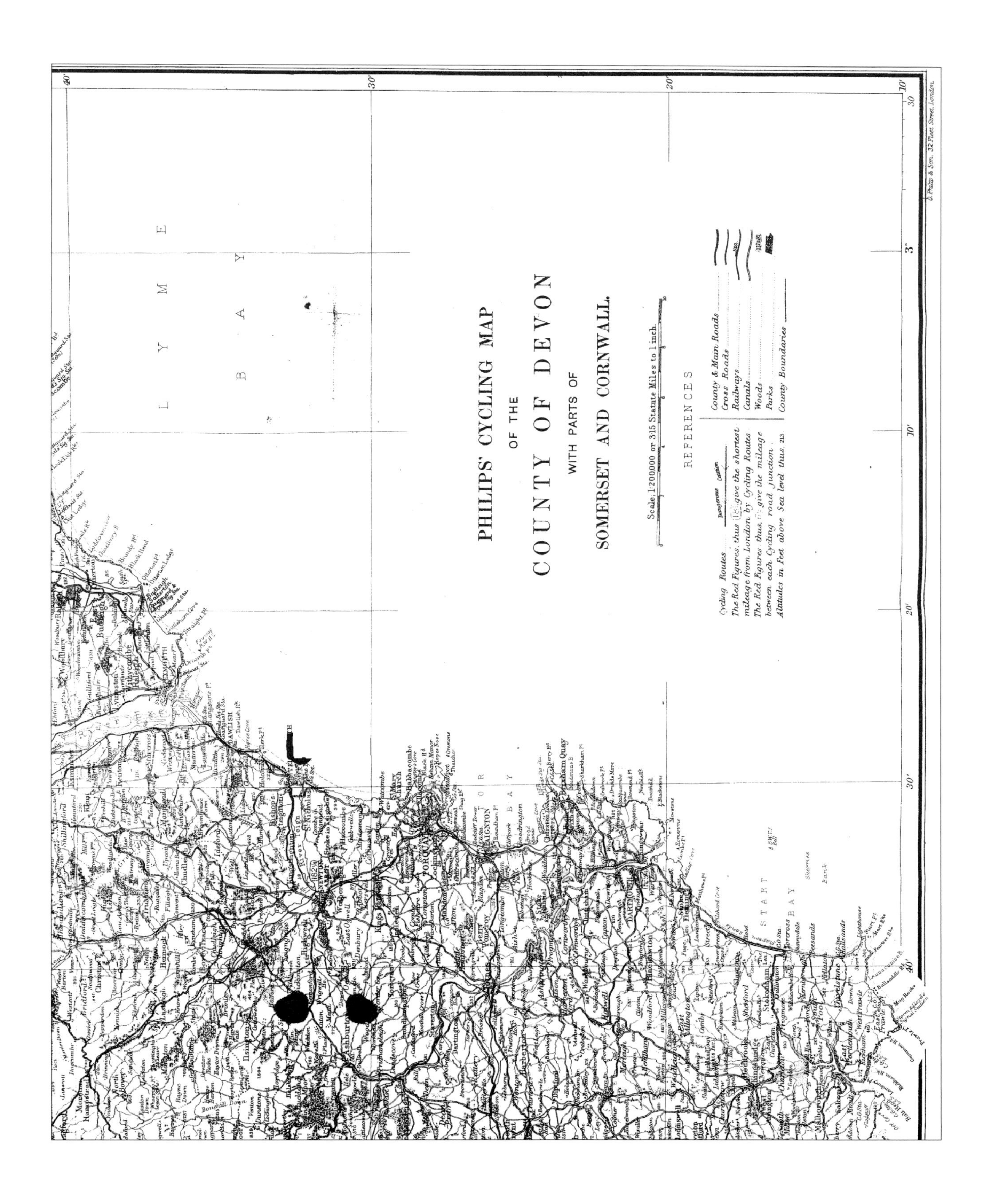

179.1 **E G Ravenstein** *County Of Devon*

180

ANON – GAZETTEER
1901

At the beginning of the twentieth century a small, illustrated guide of Devon (28pp of text and photographs) was produced with a soft cover typical of today's cheap guides. It had no publisher's imprints but the large adverts on the back show it was possibly produced for Thomas Cook. There is no printer's address.

The last page opposite the map has another unframed and untitled sketch map of only the southern part of the county showing coach roads and railways. This is part of an advertisement for Thomas Cook coach tours of the area. A second sketch map on the inside back cover advertised the Prince's Devon Tour showing a circuit Exeter-Crediton-Okehampton-Plymouth-Exeter with excursions to Princetown, Ashburton and Dartmouth and by boat to Exmouth.

A note in the guide refers to a cricket club, (re)formed in 1900, noting that the ground would be ready in 1902.

Size: 110 x 130 mm. No scale.

1. 1901 *Gazetteer Of Devon For Tourists, Travellers And Sportsmen*
Devon? Thomas Cook? (1901). NLS.

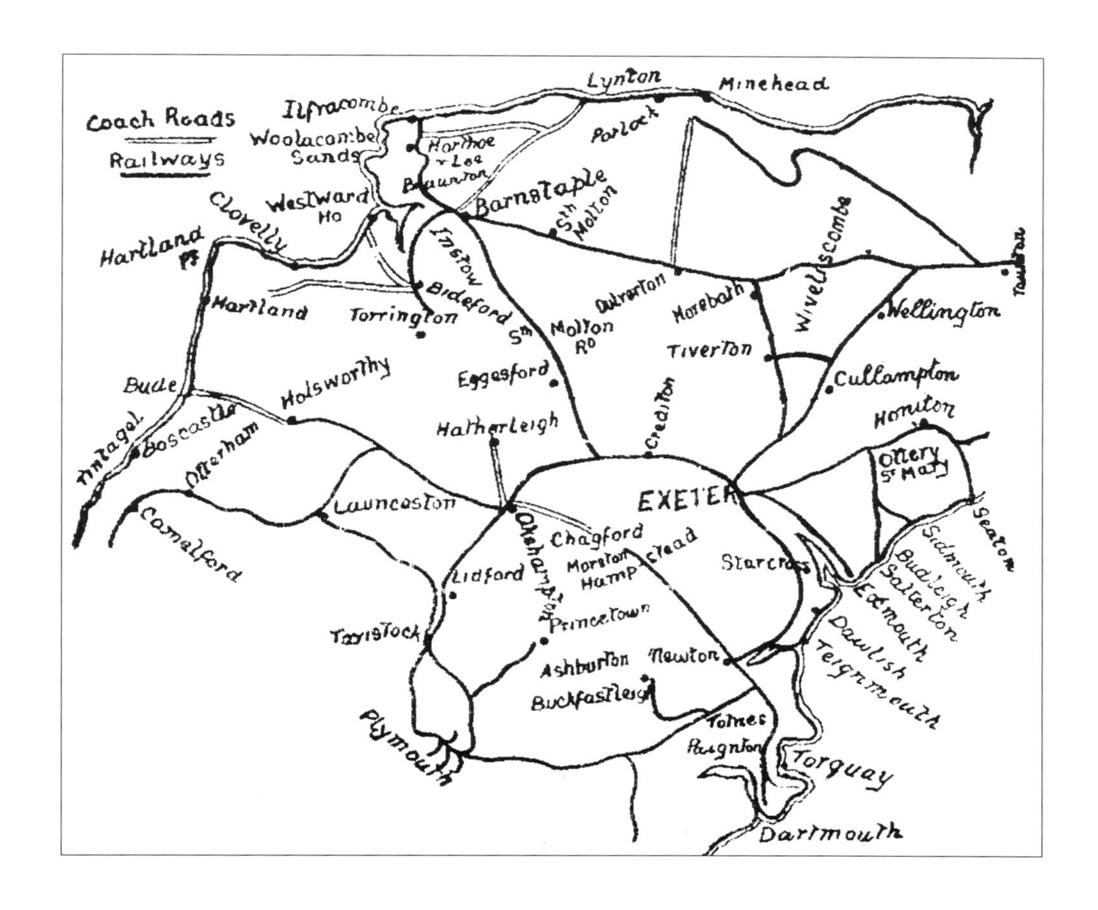

180.1 Anon *Gazetteer Of Devon For Tourists, Travellers And Sportsmen*

181

ORDNANCE SURVEY

1901

As well as the index maps produced by the Ordnance Survey during the latter years of the nineteenth century one county map was also published at the beginning of the twentieth century. In 1901, using information gathered during the re-survey of Devon, a one-sheet map of *Devonshire* was printed and published at a scale of four miles to an inch.

Devon was one of the last counties to be re-surveyed, somewhat ironically as it had been one of the first to be surveyed. This new map made use of the latest research and was published by D A Johnston, the then Royal Engineers Director General. All OS sheet maps would have been available singly, including the index maps. This particular single sheet map was available at 6d for the sheet, but could be bought folded in a cover for a further 3d.

The map was the same size as those produced in 1885 and 1888 (**162** and **164**) and the coast was similar. However, far more detail has been added including minor roads, cliff shading with an attempt to show parks and gardens or forested areas. Whereas all earlier maps only showed Devon, this map has a similar amount of detail in surrounding counties as in Devonshire (printed centrally across the county) and is probably taken from a transfer of England and Wales. The map has a standard rectangular frame and all references are below this frame.

The 6" sheets were revised about 1901: these had been published by the then Director General of the OS, General Sir C W Wilson in 1893; revised again in 1897 and republished by the Director General, Col. J Farquharson in 1898.

Size: 490 x 590 mm. **Scale of Four Miles to One Inch** (10+20=190 mm) **Miles**.

DEVONSHIRE (CeOS) with imprints **ORDNANCE SURVEY OF ENGLAND AND WALES** (AaOS) and **PARTS OF SHEETS 17.18.21.22.** (EaOS). From left to right below the map are a key, Price, Scale Bar with note on altitudes below it and copyright (centrally), Note on rights of way. Three signatures: **Revised in 1893-98 and Published by Colonel D A Johnston, RE Director General. 1901** and **Printed at the Ordnance Survey Office, Southampton in 1901** and **Reduced from the One Inch Map**. Railways to Plymouth, Exeter to Torrington, through Okehampton, both lines to Tavistock, Launceston to Bodmin, LSWR through Holsworthy and Bude, Exmouth, Seaton, Sidmouth, Dartmouth, Brixham, Hemyock, Taunton to Barnstaple, Lynton, Launceston to Halwill, Ilfracombe, Ashton, Exeter to Morebath (Bampton), Tiverton, Ashburton, Kingsbridge, Yealmpton, Budleigh Salterton and Moreton Hampstead.

1. 1901 *Ordnance Survey Of England And Wales - Devonshire*
Southampton. Ordnance Survey. 1901. BL.

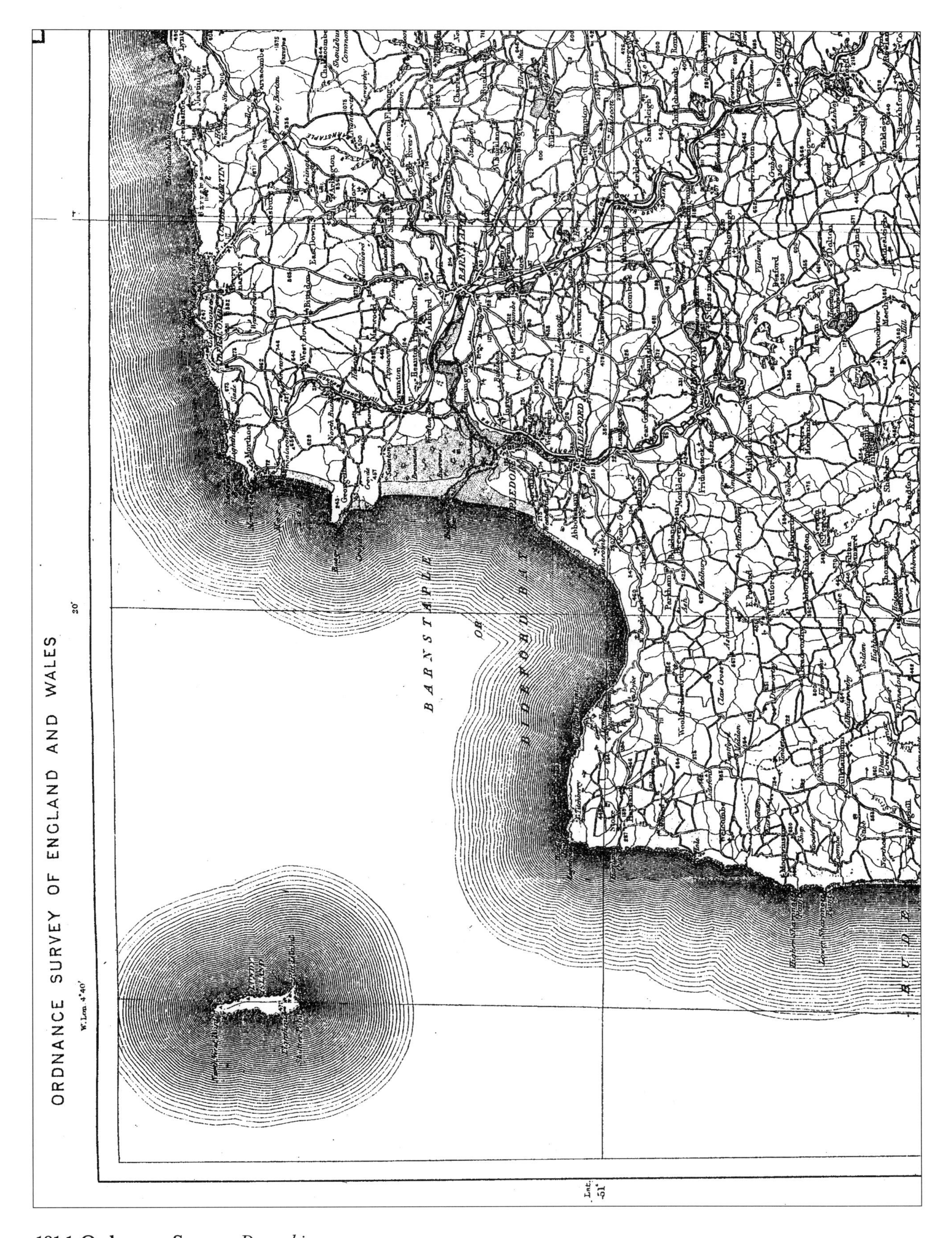

181.1 Ordnance Survey *Devonshire*

182

HARRY INGLIS

1901

James Gall (1783-1874) set up his own printing business in Edinburgh in 1810 concentrating on religious works (and publishing the St John's gospel for the blind!). His son, also James, joined the business in 1838 and the firm traded as James Gall & Son, until James junior left the business to become a minister for the Church of Scotland in 1847. Robert Inglis joined the company in the same year and the variety of works, both published and printed, increased. Robert married James' daughter, cementing the relationship. His brother, Henry Robert Gall Inglis, joined the firm in 1887. James Gall senior died in 1874, at the age of 91, and the two brothers died in 1939.

One of Gall and Inglis' innovations was the reintroduction of the strip road map. This represented roads as straight lines and in their *Contour Road Book* series these were seen in elevation, thus making clear to cyclists the hazardous stretches of road. They produced their *"Contour" Road Book of England* in three parts, each covering a *division.* The third volume, covering the Western Division, appeared in 1901 with a preface dated 1900; This volume covering Wales, the Midlands, and South-West, completes the "Contour" Road Book of England in three volumes. Sectional maps of the division preceded the strip maps with their elevations. Although sheets 39 and 40 do not show Devon completely (everything east of Cullompton is on sheet 37, they have been included as they do cover the majority of the county.

Further issues appeared as the complete *Contour Road Book of England,* the *Western Division* and in an abridged form *The Royal Road Book Of Great Britain.* Although not great innovators Gall & Inglis provided the hobby traveller with cheap and affordable maps and they became important suppliers of mass-produced maps.

Size: 140 x 170 mm. No Scale.

No title. Sheet no. **39** (AaOS) and **40** (EaOS). Note at top of sheet **Continuation North 34**. Map of Devon and Cornwall from Lands End to Tiverton with the southwest promontary from Truro in inset (Aa). Railways are shown to date, ie no line to Budleigh Salterton or Ashton to Exeter.

1. 1901 *The "Contour" Road Book of England. Western Division. By Harry Inglis*
London & Edinburgh. Gall & Inglis. (1901). KB.

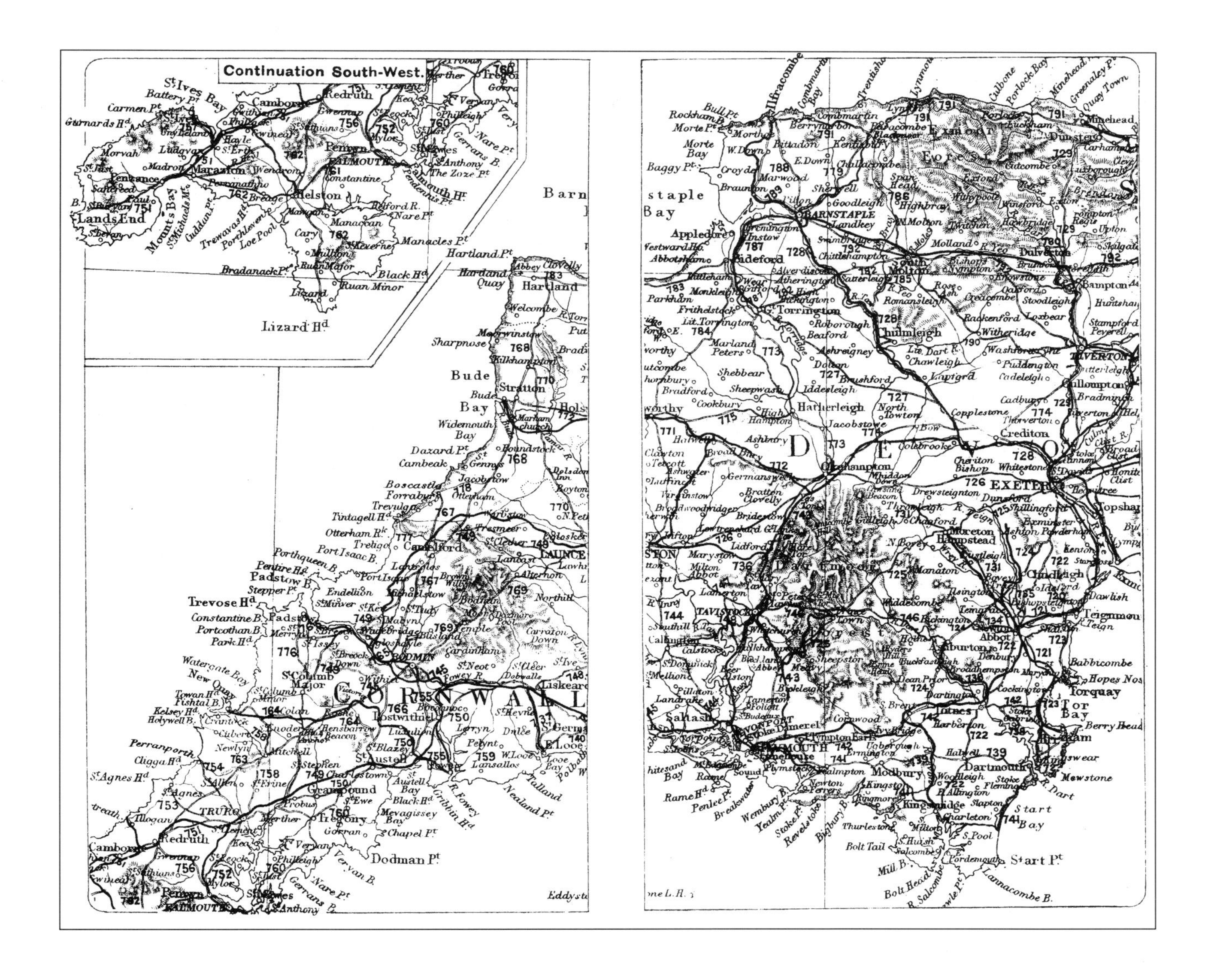

182.1 **Inglis** *"Contour" Road Book*

APPENDIX I – Aaron Arrowsmith

80A
AARON ARROWSMITH
1814

Aaron Arrowsmith (1750-1823) was a well-known engraver of maps, who together with his sons, Aaron and Samuel, and later a nephew John, established a reputation for their maps of the Americas and the new discoveries. The Arrowsmiths also published a number of maps, including Devon, for government publications: Aaron Arrowsmith engraved this map of Devon and Cornwall in 1814; a map of Devon and Somerset was completed for a report in 1833, probably by Samuel (**110**); and a map of Devon and Cornwall in the *Third Report from His Majesty's Commissioners* was signed by Samuel (**117**). The earlier map was a map of the diocese. It is more detailed than Samuel's later map, and it is surprising that Samuel did not copy it for his later report. The map was one of a set of 15 maps drawn by Aaron and his son Samuel to accompany the 5 volume *Eccliasticus, temp. Henrici VIII, auctoritate regia institutus* edited by J Caley and the Rev. J Hunter.

Aaron's original engraved map was lithographically reproduced, reduced in size and included in a late nineteenth century work on the state of the World Church in the *Historical Church Atlas.* This was published in London by the Society for Promoting Christian Knowledge (founded in 1698) and printed by Stanford's (many maps are signed *Stanford's Geographical Establishment*). A Christian atlas of the world, it included various chapters and maps on such topics as the Roman Empire, Gothic Invasions, Rise and Spread of Mohammedanism, and the Crusades. The author, Edmund McClure had associations with both the SPCK and the West Country. He was secretary of the SPCK from 1875 for 40 years and an honorary Canon of Bristol Cathedral. He died in 1922. Although the map of Devon and Cornwall is a double-page map, it has been split and there is a margin down the middle of the page. This map of 1814 was copied in 1854 by the Reverend George Oliver (**129**).

In 1815 Aaron Arrowsmith completed a *Map of England and Wales* on 15 full and 3 half sheets. This was published both as book and individual sheets and reissued 1816 and 1818[1]. The plates were reused in the 1870s by Stanford to produce a *Railway & Station Map of England & Wales* in 24 Sheets. Central and west Devon was on sheet 19[2], with south Devon below Torquay on sheet 24 and east Devon (from Exmouth) on sheet 20.

Size: 392 x 552 mm. No scale bar.

DIOC' EXON'. Imprint: **Tabula Juxta Valerem Eccliasticum XXVI; Henrici VIII Institutum Geographica** (CaOS). Signature: **A Arrowsmith** (EaOS) and dated **AD MDCCCXIV** (EeOS). In addition it has two inset tables: **NOTAE EXPLAN**. and **Civitas Exon**. Graticulated border with degree lines. Explanation (Ee).

1. 1814 *Eccliasticus, temp. Henrici VIII, auctoritate regia institutus*
London. Ed. J Caley and Rev. J Hunter. (1810-34). RGS.

2. 1897 Map is reduced to 230 x 320 mm and printed in two parts, on two pages. The earlier imprints and signature are erased, leaving the title and the inset tables. Left page has page number **66** (AaOS) and imprints: **HISTORICAL CHURCH ATLAS**. (AcOS) and **EXETER WESTERN SECTION**. (EeOS). Right page has page number **67** (AaOS) and imprints: **HISTORICAL CHURCH ATLAS**. (AcOS) and **EXETER EASTERN SECTION**. (EeOS). Place names have been added.

Historical Church Atlas by Edmund McClure M A
London: SPCK. New York: E & J B Young and Co. 1897. BL, RGS.

1. British Library has various copies: BL K 5-74 England Tab II; Maps STE (3); and Maps 148.d.25.
2. *Large Scale Railway & Station Map of England & Wales* (CaOS) with *Scale of Miles* (10=90) (AeOS) 630 x 470 mm. DEVON printed across county. Editions of sheets are known dated from 1876 to 1884.

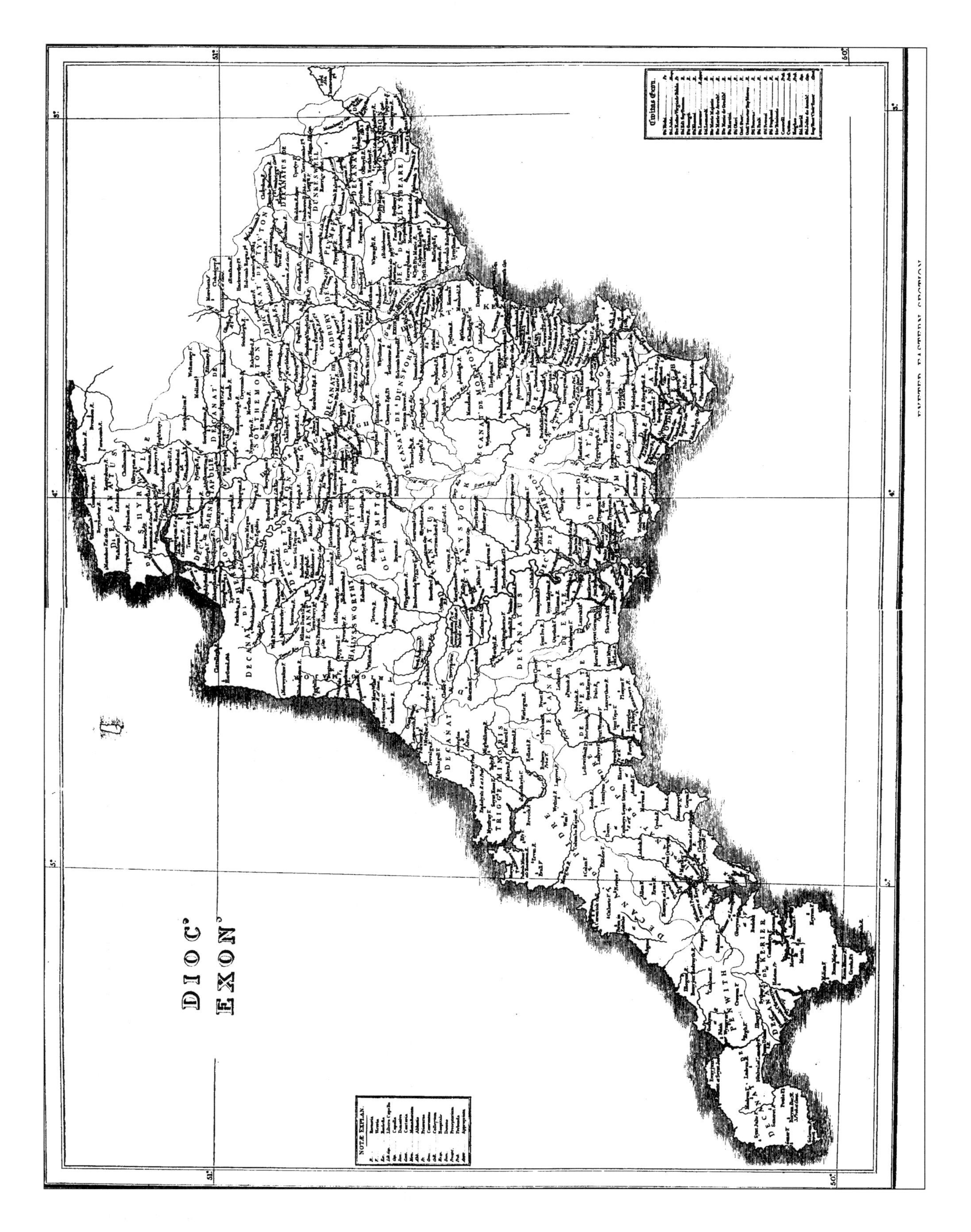

80a.1 Arrowsmith *Eccliasticus, temp. Henrici VIII, auctoritate regia institutus*

APPENDIX II – John Cary

John Cary (1755-1835) came originally from Wiltshire. He was the second son of George, a maltster, and Mary Cary. He had three brothers, Francis (1756-1836) also an engraver, William (1759-1825) a map publisher and globe maker with whom he collaborated, and George the elder (*d.*1830) who was a partner at 86 St. James Street, London (1820). His two sons, George the younger (*d.*1859) and John, joined the firm in 1820. They continued the business for a time after John's death reissuing atlases. However, they soon gave up the business and stock and plates were sold at auction. Much of this property of the business passed to G F Cruchley *c.*1844, and when he went out of business Gall & Inglis purchased the effects.

John Cary was one of the finest English cartographers. Cary's first county atlas was the *New and Correct English Atlas* published in 1787 (**51** and **73**). Other county atlases followed (*Cary's Travellers Companion* and *Cary's New English Atlas*, **55**, **69** and **71**). In addition his work covered world atlases, road maps, sea charts, town and canal plans. In 1789 he produced maps for an edition of Camden's *Britannia* translated by Richard Gough (**54**). The maps from this work were used by John Stockdale in his *New British Atlas* published in 1805.

All of the above atlases included maps based on the county format. However, as a result of becoming Surveyor of Roads to the General Post Office, commissioned to survey the roads of England in 1794, Cary also produced sectional maps of England and Wales. Some of these resurfaced in the twentieth century, redesigned so as to portray specific counties.

The following two maps were taken from plates first prepared by John Cary and used as sectional maps, and were not covered in the previous cartobibliography of Devon. All copies so far seen by the authors have been post 1901 and so they do not appear in the current cartobibliography. However, there may well be earlier examples which would belong and for this reason the following maps are listed here.

1. GALL & INGLIS

In 1810 James Gall set up a printing business in Edinburgh and was joined by his son, also James, in 1838 and the firm traded as James Gall & Son. In 1847 James junior left the business and Robert Inglis joined and they traded thereafter as Gall & Inglis. Although the company produced one atlas about 1850, cartographic publications did not assume an important role until the 1870s. In 1877 the company were able to buy the left over stock, and more importantly, the plates of George Frederick Cruchley at auction. Robert's son, James Gall Inglis joined the company in 1880 and the publication of maps and atlases became increasingly important.

Gall & Inglis had acquired the plates of *Cary's Improved Map of England and Wales* amongst Cruchley's stock and began to use these plates in a variety of ways. A number of hunting maps appeared before the turn of the century including one by Swiss & Co. of Devonport with whom the company seem to have a working relationship (see **166**).

Amongst Cruchley's stock there must have been a large amount of loose maps with Cruchley's imprint. Gall and Inglis were happy to sell this stock off either with Cruchley's imprint replaced by a Gall & Inglis imprint, exploiting Cruchley's reputation, or alternatively, simply with a sticker over the Cruchley imprint. The map below had a Gall & Inglis imprint with an old Cruchley cover but the plates were from Cary's ½" map reduced lithographically to ⅓" which had originally been produced in parts from 1820 to 1832.

Size: 520 x 620 mm. **SCALE THREE MILES TO ONE INCH** (10=87 mm).
Folded Map (182 x 105 mm).

DEVONSHIRE (Title in centre of map). Imprint: **Gall & Inglis, Edinburgh** (Ee OS). Inset map of **Lundy** (Ae). Note at top of sheet **For Continuation North see Foot of Map** referring to inset of Morte Bay to the Foreland (Ee). Grid lines off the coast of Cornwall, Plymouth Sound and Bideford Bay; depths, rocks and banks off the coasts; hachured hills; principal roads are coloured red; railways are coloured black and include Kingsbridge and Yealmpton, Lyme Regis and Budleigh Salterton to Exmouth, Ashton to Exeter (but no line to Appledore).

1. 1903 *Cruchley's County Maps of England. Devon. For Cyclists, Tourists &c*
London & Edinburgh. Gall & Inglis. (1903).

2. G W BACON

Towards the end of hte nineteenth and beginning of the twentieth centuries, the company of G W Bacon was successfully exploiting a large number of maps, that has been produced much earlier, in order to cater to the growing needs of a socity becoming increasingly mobile. He was still issuing edtions of the *Commercial and Library Atlas* utilising the plates of the *Dispatch Atlas* (**136**) for nearly forty years old. A few folding maps are known.

George Bacon used an even older map as the basis of one cycling and motoring map: a folding map is known which was a reissue of Benjamin Baker's copy of the smaller Donn map of Devon of 1765. Baker's map, published in 1799 by William Faden (**62**), had been reissued by the Wyld family in the 1850s and resurfaced in the late 19th century with Bacon's imprint as *Bacon's County Map And Guide Devonshire with parts of adjoining counties.*

In addition, Bacon must have good connections with John Bartholomew as two of his maps were produced as folding maps with Bacon covers. Bartholomew's 1872 map (**150**), extensively used by Black after 1879, has been seen with a Bacon cover: *Bacon's County Map of Devonshire with parts of adjoining counties*; and his 1895layer-coloured maps (**174**) appeared as folding maps, but without the typical colouring, as *Bacon's New Half-Inch Maps Cycling and Motoring Devon* and *Bacon's New Library map of Devonshire And Part of Somerset.*

Although George Frederick Cruchley was able to acquire most of the Cary plates and produced a number of atlases from them (see **71, 73** and **92**) it seems that George Bacon procured the plates of *Cary's New Map of England.*[1] The full revision of this map, *The Improved Map* (1832), was acquired first by Cruchley and susequently by Gall & Inglis. Bacon also exploited old Cary plates for his *Bacon's Waistcoat-pocket Maps c.* 1887 and was once advertised as *Exeter and District.*

Two locally sold maps can be found with G W Bacon's imprint. The first is a cycling map of central Devon. This covers an area from Dartmouth and Plymouth north as far as Barnstaple and Bridgwater, ie it covers approximately 80% of the county. With various cover titles, usually town (eg *Dawlish* or *Teignmouth*) *and District*, the maps are 460 x 580 mm, have a selection of routes (Ee) and are divided into circles around Exeter. They were produced about the turn of the century. One example in the author's possession was sold by F P Davies of Brunswick Place, Dawlish. It was sold as *Davies' Cycling Road Map Dawlish And District* (cover title), but Bacon's imprint appears below the map. Davies was a photographer who sold views and guides. The railways are confusing as Exmouth-Budleigh is not included although the line to Westward Ho1 has been included.

The Second, the map below, (folded to 165 x 90 mm), was most likely taken from the second state of *Cary's New Map of England* (five miles to the inch) which had originally appeared in sections in 1792 - nearly 100 years earlier! After re-issues in 1794, re-engraving in 1816 with corrections and re-issue in 1822 (all by the Carys) the map reappeared in 1885 as *Bacon's Road Map of England in Seven Sheets.* What had happened to the sheets in the intervening 50 years is not known.

Size: 350 x 465 mm. **Scale of English Miles** (20=100 mm).

No Title. Imprints: **Copyright** (AeOS) and **G W Bacon & Co., Ltd., 127 Strand, London** (EeOS). Notes: **Hills to be ridden** (CeOS) and **Divided into 10 mile sq**. (DeOS). Plain border broken for St Austell. Inset (Ee) shows the south continuation below Modbury. Railways to Yealmpton, Bude and Lynton (all *c.*1898) as well as the post-1900 lines Budleigh Salterton to Exmouth, Ashton to Exeter, Bideford to Northam.

1. 1903 *Bacon's Road-Map For Cyclists & Motorists - Newton Abbot District ...*
London. G W Bacon & Co. Ltd. (1903).

1. See Eugene Burden's letter to *The Map Collector*, Winter 1990; Issue 53; p. 52.

APPENDIX III – John Bartholomew

After an apprenticeship with the Edinburgh company W & D Lizars, George Bartholomew (1784-1871) established himself as an engraver in Edinburgh. George's son, John I (1805-61), set up his own business also producing work for Lizars. His son, John Bartholomew II (1831-93), and another John George Bartholomew (1860-1920), continued the work of the company. The Bartholomew family also produced maps for other publishers (**150, 157, 175**) as well as maps for the *Encyclopaedia Britannica* (**153**).

In 1895 John Bartholomew began his famous series of layer coloured maps (see **174**). However, he had long experimented with ways of depicting heights and in the mid-1880s produced a few maps with highly accurate hill shading. One of the maps that was produced was a map of South Devon. This was the precursor to the later coloured map.

The only copies known are hachured and only South Devon has been seen; it is believed that no map of North Devon was produced. The records of the Bartholomew Printing House (now in the custody of the National Library of Scotland) reveal that the map was printed three times. The original printing took place in 1884; 1040 copies were printed 19th May 1886 and a further 1010 copies 8th September 1887.

The maps were hachured had railways in red, clear cased roads and were overprinted in a pale yellow wash. One copy extant has the cover title: *Map Of South Devon And Dartmoor* – Scale Two Miles To An Inch – Reduced from the Ordnance Survey Corrected to date and Coloured – Price One Shilling – Cranford's Library – Dartmouth and Paignton. R Cranford was a seller of guide books, maps and photographic views who also advertised bric-a-brac, antique furniture and china. Visitors wishing to see the interior of the Butterwalk in Dartmouth had to apply for admission to Cranford's Library.

In 1893 and 1894 further maps of South Devon were printed. These were titled Reduced Ordnance Map of South Devon and were printed on 13th May 1893 (655 copies) and again 29th March 1894 (525 copies) respectively. These maps had no layer colouring, no contours but were hachured as the earlier maps. No copy of North Devon is known. The Bartholomew Printing Records has copies of these.

Size: 480 x 700 mm. **SCALE 2 MILES TO AN INCH** (6=75 mm).

W. H. SMITH & SON'S REDUCED ORDNANCE MAP OF SOUTH DEVON (Ee). Imprint: **London: W H Smith & Son, 186 Strand** (CaOS). The map covers an area from Looe and Liskeard to Sidmouth and Prawle Point to Exeter and Okehampton Station. Railways are shown to Sidmouth, Exmouth, Dartmouth and Brixham, Moreton Hampstead and Ashburton, Dartmoor Loop, Launceston, Looe, Ashton (although no station): mineral lines to Princetown and Kellybray to Calstock thinner red line: proposed lines from Okehampton to Holsworthy and from Launceston north to Halwell with no black engraved edge: and proposed lines to Kingsbridge and Sidmouth faintly engraved but not coloured.

1. 1884 *W H Smith & Son's Reduced Ordnance Map Of South Devon*
London. W H Smith & Son. (1884), (1886), (1887).

APPENDIX IV
CORRIGENDA AND ADDENDA

In 1996 the same authors published *The Printed Maps of Devon: County Maps 1575 – 1837.* As with all cartobibliographies there will always be some mistakes and omissions. Since the first volume was published the following mistakes have been found or the following new states have been discovered. See Appendix I for details of the Arrowsmith map of 1814.

Page xii	1st para: 1627 (46) Ptolomy should read Ptolemy.
Page 27	All sources except *Geographiae Blavianae*: Delete C as source.
Page 40	The footnote 3 should read the Whitaker Collection not Walker.
Page 42	State 2, second entry (1815) is at BRL.
Page 98	See foot note 3 - A copy of this map with the revised imprint was offered for sale by R Collicot 9.4.99
Page 104	State 4: Delete C for 1793 (1804) atlas.
Page 104	State 6: Delete BL for 1793 atlas.
Page 130	In 1840 Hansard published *The Charities in the County of Devon Vol.I,* this included a copy of the Neele map in State 1. This would suggest, but not prove, that State 2 was ssued later than 1840, not (1830).
Page 138	A first state without the Note *Roads...* has been found in another and later edition published by C.Cooke but titled *Topographical and Statistical Description of the County of Devon.* This has a different text referring to the 1811 census figures.
	State 4: Second part of imprint removed (Note 1). A copy has been seen and guide refers to 1831.
	Topographical and Statistical Description of the County of Devon. London. Sherwood & Co. (1831). KB.
Page 150	Ordnance Survey 1815 - An INDEX map of the County showing the relative sheets [XX-XXVII] was fixed to the cover of PART the IID....DEVONSHIRE...Plates No.20....27 Size: 75 x 60 with INDEX (CaOS) in a raised part of the border. (See Fig.1 p.xxviii)
Page 154	State 2 becomes State 3: State 2: Date changed to **Jan 1st 1823** (E).
Page 160	State 1 map imprint date is 1812 (not 1815).
Page 160	State 3: Sources are **CCCLXII**, BL, C and **CCCLXIII**, BL, BCL.
Page 162/3	**80A** Aaron Arrowsmith - 1814 - see APPENDIX I (p.176 *Victorian Maps)*
Page 166	State 4: *Official Map Of Devon* with vignette deleted and many changes in *Devonshire The Official County; And Guide* published by Simpkin, Marshall & Co. Copy at BL.
Page 174	State 1: Date of publication was 1819.
Page 186	State 1: Delete BL for 1828 entry.

Page 186 State 2: Change date to 1828 (1835).

Page 192 State 2 becomes State 3[1]: States 1 and 2 as follows:

1. 1825 Incomplete set .... W.

2. 1828 Imprint now reads:
London. Published Septr 20 1828 by S MAUNDER. 10. Newgate Street. (E).

Page 194 Additions noted in State 3 are included in State 1; ie States 2 and 3 are identical.

Page 198 Amend all 'state' entries to read:

1. 1829 *Pigot & Co's British Atlas of the Counties of England*
London & Manchester. J Pigot & Co. 1829. W.

2. 1830 Imprint address: **17 Basing Lane London & 18 Fountain St. Manchester.**
Addition of **Long West from Greenh** between borders (BeOS).
(E), (NDL).

Pigot & Co's British Atlas of the Counties of England
London & Manchester. J Pigot & Co. 1830, 1831. GL, B; **CCCCXXV**, C, W.

Pigot & Co's National Commercial Director
London & Manchester. J Pigot & Co. 1830. GL, WM.

3. 1831 Polling places added. **North Division** and **South Division** added. (NDL).

Page 210 The printer was Joseph Netherclift. (not Netherclist)

Page 230 *Pigot & Co's Pocket Atlas* was first published in parts in 1838 and Devon appeared in April 1839. That would make this map the first Victorian map of Devon.[2]

Page 236 State 2: The notes ***Bishops Residence*** and ***Cathedral Church*** are omitted and new notes added slightly higher (one line below *Do......Deaneries*). Provenance unknown - a single sheet but still with guard showing binding into a publication. (E)

1. New State 4 appeared *c.*1855 with new title and imprint. The Hundreds listing has been altered and railways added. *The Hand-Book To South Devon*; Devonport; W Wood.
2. Eugene Burden; A Pocket Topography of England; *IMCoS Journal*; Issue 80; Spring 2000.

APPENDIX V
EARLIER MAPS ISSUED AFTER 1837

The following maps were engraved before 1837 (See Batten and Bennett, 1996) but were still being updated and reissued during Victoria's reign. The number on the left is the corresponding entry in *The Printed Maps of Devon, 1575-1836*. Principal later works are named, but not various editions, which may sometimes have other names. The maps are numbered in chronological sequence of first publication. The chosen order of preference throughout the catalogue is surveyor, draughtsman, engraver, author, publisher. Names joined by a slash indicate that both/all were connected with the original issue. Where maps are commonly associated with another mapmaker or printer their name is in brackets.

57.	1791	B Baker	*The Universal Magazine*
62.	1799	Baker/Faden	*Bacon's County Map And Guide Devonshire*
63.	1801	Smith/Jones/Smith	*Smith's New English Atlas*
64.		John Wilkes	*The Charities of Devon*
67.	1805	Cole/Roper	*The British Atlas*
71.	1807	J Cary	*Cary's New English Atlas*
			Cruchley's Railway Maps
			Cruchley's County Maps
			Deacon's Court Guide 1882
			White's History, Gazetteer 1889
72.	1808	Cooper/Capper	*A Topographical Dictionary of the U. K.*
73.	1809	J Cary	*New and Correct English Atlas* II
			Cruchley's County Atlas
74.	1809	Mudge/OS	*The Second Part of the General Survey*
78.	1812	J Wallis	*Wallis's New British Atlas*
81.	1816	R Rowe	*The English Atlas*
			The British Gazeteer 1860-Collins
82.	1816	Dix/Darton	*A Complete Atlas of the English Counties*
85.	1819	Crabb/(Ramble)	A Set of Cards
			Reuben Ramble's Travels -1846
87.	1820	Hall/Leigh	*Leigh's New Pocket Pocket Road Book*
90.	1822	J Walker	*Crosby's Complete Pocket Gazetteer*
			Camden's Britannia
91.	1822	Gardner/Smith	*Smith's New English Atlas*
92.	1822	G & J Cary	*Cary's Traveller's Companion* III
			Cruchley's Railroad Companion
93.	1822	Smith/Davies	*The Exeter Pocket Journal*
			Exeter Journal & Almanack
95.	1825	Ebden/Duncan	*Ebden's New Map of the County of Devon*
			Wood's Handbook
98.	1829	J Pigot	*Pigot & Co's British Atlas*
			Slater's British Atlas
			Slater's Directory
99.	1830	H Teesdale	*A New Travelling Atlas*
			The Tourist's Atlas
100.	1830	T L Murray	*An Atlas of the English Counties*
			Commercial Directory
101.	1830	S Hall	*A Topographical Dictionary of Great Britain*
			A Travelling County Atlas
			The New County Atlas
103.	1831	Creighton/Lewis	*A Topographical Dictionary of England*
106.	1832	W Cobbett	*A Geographical Dictionary*
107.	1833	Scott/Fullarton	*A New and Comprehensive Gazetteer*
			Parliamentary Gazetteer
108.	1833	Archer/Pinnock	*The Guide to Knowledge*

			Descriptive County Maps
			Johnson's Atlas of England
111.	1834	Dower/Moule	*Moule's English Counties*
			Complete & Universal English Dictionary
			The History of England - Smollet
113.	1835	Creighton/Lewis	*A Topographical Dictionary of England*
114.	1835	J Pigot	*Pigot & Co's Pocket Atlas*
			Pocket Topography & Gazetteer of England
116.	1836	J & C Walker	*This British Atlas*
			Hobson's Fox Hunting Atlas
			White's Gazetteer 1878
			Murray's Handbook for Devon & Cornwall
			Lett's Popular County Atlas
			Walker's Fox Hunting Atlas

APPENDIX VI
TITLES AND SIZES OF MAPS

Year	Maker	Title	Size
1839	De la Beche	*INDEX to the ORDNANCE GEOLOGICAL MAPS ...*	305 x 410 mm
1842	Archer/Dugdale	*DEVONSHIRE*	180 x 230 mm
1845	Becker/Fisher	*DEVONSHIRE*	355 x 505 mm
1845	Becker/Besley	*ROUTE MAP OF THE ROADS OF DEVON*	235 x 285 mm
1846	Becker/Besley	*DEVONSHIRE*	238 x 288 mm
1848	Emslie/Reynolds	*DEVONSHIRE*	175 x 235 mm
1850	Rowlandson	*Illustrated London News ... Part II DEVONSHIRE*	65 x 60 mm
1851	Rock	*DEVONSHIRE*	185 x 245 mm
1852	Walker/Knight	*DEVONSHIRE*	210 x 160 mm
1852	Archer/Collins	*DEVON*	76 x 53 mm
1854	Becker/Besley	*DEVONSHIRE*	315 x 380 mm
1854	Oliver	*DIOC' EXON'*	340 x 477 mm
1855	Schenk,McFarlane/Black	To Accompany BLACK'S GUIDE to *DEVONSHIRE ...*	225 x 210 mm
1856	Brooks	*MAP of DEVONSHIRE*	149 x 174 mm
1856	Becker/Kelly	*THE POST OFFICE DIRECTORY...*	220 x 270 mm
1857	Billing	*MARTIN BILLING'S MAP OF DEVONSHIRE*	630 x 448 mm
1857	Becker/Besley	*NORTH DEVON*	310 x 400 mm
		SOUTH DEVON, AND DARTMOOR	305 x 405 mm
1858	Johnston	*HANDBOOK MAP OF DEVON AND CORNWALL*	335 x 485 mm
1889		*DEVON*	245 x 340 mm
1896		*Deacon's Devon And Cornwall* (four maps, each)	245 x 340 mm
1858	Weekly Dispatch I	*DEVONSHIRE (SOUTH DIVISION)*	300 x 425 mm
		DEVONSHIRE (NORTH DIVISION)	300 x 425 mm
1864	Weekly Dispatch II	*DEVONSHIRE (SOUTH DIVISION)*	300 x 425 mm
		DEVONSHIRE (NORTH DIVISION)	300 x 425 mm
1869	Weekly Dispatch III	*BACON'S MAP OF DEVONSHIRE*	680 x 615 mm
1876		*BACON'S MAP OF DEVONSHIRE* (two sheets, each)	305 x 430 mm
1883		*DEVONSHIRE (SOUTH & EAST DIVISIONS)*	330 x 430 mm
		DEVONSHIRE (NORTH DIVISION)	305 x 430 mm
1884		*DEVONSHIRE (SOUTH & EAST DIVISIONS)*	310 x 430 mm
1885		*DEVONSHIRE (SOUTH SHEET)*	450 x 650 mm
		DEVONSHIRE (NORTH SHEET)	425 x 610 mm
1886		*DEVONSHIRE (SOUTH SHEET)*	310 x 425 mm
		DEVONSHIRE (NORTH SHEET)	305 x 425 mm
1859	Walker/Stanford	*COUNTIES OF CORNWALL AND DEVON*	226 x 298 mm
1859	G F Cruchley	*CRUCHLEY'S MAP OF DEVONSHIRE* (2 sheets)	615 x 950 mm
1860		*CRUCHLEY'S MAP OF DEVONSHIRE*	1230 x 960 mm
1868		*CRUCHLEY'S MAP OF DEVONSHIRE (North)*	500 x 925 mm
		CRUCHLEY'S MAP OF DEVONSHIRE (South)	510 x 925 mm
1875		*CRUCHLEY'S MAP OF DEVONSHIRE*	1230 x 960 mm
		CRUCHLEY'S NEW MAP OF NORTH DEVON	475 x 640 mm
1860	Becker	*AN ECCLESIASTICAL MAP OF THE DIOCESE*	394 x 563 mm
1861	McLeod	*DEVONSHIRE*	131 x 155 mm
1862	G Philip & Son	*DEVONSHIRE*	410 x 335 mm
1876		*THE PICTORIAL WORLD MAP ...* (this issue only)	410 x 335 mm
1862	Bartholomew/Black	a) *DEVONSHIRE*	280 x 260 mm
1862		b) *DORSET, DEVON & CORNWALL*	270 x 540 mm
1869		c) *DEVON & CORNWALL*	277 x 400 mm
1873		d) *DEVONSHIRE (reduced size)*	255 x 263 mm
1864	Sackett	*Exeter:*	365 x 500 mm
1864	Hughes	*DEVONSHIRE*	240 x 300 mm
1865	Spargo	*PLAN OF THE COUNTY OF DEVON*	232 x 353 mm
1868	James	*DEVONSHIRE (NEW DIVISIONS OF COUNTY)*	480 x 335 mm
1868	Jackson & Partridge	*HIEROGLYPHICAL COUNTY READINGS No. IV*	30 x 145 mm
1870	Barnes & Home	*A PAROCHIAL BOUNDARY MAP*	587 x 461 mm
1872	Bartholomew/Heydon	*JOHN HEYDON'S MAP OF ... PLYMOUTH, DEVONPORT AND STONEHOUSE*	473 x 568 mm

1873		*NORTH DEVON*	490 x 500 mm
		SOUTH DEVON	490 x 500 mm
1879		*DEVONSHIRE*	460 x 455 mm
1881		*DEVONSHIRE (SOUTH SECTION)*	235 x 450 mm
		DEVONSHIRE (NORTH SECTION)	235 x 450 mm
1892		*DEVONSHIRE NORTH EAST SECTION*	240 x 250 mm
		DEVONSHIRE NORTH WEST SECTION	240 x 250 mm
		DEVONSHIRE SOUTH EAST SECTION	240 x 250 mm
		DEVONSHIRE SOUTH WEST SECTION	240 x 250 mm
1895		*DEVONSHIRE*	460 x 455 mm
1895		*THE COUNTY OF DEVON ... ORDNANCE SURVEY*	480 x 490 mm
1898		*DEVONSHIRE*	480 x 505 mm
1874	Murby	*THE COUNTY OF DEVONSHIRE*	100 x 135 mm
1875	Collins	*DEVONSHIRE*	206 x 158 mm
1877	Bartholomew/Black	*DEVON*	250 x 200 mm
1878	Stanford/Worth	*MAP TO ... THE GUIDE TO SOUTH DEVON*	245 x 300 mm
		MAP TO ... THE GUIDE TO NORTH DEVON	245 x 300 mm
1881	Stanford	No title: *DEVON* across map halves (two maps, each)	139 x 77 mm
1886		*DEVON, WEST* and *DEVON, EAST* (two maps, each)	139 x 77 mm
1882	Bartholomew/Black	*KEY MAP OF DEVONSHIRE*	145 x 190 mm
1895		*SKETCH MAP OF DEVONSHIRE*	145 x 190 mm
1882	Bartholomew/Dulau	*INDEX MAP SHOWING THE MAPS IN THE GUIDE*	145 x 199 mm
1896		*INDEX MAP Shewing The SECTION MAPS ...*	145 x 199 mm
1883	Anon/J P	*PUZZLE PICTURES. No. IX. - DEVONSHIRE*	25 x 33 mm
1883	Bryer/Kelly	*KELLY'S MAP OF DEVON*	482 x 510 mm
1885	W. Morning News	*DEVON & CORNWALL ELECTORAL DIST.*	540 x 737 mm
1885	Stanford	*DEVON (NORTH)*	177 x 238 mm
		DEVON (SOUTH)	177 x 238 mm
1892		*COUNTY ... DEVON to illustrate BIRDS OF DEVON*	253 x 276 mm
1885	Jones/O S	*DEVONSHIRE, NEW DIVISIONS OF THE COUNTY*	470 x 470 mm
1885	Bacon	No title: Sh.11 (adjoining sheet 7, 10, 12)	145 x 170 mm
1888	Ordnance Survey	*DIAGRAM of the Sanitary Districts*	530 x 630 mm
1891		*INDEX to the ORDNANCE SURVEY ... 1891*	530 x 630 mm
1888	Cassell/L&SWR	ROUTE MAP - IV	123 x 80 mm
1890	Swiss (Cary)	*SWISS & CO.'S NO. 1 HUNTING MAP*	976 x 753 mm
1890		*... ROAD & RAILWAY HUNTING MAP ... SOUTH*	620 x 872 mm
1890		*... ROAD & RAILWAY HUNTING MAP ... NORTH*	520 x 872 mm
1890	Jaques & Son	*DEVONSHIRE* (one card only, no title on 2nd card)	58 x 77 mm
1891	Southwood	*DEVON. ... PIDSLEY'S BIRDS OF DEVONSHIRE*	460 x 510 mm
1892	Ordnance Survey	*ORDNANCE SURVEY OF DEVONSHIRE*	650 x 670 mm
1892		*INDEX to the SIX INCH SCALE ... ORDNANCE SURVEY*	650 x 670 mm
1900		*ENGLAND & WALES. DIAGRAM OF DEVONSHIRE*	650 x 740 mm
1893	Page	*MAP OF THE RIVERS OF DEVON*	245 x 225 mm
1894	Weller	*DEVONSHIRE*	216 x 278 mm
1895	Page	*MAP OF THE COASTS OF DEVON*	345 x 325 mm
1895	Firks & Son	*G FIRKS & SON'S `WESTWARD HO!' MAP OF DEVON & CORNWALL*	1375 x 1197 mm
1895	Bartholomew	*NEW REDUCED OS ... SOUTH DEVON*	480 x 705 mm
1896		*BARTHOLOMEW'S REDUCED OS ... SOUTH DEVON*	480 x 705 mm
1896		*BARTHOLOMEW'S REDUCED OS ... NORTH DEVON*	480 x 730 mm
1897	Bartholomew/Pattisons	*DEVONSHIRE*	225 x 170 mm
1897	Cassell & Co.	*THE RIVERS OF DEVON*	104 x 102 mm
1897	Jervis	TORRINGTON & OKEHAMPTON RAILWAY	245 x 202 mm
1897	Clark/Norway	No title: *Highways and Byways*	182 x 250 mm
1899	E G Ravenstein	*Philips' Cycling Map Of The County of Devon*	600 x 700 mm
1901	Anon – Gazeteer	No title: *Gazetteer of Devon ...*	110 x 130 mm
1901	Ordnance Survey	*DEVONSHIRE*	490 x 590 mm
1901	Harry Inglis	No title: Sheets 39 and 40	140 x 170 mm

SELECT BIBLIOGRAPHY

Barker, K & Kain, R (Eds)	*Maps and History in S.-W. England*; Univ. Exeter Press; 1991.
Batten, K & Bennett, F	*Printed Maps of Devon 1575-1836*; Devon Books; 1996.
Baynton-Williams, R	*Investing in Maps*; London; 1969.
Beresiner, Yasha	*British County Maps*; Antique Collectors' Club; 1983.
Booker, Frank	*Industrial Archaeology of the Tamar Valley*; David & Charles; 1967 (1974).
Burden, Eugene	*County Maps of Berkshire 1574-1900*; (1988) 1991.
Carroll, Raymond	*County Maps of Lincolnshire 1574-1900*; Lincs Record Soc. 1996.
Chubb, Thomas	*The Printed Maps in the Atlases of G. Britain and Ireland*; London; 1972.
Fordham, Sir H G	*John Cary, Engraver and Map Seller*; Cambridge; 1910.
Gardiner, Leslie	*Bartholomew 150 years*; J Bartholomew & Son Ltd; 1976.
Hodgkiss, A G	*Discovering Antique Maps*; Shire Publications Ltd; 1988.
Hodson, Donald	*The Printed Maps of Hertfordshire*; 1974.
IMCoS JOURNAL	Magazine of the International Map Collectors' Society. Quarterly.
Keir, David	*The House of Collins*; Collins; 1952.
Kingsley, David	*Printed Maps of Sussex*; Sussex Record Society; 1982.
Lister, W B C	*A Bibliography of Murray's*; Dereham Books; 1993.
The Map Collector	Published Quarterly; Issues 1978-1995.
Mackenzie, Ian	*British Prints*; Antique Collectors' Club; 1988.
Moreland, C & Bannister, D	*Antique Maps*; Phaidon; 1986.
Nowell-Smith, Simon	*The House of Cassell*; Cassell & Co. Ltd; 1958.
JoPHS	*Journal of the Printing Historical Society No. 27*; 1998.
Sellman, R R	*Aspects of Devon History*; Devon Books; (1962) 1985.
Smith, Martin	*The Railways of Devon*; Ian Allen Publishing; 1993.
Smith, David	*Victorian Maps of the British Isles*; Batsford Books; 1985.
Somers Cocks, J V	*Devon Topographical Prints*; 1977.
St. John Thomas, David	*Regional History of The Railways, Vol I*; David & Charles; 1988.
Thompson, Victor	*Back Along The Lines - North Devon's Railways*; Badger Books; 1983.
Todd, William B	*A Directory of Printers*; Printing Historical Society (PHS); 1972.
Tooley, R V	*Maps and Mapmakers*; Batsford Books; 1987.
Tooley, R V & Bricker, C	*Landmarks of Mapmaking*; Wordsworth; 1989.
Twyman, Michael	*A Directory of London Lithographic Printers*; P H S; 1976.
Twyman, Michael	*The British Library Guide to Printing*; British Library; 1998.
Vaughan, John	*The English Guide Book c.1780-1870*; David & Charles; 1974.

In addition to the work quoted above, David Smith has done a lot of research into the major printers, publishers and engravers of the nineteenth century and much has been published in magazines and journals. The following is a brief list of his publications concerning figures mentioned in this cartobibliography:

The Philip Family Firm;	*The Map Collector* 38;	1987.
George Frederick Cruchley;	*The Map Collector* 49;	1989.
200 Years of W H Smith;	*The Map Collector* 60;	1992.
George Washington Bacon;	*The Map Collector* 65;	1993.
William Blackwood & Sons;	*Mercator's World* Vol. 1.5;	1996.
Cassell & Co.;	*IMCoS Journal* 70;	1997.
A & C Black;	*IMCoS Journal* 71;	1997.
Gall & Inglis;	*IMCoS Journal* 73;	1998.
The Business of the Bartholomew Family Firm;	*IMCoS Journal* 75;	1998.
The Cartography of the Bartholomew Family Firm;	*IMCoS Journal* 76;	1999.
The Business of W & A K Sohriton	*IMCoS Journal 82*;	2000

INDEX

C

D

N

O

P

Q

R

U

V

W

Y

Z